Sustainable Textiles: Production, Processing, Manufacturing & Chemistry

Series Editor

Subramanian Senthilkannan Muthu, Chief Sustainability Officer,
Green Story Inc. Canada

This series aims to address all issues related to sustainability through the lifecycles of textiles from manufacturing to consumer behavior through sustainable disposal. Potential topics include but are not limited to: Environmental Footprints of Textile manufacturing; Environmental Life Cycle Assessment of Textile production; Environmental impact models of Textiles and Clothing Supply Chain; Clothing Supply Chain Sustainability; Carbon, energy and water footprints of textile products and in the clothing manufacturing chain; Functional life and reusability of textile products; Biodegradable textile products and the assessment of biodegradability; Waste management in textile industry; Pollution abatement in textile sector; Recycled textile materials and the evaluation of recycling; Consumer behavior in Sustainable Textiles; Eco-design in Clothing & Apparels; Sustainable polymers & fibers in Textiles; Sustainable waste water treatments in Textile manufacturing; Sustainable Textile Chemicals in Textile manufacturing. Innovative fibres, processes, methods and technologies for Sustainable textiles; Development of sustainable, eco-friendly textile products and processes; Environmental standards for textile industry; Modelling of environmental impacts of textile products; Green Chemistry, clean technology and their applications to textiles and clothing sector; Eco-production of Apparels, Energy and Water Efficient textiles. Sustainable Smart textiles & polymers, Sustainable Nano fibers and Textiles; Sustainable Innovations in Textile Chemistry & Manufacturing; Circular Economy, Advances in Sustainable Textiles Manufacturing; Sustainable Luxury & Craftsmanship; Zero Waste Textiles.

Subramanian Senthilkannan Muthu
Editor

Natural Dyes and Sustainability

Springer

Editor
Subramanian Senthilkannan Muthu
Chief Sustainability Officer
Green Story Inc.
Kowloon, Hong Kong

ISSN 2662-7108 ISSN 2662-7116 (electronic)
Sustainable Textiles: Production, Processing, Manufacturing & Chemistry
ISBN 978-3-031-47470-5 ISBN 978-3-031-47471-2 (eBook)
https://doi.org/10.1007/978-3-031-47471-2

This Springer imprint is published by the registered company Springer Nature Switzerland AG
The registered company address is: Gewerbestrasse 11, 6330 Cham, Switzerland

Paper in this product is recyclable.

This book is dedicated to:
The lotus feet of my beloved Lord
Pazhaniandavar
My beloved late Father
My beloved Mother
My beloved Wife Karpagam and Daughters –
Anu and Karthika
My beloved Brother
Last but not least
To everyone working in the textile sector to
make it "SUSTAINABLE"

Contents

Natural Dyes in Traditional Textiles: A Gateway to Sustainability 1
R. Surjit, B. Anusri Mahalakshmi, E. Shalini, and M. Shubha Shree

Sustainable Extractions and Applications of Natural Dyestuffs......... 25
Elçin Emekdar Karaman and Umut Kıvanç Şahin

Dyes and Pigments from Agricultural Wastes in the Coloration of Textiles.. 45
Noureddine Baaka, Shahid Adeel, Fozia Anjum, Aamir Ali, and Rehab Ahmed Abdelghaffar

Recent Approaches and Advancements in Natural Dyes............... 63
Silpinwita Das and Sankar Roy Maulik

Mushroom-Based Natural Dyes for Sustainable Color in Textiles....... 79
Muhammad Zulqurnain Haider, Shahid Adeel, Naveed Sultan, Noman Habib, Muhammad Arslan Ashraf, and Mozhgan Hosseinnezhad

Extraction and Application of Plant-Based Tannins as Sources of Natural Colourants 103
Shahid Adeel, Muhammad Yameen, Fariha Asghar, Nimra Amin, Meral Ozomay, and Somayeh Mirnezhad

Social and Environmental Impact of Natural Dyeing 127
Shanthi Radhakrishnan, Radhika Rajangam, and Praveena Peruran

Classification of Natural Dyes for Sustainable Exploitation............ 153
Jamiu Mosebolatan Jabar

Natural Indigo Dyes: A Potential Dye for Sustainability 193
R. Surjit, H. Shanruthi, V. Sruthi, and K. S. Tarunvalavan

Natural Dyes: Coloring Tool for Textiles 211
Shanthi Radhakrishnan, Yamunadevi Ayyaswamy, and Jasirabanu Ahamed Basheer

Extraction of Natural Pigments Using Supercritical Carbon Dioxide and Its Application on Different Fabrics 231
Tarek Abou Elmaaty and Heba Sorour

Natural Dyes as Corrosion Inhibitors 265
Priya Vashishth, Bindu Mangla, Himanshi Bairagi, Rajni Narang, and Harish Moudgil

Evaluation of Colorimetric and Color Fastness Properties of Marble-Printed Silk Fabrics with the Use of Various Mordants and After-Treatments 287
Habibe Kahvecioğlu Sarı, Fatma Filiz Yıldırım, Ozan Avinc, and Arzu Yavas

Natural Dyeing of PA 6, PTT, PBT, PLA, Silk, and Soybean Textile Fibers with Black Mulberry Fruit Extract by Conventional and Microwave Dyeing Techniques 329
Ece Kalayci, Arzu Yavas, and Ozan Avinc

Index 353

Natural Dyes in Traditional Textiles: A Gateway to Sustainability

R. Surjit, B. Anusri Mahalakshmi, E. Shalini, and M. Shubha Shree

Abstract Fabrics constructed using ancient manufacturing and dyeing techniques around the world are termed traditional textiles. Over centuries, most regions developed specialities based on local resources: the golden silk of Assam, the fine cotton of Bengal and the Bagru prints of Rajasthan. The chief source of colouring matter, until about a century ago, had been nature in general and vegetation in particular and now it has expanded to invertebrates, insects, fungi and minerals. The development of natural dyes took place at the same time after the technique of weaving had been discovered in about 5000 BC. From the ancient period, the usage of natural dyes has played a major role in traditional textiles. Natural dyes have been used since ancient times for colouring and printing leather as well as common textile fibres like cotton, wool and silk. Natural dyes are extracted from a variety of substances that are in nature such as plants (e.g. woad and turmeric); insects (e.g. cochineal beetles and lac scale insects); animals (e.g. some species of snails or cuttlefish) and minerals (e.g. iron ore and clay) without any chemical treatment. When these are used for dyeing fabrics, they not only impart colour to the fabric but also act as antifungal agents whereby they impart protection to the fabric against bacterial or fungal infections or act as moth repellents. Some dyes like indigo have a cooling sensation too Among most natural dyes, plant-based dyes/pigments have a wide range of applications in fabric. Fixing the colour is the great challenge of dyeing cloth. Indian dyers' use of mordants was key to their mastery, which was unrivalled until the invention of Western chemical dyes in the nineteenth century. One kilogram of dye plant is required for dyeing 1 kg of fibres as the natural dyes are contained in plants in small quantities, and this creates a need to handle a large amount of biomass waste which can be reused again into derivative products to support the 3R principle (reduce, reuse and recycle). Man learnt to use different kinds of natural colouring matters to dye cloths, but development in technology has resulted in

R. Surjit (✉) · B. Anusri Mahalakshmi · E. Shalini · M. Shubha Shree
Department of Fashion Technology, PSG College of Technology,
Coimbatore, Tamil Nadu, India
e-mail: rst.fashion@psgtech.ac.in

S. S. Muthu (ed.), *Natural Dyes and Sustainability*, Sustainable Textiles: Production, Processing, Manufacturing & Chemistry,
https://doi.org/10.1007/978-3-031-47471-2_1

shifting to modern techniques like the shift from natural to synthetic dyes. Despite its advantages, the employment of synthetic dyes is hazardous to the environment, and the achievement of sustainability has become difficult. Thus there are slow fashion movements and sustainable living principles to switch to using natural products in the textile industry. Production of natural dyes from indigo, madder and weld plants is economically feasible, environmentally friendly and hence sustainable. In this chapter, the role of natural dyes in traditional textiles and their aspects of sustainability have been discussed.

Keywords Natural dyes · Traditional textiles · Indigo · Madder · 3R

1 Introduction

Textiles are one of the basic human needs, and it has been an essential requirement for mankind since the beginning. During the early days, people made use of leaves and barks for making clothes. The creation of textiles evolved gradually. The methods and techniques adopted for fabricating textiles varied across people of different regions. Natural fibres from plant and animal sources remained as their basic raw materials for a very long time until the development of synthetic fibres. These ancient techniques were passed on through generations through which the tradition of the textiles was preserved. Traditional textiles are recognized as the cultural identity of people of the respective locality. There are numerous traditional textiles across the world among which a few traditional textiles have been briefly discussed in this chapter. Natural dyes have been an important element in traditional textiles. People made use of the natural resources from their surroundings to dye the fabric and enhance the look of their textiles. Dyes extracted from plants, animals and mineral resources are termed natural dyes. Though they have low fastness properties, they have many advantages like antibacterial properties, medicinal values and sustainability. The dyes being sustainable, and leading to lower carbon footprint, is the most important factor for using them. In the ancient period, the extraction of dyes from natural sources remained a manual process. With the development of technology, the efficiency of extraction has widely improved. Mordants were incorporated along with the natural dyes during its application for improvement in bonding between the dye and the fabric. The extraction, classification, properties and application of natural dyes in traditional textiles are discussed in this chapter. In the present state, a gradual shift is observed towards functional garments and a healthy lifestyle rather than fashion in textile industries. Natural dyes are considered to be sustainable because the waste produced during the dyeing process inflicts minimal or no harm to the environment. In this chapter, the history of traditional textiles and sources and extraction of natural dyes are discussed. The role of natural dyes in traditional textiles is detailed along with their applications. The scope of natural dyes in promoting sustainability is discussed elaborately.

2 Traditional Textiles

2.1 *History*

Textiles fabricated using the ancient form of techniques, dyed using natural sources, and designed using traditional methods are referred to as traditional textiles. Textiles have existed since the 3rd millennium BCE. Flax and hemp are known to be the most ancient plant fibres used for creating textiles. Linen finds the most important place in ancient Egyptian textiles [32]. The existing studies show the early existence of silk cloth in China region during the Neolithic period (5000–1700 BC) [24]. It is found that people have had knowledge on improvising the bonding between dye and fibre with the help of mordants during 2500–1500 BC. During the Vedic period, there were textiles of various colours and the prevalence of embroidery during that time is known through Vedic texts. Further references show that the fabrics were also printed in this era. Post Vedic period, the sources of materials for making textiles were different for two divergent groups. One group used skin as sourcing material while the other used cotton. The epic Mahabharata mentions the usage of silk fabrics and the epic Ramayana mentions the usage of fur, woollen clothing and silk fabrics of various colours. During the fourth century BC, fibres like cotton, silk, linen, white bark, skin, and wool and plant fibres were used in making textiles. During the seventh century, the tie and dye process was employed in manufacturing costly fabrics. The discovery of printed resist dyed cotton fragments from the fifteenth to the sixteenth century shows the evolution of dyeing textiles.

2.2 *Traditional Textiles Across India*

The traditional textiles in India are not particular to a single region and are spread over the entire landscape of India. In this section, the traditional textiles across India are detailed.

2.2.1 North

2.2.1.1 Tribal Textiles

The clothing of tribal people is a great reference for traditional textiles because of their total dependence on natural sources for satisfying their needs over man-made developments. The tribal textiles of India are considered to be one of the oldest in the world. The design and craftsmanship of Indian tribal textiles have remained strong and beautiful for thousands of years [38]. Various techniques and practices were followed by various groups of tribes across the country depending upon their culture, climate in their geographical locations and available resources. There are various tribe groups present in Northern India, and the textiles of a few among the tribes are explained below:

Karbi is one of the main tribal groups of Karbi Anglong and West Karbi Anglong districts of Assam. Based on the survey conducted by Karolia and Ladia [25], the weavers of the Karbi tribe used their traditional fly shuttle wooden looms to create textiles. They made use of cotton yarns and eri silk yarns for making textiles. For dyeing the textiles, the Karbi tribe made use of three colours in particular: white, indigo black, and red. It is to be noted that the colours used were natural dyes extracted from natural wild herbs. Generally, the tribes use motifs that are highly inspired by nature.

The Biates are one of the oldest hill tribes of Northeast India. According to the survey conducted by Karolia and Ladia [25], the Biate weavers use back strap looms for weaving textiles, and for dyeing, they extracted blue dye from indigo plants and red colour by boiling the cloth with roots of various plants in the jungle.

2.2.1.2 Sambalpuri Sarees

Sambalpuri textiles is one of the ancient textiles of Odisha. It is believed that the craft of ikat has been practised since the twelfth century. In Sambalpuri ikat sarees, initially, the yarns used as warp and weft undergo a tie and dye process and then the weaving is done. The traditional Sambalpuri sarees are thematic, and they derive inspiration mainly from the surrounding nature or from great mythological epics and folk tales of Odisha which makes this textile one of the best examples of cultural sustainability. Cultural sustainability could be described as the preservation of cultural identity. The handloom weaving technique is employed in creating Sambalpuri ikat textiles [40].

2.2.1.3 Handloom Sarees of Bengal

The traditional handloom sarees in Bengal have many varieties where Baluchari, Dhaniakhali and Santipore find a very significant place. The process of fabrication makes these traditional sarees exceptional. Baluchari sarees are silk sarees from West Bengal created by weavers experienced in operating drop box looms. The pallu of a baluchari saree is known to contain brocade figured motifs of contemporary times. The inspiration for the motifs used in these sarees is mostly based on nature, Islamic patronage, British patronage and Hindu epics. The Dhaniakhali sarees are described as having near-opaque off-white ground having borders emphasized by fine stripes above. These sarees are distinguished by their compact texture.

2.2.1.4 Chanderi Sarees

The origin of Chanderi sarees is from Chanderi district of Madhya Pradesh. Chanderi sarees have a transparent texture and hand-woven motifs. These sarees are woven using a common throw shuttle pit loom. Chanderi sarees are plain weave fabrics of degummed or raw silk warp yarn and cotton weft yarn. In earlier days, pure silver or gold were used as zari threads. These sarees are known to be delicate and elegant.

2.2.2 South

2.2.2.1 Kodalikaruppur Sarees

These sarees get their name from the village name which is Kodalikaruppur. These sarees are woven using the jamdani weaving technique using a simple throw shuttle pit loom. The weavers make use of unbleached cotton weft for producing ground fabric, and the figure formation on the fabric is done by interlacing the zari threads. This process is followed by kalamkari drawing–dyeing process. This process consists of scouring, bleaching, mordanting, outlining the design and dyeing using colours in the following order: brown, red, indigo blue and finally yellow [36]. It is to be noted that all the stages are carried out using natural ingredients which are tabulated in Table 1.

2.2.2.2 Paithani

Paithani saris of Maharashtra are tapestry woven silk saris created by indigenous weavers. The existence of Paithani sarees during the olden days is proven by the mentions of Rig Veda about a golden woven fabric and Greek records about Paithani fabrics. Paithani saris were woven using a fly shuttle frame loom. Based on the weaving process, they are classified into two types, Traditional Paithani and Kadiyal Paithani. These sarees contain nature-inspired motifs and a Narali border of contrast or matching colour [5].

2.2.2.3 Kancheepuram Sarees

Kancheepuram silk sarees are famous all over the world mainly because of their unique designs and are entirely handwoven by weavers of Kancheepuram. The weaving technique of Kancheepuram saree is unique and it is known to be passed

Table 1 Process and natural materials used for each process

Process	Natural materials used
Scouring	Reetha and shikakai
Bleaching	Sheep or cow dung, lime
Mordanting	Gall nuts, myrobalan and buffalo milk
Outlining the design	Charcoal powder Iron pieces and jaggery
Dye (red)	Manjistha
Dye (blue)	Indigo plant
Dye (yellow)	Myrobalan flowers

through generations. Kancheepuram sarees are woven with dyed mulberry silk yarns, and the designs are created using zari threads containing 57% silver and 0.6% gold [15]. These sarees are stronger than others because they are woven with three single silk yarns twisted together. Kancheepuram sarees are heavier in weight and contain wide borders and solid 'Mundhi'. The creation process of Kancheepuram sarees is long, complex and tedious and hence it requires skilled weavers. Traditionally, punching the design was carried out manually followed by weaving of the saree. But nowadays, efficiency is improved by computerizing the creation process of the punch card using punching card software known as 'Techmen' [6]. Inspiration for these sarees is derived from the architectural excellence of the temples located in this city [26]. These sarees are thereby considered to play an important part in preserving the culture and heritage of this region.

3 Natural Dyes

Natural dyes are colouring substances obtained from either animal or vegetable matter. Natural dyeing was one of the oldest methods practised by the people of the early days [39]. Indians are considered to be the ancestors of dyeing using natural sources [48]. The art and craft of dyeing textiles using natural dyes is practised by many craftsmen across the country. In most cases, the natural dyeing process requires the usage of mordant. This is due to the substantivity property being low for the majority of these dyes while dyeing cellulosic and other textile fibres [14]. Natural dyes are harmonious with nature and they pose no hazard to the surrounding nature and health; in fact, they also act as health-care substances in some cases. Natural dyes are known to possess a lot of health benefits to the wearer.

3.1 Natural Dyes from Vegetable and Animal Sources

Up until the mid-nineteenth century, only natural resources were used to extract the dyes. Natural dyes are derived from several plants and herbs like roots, bark, leaves, flowers, seeds and fruits. Around 450, dye-yielding plants are present in India. Although plants exhibit a variety of colours, all the colours cannot be used because of the lack of a few of the properties required by the dyes [48]. The term *tinctorius* or *tinctorum* or similar can be found in many systematic names of plants capable of yielding dyes which means, *the dyer's* [44]. The shade of the dye derived from plant sources varies depending on the year the plants are picked, the type and conditions of the soil used for cultivation of the plant, cultivating methods, etc.

A few sources and colours of natural dyes are given in Table 2.

Table 2 The source and colour of natural dyes

Source of natural dye	Colour of natural dye
Dyes of madder	Red
Saffron	Yellow
Woad	Blue
Bark of oaks, walnut hulls and oak galls	Brown to black
Fustic	Daffodil yellow to deep gold to orange
Coccus cacti	Red, pink and purple
Dyers' bugloss	Red
Weld	Yellow
Indigo	Blue
Logwood	Blue
Brazilwood	Red
Acacia catechu	Brown
Bixa Orellana	Red
Jackfruit	Yellow
Onion	Yellow
Barberry	Yellow

3.2 *Natural Dyes from Mineral Sources*

Here, the colorants are extracted from natural organic compounds [31]. Mineral colours are generally insoluble in water but it is made to precipitate through the fabric by means of double decomposition. A few examples of minerals used for this dyeing are cinnabar, red lead, zinc white, etc. [37]. Geru and sulphide of mercury are examples of natural red dyes derived from mineral sources.

3.3 *Natural Dyes on the Basis of Application Methods*

- Direct/substantive dye: Direct dyes can be applied on all natural fibres. The stains obtained from berries, barks, leaves and roots come under this category.
- Acid dye: Wool and silk are dyed using these dyes.
- Basic dye: Basic dyes are used to dye woollen fabrics, silk and tannic acid–treated cotton [37].
- Mordant dye: Mordant dyes can be applied onto both cellulosic and protein fibres from plants and animals. Many natural dyes come under the group of mordant dyes. One of the examples of mordant dye is 'madder root', the plant *Rubia tinctorum*. It contains alizarin and is used for creating many shades of red. Similarly, 'Kermes' from the Mediterranean scale insect *Kermes ilicis* and 'Lac' derived from the insect *Coccus lacca* are some of the examples for mordant dyes creating red colour [44].

- Vat dye: Vat dyes are suited to dye both cellulosic and protein fibres but mostly cellulosic fibres. One of the sources of the dye is the *Indigofera tinctoria* plant, and it was first introduced by Phoenician traders and people migrating to the Mediterranean area. In the process of dyeing with indigo, the glucoside indican is converted into water-soluble indoxyl and the indoxyl is made insoluble by oxidation with the help of sunlight. Tyrian purple is another example of vat dye which was extracted from Mediterranean shellfish of Purpura and Murex species. A reddish-purple natural dye is obtained [44].

3.4 Natural Dyes on the Basis of Colour

Natural dyes are also classed based on the category of the colour it imparts over a fabric. Colour of the natural dye depends upon the type of coloured constituents present in the source. For example, Indigotin present in the plant Indigo is responsible for the blue colour. Various colours of natural dyes like yellow, red, blue, brown, black and many more are present. It is to be noted that the number count of plants yielding yellow dyes is more when compared to plants yielding other colours [37].

3.5 Uses of Natural Dyes

Natural dyes are used for the purpose of dyeing and printing fabrics using various techniques like block printing and Kalamkari [14]. The development of synthetic dyes during the nineteenth century caused a revolution and led people to drift towards it due to its advantages in terms of cost, application methods, etc. But in recent days, awareness has been created of the environmental impacts caused by the dyeing process involving synthetic dyes and chemicals, and this has led to an increase in the initiation of usage of natural dyes among many worldwide clothing brands. This is because natural dyes are beneficial to the environment as they are biodegradable and eco-friendlier than synthetic dyes which have many positive impacts on the surroundings. In fact, there is a huge reduction in the involvement of harmful chemicals in textiles [33]. Various natural dyes are also known to provide UV protection. Based on the study conducted by Hou et al. [19], wool dyed using orange peel is said to exhibit about six times higher UV protection factor compared to synthetic dyed wool. Another interesting information is that the UV protection of these fabrics is good even after 30 wash cycles compared to the fabrics dyed using synthetic colourants. Many scientific researches are carried out based on natural dyes. Apart from its application in the textile industries, natural dyes are also used in the cosmetics and food industries. For example, among the well-known brands, Yves Saint Laurent developed a line of makeup using natural dyes [7].

Table 3 Natural dyes, colour produced and their medicinal importance

Natural dyes	Colour	Medicinal importance
Turmeric	Yellow	Antioxidant, anti-inflammatory, antifungal, antiviral and antibacterial
Saffron	Golden yellow-orange	Antispasmodic, eupeptic, nerve sedative, antidepressant, antitumor, etc.
Annato	Reddish orange	Antimicrobial, anti-clastogenic, anti-pruritic
Pomegranate	Red-purple	Anticarcinogenic, antimicrobial, antiviral, antioxidant
Tomato	Red	Prevents all types of cancer from the body and acts as an effective antioxidant
Tagetus	Yellow to orange	Digestive, diuretic and stomachic
Henna	Red	Anticarcinogenic, anti-inflammatory, antipyretic and analgesic

3.5.1 Medicinal Properties of Natural Dyes

Natural dyes can provide skin benefits to the wearer. Antibacterial properties are highly desirable for textiles used for medical purposes [16]. This is achieved by the usage of natural dyes for dyeing textiles. Other than the medical field, natural dyes are known to possess more medicinal benefits in the pharmaceutical industry. Many dye-yielding plants are medicinal and some of these plants also possess remarkable antibacterial and antifungal properties. Some of the dye-yielding plants and their medicinal properties are discussed in Table 3.

According to the study by Gupta et al. [16] on testing the antimicrobial properties of natural dyes, it is stated that the dyes extracted from kamala, pomegranate and gall nuts were observed to be more effective biocides when they were used for dyeing of cotton material, among which gall nuts were found to be more efficient. The antimicrobial effect of these dyes is assumed to be durable in practice as they possess good wash fastness properties.

4 Extraction of Natural Dyes

4.1 Evolution of Extraction of Natural Dyes

Since ancient times, dyes and colorants extracted from natural sources have been used for the coloration of hair and skin, festive food, clothing and furnishing textiles, etc., in our day-to-day life. During the olden days, the most commonly used dyes in India were madder, indigo and saffron [7]. Natural dyes are obtained from different parts of the plants like roots (e.g. beetroot, turmeric, onion); bark/branches (e.g. purple bark, sampan wood, sandalwood); leaf (e.g. indigo, henna, eucalyptus, tea, cardamom, coral jasmine, lemon grass); flowers (marigold, kusum) and fruits/seeds (e.g. pomegranate rind, bell nut, myrobalan). Natural dyes are also extracted from insects, animals and minerals with no chemical treatment. Microorganisms like fungi, algae and bacteria also act as sources for the extraction of natural dyes. These dyes are environmentally friendly and are known for coloration in a single stage which makes the treatment process less expensive [27]. Indigo dyes that are obtained from natural sources (e.g. shellfish and lichens) produce shades of blue and purple. The red colour was imparted using derivatives of anthraquinone that are obtained from plants or insects. Always derivatives of flavonoids obtained from a wide range of plant species produced yellow shades. Over-dyeing of primary colours was widely used in obtaining the secondary and tertiary colours, for example, over-dyeing of blue and yellow dyes resulted in green hues. New plants are discovered to obtain new colours to interpret the dyes obtained synthetically.

The demand for indigo is increasing as it is used as a colorant for denim. Green, blue and violet shades can be obtained from indigo when it is mixed with other substances. Synthesis of indigo can be done both naturally and chemically. *Couroupita guianensis* tree contains cannon ball fruit which is cut in half and fermented. After fermentation the powder is dried and crude dye powder is obtained. The powder is used to dye the fabric at room temperature for 1 hour [50].

4.1.1 Methods of Extraction of Natural Dyes

The process involving extraction of colorants from natural sources is more complex. The extraction techniques of every dye are specific. A study by Chungkrang et al. [8] reported that there were different techniques employed in deriving dyes from natural resources. In aqueous extraction, if the raw materials are in a wet form then they are ground into a fine paste, then to remove the non-colouring matters they are boiled and filtered. When the materials used are in a dry state, they are ground into powder form and immersed in water to loosen up their structure. They are used to apply colours to various things. Some disadvantages are they are time-consuming and need more water, high temperatures, etc. The extraction of yellow dye from extracts of mango trees by using the aqueous extraction method is one of the examples of sustainable practices as this method is found to cause minimal pollution to the environment [4]. The Soxhlet extraction method is found to be conventional and is known for producing value-added natural dyes. For example, the dye from Gunda Gundo Orange peels is extracted using this method. Here the dried samples of orange peel and ethanol (organic solvent) are raw materials. Based on the findings, the dye imparts good colour strength, and it is considered to be sustainable as it consumes less energy and reduces environmental pollution [52]. Dye from weld is also extracted using the Soxhlet method and is commonly used for dyeing fibres like wool. This dye is found to be non-toxic than synthetic dyes and thereby it is considered to be sustainable [28]. In acid or alkaline extraction, the dyes are extracted under acid and alkali medium. Dilute alkali and acids are used here for hydrolysis of glucosides for better extraction. Tesu natural dye from the tesu flower is extracted using the acid method. The alkaline method is used to extract dyes from the compounds containing phenolic groups. In the ultrasonic microwave extraction method, a faster rate of extraction in less time can be achieved. Here the raw materials are treated using microwave or ultrasonic rays to extract the dyes.

Indigo dyes from the *Couroupita guianensis* tree and colour from turmeric are extracted by the fermentation method. In this method, microorganisms present in the raw materials carry out the fermentation process and the dye is extracted. This method is time-consuming and creates a foul smell due to microbial extraction. In the enzymatic extraction method, the binding materials in the plants are loosened with the help of enzymes like pectinase, cellulose and amylase. It is one of the most sustainable ways of extracting natural dyes from plants. The aqueous extraction method is considered to be more effective than the solvent extraction method. Organic solvents like acetone and petroleum are used to extract the dyes. Low temperature and less amount of water are required and the dye extracted is of good quality, which are some of the advantages of this method.

5 Properties of Natural Dyes

5.1 Dye Uptake

Dyes extracted naturally from plants tend to show antimicrobial and antioxidant properties. These properties are said to enhance their role in textile processing. Though the usage of natural dyes is considered to be more sustainable than synthetic dyes, the low affinity of these dyes restricts their usage in the textile industry. Various techniques for modifying the surface of the fabric which increases their affinity towards the dyes have been developed. Plasma treatment, gamma radiation, microwave radiation, ultraviolet radiation and enzyme treatment are some of the examples of the techniques employed in improving affinity, which are explained below:

5.1.1 Plasma Treatment

Plasma treatment is an environmentally sustainable approach to improve the dye uptake of natural dyes with fibres. Plasma treatment is considered one of the best alternatives for textile wet processing since it reduces energy, water, and chemical consumption and generates less pollution. Plasma modifies the textile surface through physical and chemical changes. Depending on the working temperature, plasma can be classified as hot and cold plasma. Dave et al. [9] suggested that the processing conditions and careful selection of the plasma gas lead to various surface modifications including the elimination of contaminants, bond breaking, crosslinking, functionalization, post-irradiation grafting, etc. Based on the studies done between 1998 and 2020, the outcome of the improvement of natural dyeing through plasma treatment has increased up to 19.1% for cotton fibres, 16.7% for polyester, 9.5% for nylon, 45.2% for wool and 9.5% for silk [17].

5.1.2 Gamma Radiation

Gamma radiation has a high frequency and shorter wavelength. It has more importance in textile dyeing as it modifies the fabric surface to improve the dye uptake without altering the chemical structure. The internal structure of the plant cell is modified by gamma radiation which enhances the penetrating capability of the plant cell and increases the transfer of colour to the liquid media. At a very low dose, gamma radiation improves the dye uptake and colour fastness of the naturally dyed fabric [1]. The gamma irradiated natural dyed cotton fabric shows better colour strength and fastness. Gamma radiation resists shrinkage, pilling, water repellency, abrasion, wrinkling and microorganism attack.

5.1.3 Microwave Radiation

This is one of the advanced methods adopted in the treatment of fabrics using microwaves. With microwave treatment, the extraction of dyes has not only increased but the colour characteristics of the dye have also improved. When microwave treatment is used for separating the tannins from the neem bark, extraction of colorant becomes more efficient. This is due to the thermal and dielectric properties of matter where microwave heating greatly influences the vibrational energy between solvent and colorant [12]. It has many advantages like inexpensiveness, less energy consumption, reduced exposure time and eco-friendly character. Microwave radiations are capable of enhancing the coloration capability or colour retention capabilities of natural fabrics.

5.1.4 Ultraviolet Radiation

Altering the textile surface with ultraviolet (UV) radiation is one of the distinguished physical modification techniques. Reduction in the mechanical loss of fabric can be achieved by treating the fabric using ultraviolet radiation. UV radiation helps enhance the bond between the dye molecule and the carbonyl and hydroxyl groups of the cellulose, thereby increasing the dye uptake of the fabric [46].

5.1.5 Enzyme Treatment

Enzyme treatment can be performed during various phases of textile processing, and during enzyme treatment various functional groups are introduced in the fibre and the surface of the fibre is imparted to physical and chemical changes. The complete process is considered to be eco-friendlier and sustainable in nature, and it is known to consume low energy, require less water and increase the final quality of the product [46].

5.2 Fastness Properties

Natural dyeing is economical, widely accessible and sustainable in nature. This increases the importance of research in this sector in the textile and clothing industry. Dyeing the fabrics using natural dyes is always challenging because of the limitations in their fastness properties. Much research is being done to increase the fastness properties of natural dyes. Ford et al. [13] reported that during dyeing better colour fastness property was exhibited by dyestuffs like indigo and lichen. Mostly mordants are incorporated in the dyeing process to get a strong bond between the dye and textile fibre. They are done in three ways which are pre-mordanting, simultaneous mordanting and post-mordanting. Pre-mordanting is the process done before dyeing whereas simultaneous mordanting is the process that is carried out while dyeing. Post-mordanting is the process that is carried out after dyeing. The process is discussed in the later sections. Generally, natural dyes have poor colour fastness. The colour fastness of the dye is directly related to the dye molecule and the fibre interaction. A study by Silva et al. [10] reported that wool fabric tends to weaken and become yellower on exposure to light when dyed with natural dyes and the colour fastness of cotton fabric decreases by using aluminium mordants. Reasonable colour fastness properties were achieved when the bleached cotton fabric was dyed using natural dyes extracted from *Acacia nilotica* [3].

The wash fastness can be enhanced by adding mordants. A covalent bond is formed between the dye and the fibre for wool and silk fibres with the help of tannins. These tannins form metal chelation with different mordants as they have a phenolic structure. Tannins are insoluble in water after mordanting, thereby improving their wash fastness. Pre-mordanting increases the dye adsorption on cotton, but the surplus amount of dye adsorbed could be removed by washing. A study by Jothi [21] reported that as the dye extracted from marigold aggregates inside the fibre, it exhibits excellent wash fastness.

In the case of rubbing fastness, natural dyes do not require any aftertreatment as most of the natural dyes are said to have moderate to good rub fastness [41]. Rubbing fastness also depends on the types of mordants being used. A study by Umbreen et al. [51] reported that cotton fabric when dyed with turmeric dye is found to have good rubbing fastness without the use of any mordant, and when dyeing is done with the use of mordants, the rubbing fastness of cotton fabric decreases while the other fastness property increases. Also, natural dyes from Cochineal, Indigo, Prussian Blue, Madder and related dyes show adequate light fastness by modern standards [34]. The light fastness of a dyed fibre usually increases with an increase in dye concentration. Light fastness is improved by using mordants.

5.3 Antimicrobial Properties

Generally, natural dyes inhibit antimicrobial properties and are more suitable for medicinal purposes. Apart from dyeing, dyes extracted from pomegranate rind are used for bactericidal treatment of cotton fabric. Based on research on the application of chitosan on fabrics, it has been observed to exhibit antimicrobial properties and better functionality [27]. Natural dyes containing naphthoquinones or a large quantity of tannins are known for their antimicrobial properties [45].

6 Advantages and Disadvantages of Natural Dyes

6.1 Advantages of Natural Dyes

Natural dyes are mostly free from toxic substances and are non-allergic because they are obtained from naturally available sources (plants and animals) without the addition of any chemicals during their processing. Natural dyes are skin-friendly and non-carcinogenic. Some natural colours also have medicinal effects on the skin. These dyes are biodegradable thereby not causing any damage to the environment and are renewable. Few natural dyes exhibit antibacterial properties, resistance to UV light and non-allergic properties, making them more advantageous compared to synthetic dyes. Natural dyes can be extracted and purified easily. They are economically feasible and sustainable. An extensive range of colours is obtained from natural dyes by mix and match system. The colours produced from natural dyes give a soothing feel to the eyes. There is no problem in disposing of natural dye waste as the waste generated from the process is being used as fertilizers for agriculture. As the production of natural dyes is a labour-intensive process, it creates more employment opportunities in rural and sub-urban areas. Usage of dyes from natural resources has reduced the utilization of fossil fuels. Natural dyes can also be used to dye some synthetic fibres [1]. In general, dyeing using synthetic colorants produces a large amount of wastewater which is required to undergo a purification process. This purification process would not be required in natural dyeing due to the production of minimal or low wastewater which is an additional advantage of natural dyes [33]. Ancient and traditional dyeing technologies involve the usage of natural dyes. Natural dyes are suitable for the preservation and re-establishment of the heritage of old textiles.

6.2 Disadvantages of Natural Dyes

Natural dye resources are limited worldwide. The available quantity of natural dyes is still 1% of world demand [29]. The application of the dyes requires skilled workmanship, so they are costly. Dyes obtained from wild plants are not suitable for commercial textile processing. The dyes are mostly obtained from plants. The process of extracting dyes without inflicting damage to the plant still remains difficult. The natural pollination and reproduction cycle of plants are affected by the use of flowers in obtaining natural dyes. The affinity of natural dyes towards the fibre is low, thus after dyeing most of the dyes remain in the bath. Some of the mordants used to increase the fastness properties may be harmful. For obtaining required shades or darker shades more quantity of natural dyes is required. Plants contain a low amount of natural dye pigments. It is said that for dyeing 1 kg of fabric, 1 kg of dye plant is increased. This increases the need for more biomass for obtaining them which increases the quantity of waste. To eliminate the unfixed dyes after dyeing, a large amount of water and heat are required. Only the quality and stability of some of the natural dyes and pigments such as indigo, tannins and alizarin dyes can compete with synthetic dyes.

7 Role of Natural Dyes in Traditional Textiles

7.1 Natural Dyes in Traditional Textiles

Till the nineteenth century, only natural dyes were used for colouring textile materials. Based on the availability of resources in that particular region, different regions adopted different natural dyeing traditions. In India, the origin of Indian textiles was from the Indus Valley Civilization, and the people of this civilization utilized homespun cotton and applied indigo to bring out the colour in the fabric. Natural dyes played an important role in dyeing these traditional textiles due to their easy extraction of colour and their biodegradable and antimicrobial properties, and they are also known to provide higher UV absorption, thereby protecting the skin from the sun's harmful rays [43]. As seen in the previous parts of the chapter, there are many traditional textiles used across the world, which include khadi also called khaddar, kalamkari, ajrakh, etc. These traditional textiles have their own characteristics, different sources and modes of dyeing. Dyeing of different traditional textiles in earlier periods was mostly done with natural dyes obtained from various resources that were available in that particular region. The sources included indigo, saffron, sandalwood, sapanwood, Brazilwood, jackfruit, pomegranate, etc.

7.1.1 Dyeing of Khadi

Khadi is a handspun versatile fabric of India, which is symbolized to be the Indian textile heritage. Khaddar is also considered to be the most sustainable fabric since its manufacturing process does not consume any electrical energy or fossil fuel and does not produce any toxic waste; additionally, for producing a meter of fabric, only three litres of water is consumed.

When it comes to dyeing this fabric, to keep the process natural and sustainable, it is dyed with organic vegetable dyes without the use of chemicals. One such example is using pomegranate peel for dyeing, which is considered to be a more effective colorant on khadi fabric. A study was conducted by Sinnur et al. [47] on using pomegranate peel as the natural colorant as well as a UV protective and antimicrobial compound; the extract of pomegranate peel was applied on the khadi fabric along with mordants to obtain the required shade with good colour fastness and UV protection in a natural and eco-friendly method. The study concluded that the khadi fabric dyeing with pomegranate rind seed increased the fabric UPF.

7.1.2 Dyeing of Ajrakh

Ajrakh is known to be the famous textile of the Khatri communities of Dhamadka and Ajrakhpur in Kutch. 'Ajrakh' in Arabic means blue sky and it has got this name due to its printing process. It is a block-printed double-sided fabric which is created by craftsmen of this community, and it is known for the arrangement of geometrical and floral motifs, mostly dyed with indigo and madder and usually worn by the people residing in north-west India and Pakistan.

In the mid-1970s ajrakh was utilized as soft furnishings and it became one of the popular textiles in urban India; later in the new millennium, it gained popularity as a fashion accessory in both national as well as international markets [11]. Various stages of dyeing and resist printing using natural dyes and mordants are involved to produce a traditional ajrakh fabric (khatri). Once the fabric undergoes preparatory processes, using resist printing, the dye is applied on the fabric. The outline of the design is prepare dusing a natural resist paste that consists of lime and gum arabic. The inside black design is obtained by using a ten day old solution of water, jaggery and rusted iron. The solution is boiled to make a paste along with tamarind seed powder. The red colour is obtained by using a paste of alum, multani clay, gum arbic and millet flour. Once the required white areas are covered using the resist paste, the whole fabric is dyed in indigo and dried.

7.1.3 Dyeing of Batik

Batik is an Indonesian technique of wax dyeing, and the main material that is used in the manufacture of this fabric is wax, which is used to resist the colour in the design area. The fabric undergoes many processes, including designing the pattern on the fabrics, applying wax, colouring, fixing, wax removal, rinsing and drying of the fabric. The natural Batik is said to have a faded and dull effect and is also considered to have a soft feel; it also requires repeated dyeing (about 8–10 times) to achieve the desired colour, but this is considered to be more sustainable as the dye is being reused for dyeing the next set of fabric and thus does not contribute waste to the environment [49]. The dyeing process may take place in two ways: either natural dyeing or reactive dyeing can be used. In the early days, to produce batik they used materials from the surroundings, for example, naturally available dyes and natural colorants (indigo, vegetable dyes) were used and were stabilized and fixed on the fabric using mordants [20]. The processes that are carrie dout are pre-mordanting, wax application, colouration, dye interlocking in the fabric and wax removal.

7.1.4 Dyeing of Paithani

Paithani is one of the traditional Indian textiles that is fully handwoven in the rural areas of Maharashtra. This type of fabric is produced with intricate patterns and contrast borders on silk. Like other traditional textiles, these fabrics are coloured using extractions of natural resources. The bark of peepal (*Ficus religiosa*) is responsible for the brown-maroon colour. In a study by Kale et al. [22], the dye was extracted from the flowers of flame of the forest and the bark of peepal. These flowers were dried and powdered to make the dyestuff and then the dyeing of the silk fabric was carried out in the presence of an alkaline medium for about 30 minutes in each extract along with the usage of mordants. The fastness properties were enhanced in the case of post-mordanting of the silk fabric [22].

7.1.5 Dyeing of Bandhani

Bandhani is the traditional textile of Rajasthan and is known for its tie and dye method of coloration that imparts fine figurative patterns on cotton fabric. Different patterns can be obtained by varying the method of tying the fabric. In the olden days, these textiles were coloured using natural dyes, and mainly indigo was preferred. In the traditional method of dyeing, the cotton fabric (soaked in water for 5–10 minutes) was added into the dye bath that consisted of the dyestuff, soda ash, and salt and treated at a particular temperature, and then the temperature was raised gradually. The fabric is left in the bath for 10–15 minutes. After that, the fabric was taken out, rinsed and dried.

7.2 Application of Natural Dyes in Traditional Textiles

Dyes can be applied in numerous ways and methods on different types of fabric. The evolution of synthetic dyes and their availability at lower costs led to the ignorance of natural dyes in the early days. Later in modern days, due to increasing awareness and consciousness about the environment, the usage of natural dyes for the coloration of textiles is again gaining traction. Natural dyes can be applied on textiles either by means of dyeing or printing.

7.2.1 Natural Dyes in Dyeing

Natural fibres are said to have mainly two different resources from which they are extracted, plant and animal origin. The application of natural dyes on animal-originated fibres that are protein-based (wool, mohair, silk) has greater affinity, whereas on plant-originated fibres (cotton, linen, jute) it has lesser affinity. Based on the affinity of the textiles towards the natural dyes, different mordanting techniques are used for incorporating colour to the fabric.

Natural dyes are said to have poor fastness and to improve this mordants are used. Mordants, also known as dye fixatives, enhance the binding of the dye onto the fabric by forming a coordination complex with the dye, improving the dye uptake, light and colour fastness. When mordants are used in different ratios they produce different shades of colour on the fabric. Apart from the mordants, the colour strength also depends upon the metal salt used. There are three methods of mordanting:

- Pre-mordanting
- Post-mordanting
- Simultaneous mordanting

In pre-mordanting, mordants are treated along with the sample before the application or addition of dyestuff into the bath. In post-mordanting, the dyed fabric is taken from the dyebath, then squeezed and left to dry; once it is dried, the fabric is then treated with natural mordants. Whereas in simultaneous mordanting, both the dyeing of fabric and treating the fabric with mordants take place simultaneously. Hence, mordanting remains an essential process in natural dyeing. Eco-friendly mordants like alum and iron can be preferred for mordanting, making this process more sustainable [33].

In a study conducted by Kanchana et al. [23] on dyeing with marigold, clitoria, and pomegranate peel, the light and rubbing fastness of the marigold-dyed fabric was found to be higher. In the conventional application of natural dyes, natural mordants like wood ash (akali mordant), acidic fruit leaves (acid mordant), myrobalan and cow dung were highly used. For the application of natural dyes, there are numerous methods, but the highly preferred method is vat dyeing which is widely known as indigo dyeing [2].

7.2.2 Natural Dyes in Printing

Printing involves the application of colour pigments in the form of design. Usage of synthetic dyes could be replaced with natural dyes to produce environment-friendly printed fabrics [42]. Soft tones of colours can be achieved by using natural colorants in printing. Natural dyes are printed on textiles using various techniques. The fabrics printed using digital techniques are said to possess good fastness properties as compared to the fabrics printed using traditional methods [53]. Inkjet printing is considered to be one among the sustainable methods as there is a considerable reduction in the emission of effluent waste and consumption of energy [42]. Natural dyes are used for printing in various traditional textiles. For example, the print in Adinkra garments involves the usage of natural dye extracted from the Badie plant. In the current textile industry, various natural thickeners are used in the printing process. Starches from natural sources like wheat, rice and aloe vera are some examples of natural thickeners used. In a study on printing cotton fabrics using extracts of lacquer and harda, the observations state that the wash fastness of the printed fabric is good. Another water-soluble natural dye, annatto is used in printing fabrics, and the dry cleaning and rubbing fastness of these fabrics are said to be excellent [53]. Bagru printing is one such printing technique that originated from Bagru in Rajasthan where hand block printing is done using natural colours. The blocks used are of wooden type where the design is first engraved on the blocks and then the blocks are dipped in the printing paste. The dipped blocks are then placed on the fabric to get the required print design. The dyes used for the printing paste are from natural materials like dabu (mixture of mud) and iron. They use vegetable dyes mostly for which the recipe preparation has been preserved and passed on for generations now. These materials are sustainable and do not pollute the environment.

8 Sustainability of Natural Dyes

8.1 Sustainability

Sustainability is mostly used to indicate the practices that are said to be environment-friendly. In the textile industries, sustainability is achieved by following the 3R's principle (reduce, reuse and recycle) which has now extended and become the 5R's that additionally includes re-design and re-imagine that are taken to be the major strategy for eco-friendly textile processing. Among these, the reduce concept plays an important role in reducing the waste due to the textile processes, thus decreasing the various consequences that arise because of the waste emitted. The concept of recycling or upcycling is also said to have the top role in the circular economy which is also known as the cradle to cradle economy, where the resources are obtained and are used to make the required product and that is recycled again to produce a new product which in turn reduces the amount of raw material extracted. Sustainability consists of three mutually dependent goals, that is, to achieve a life-style that is environmentally, economically and socially sustainable [18]. A sustainable society is said to be able to maintain the environment the same way for some centuries from now. This brings us to a more particular term—'sustainable development'.

8.1.1 Sustainable Development

In 1987, the term 'sustainable development' was defined by the United Nations' World Commission on Environment and Development as 'the development that meets the needs of the present generation without compromising the ability of future generations to meet their own needs'.

Heinberg and Lerch formulated five axioms of sustainability [18], which included:

- There is a tendency for a society to collapse when the vital resources available to the particular society are used in an unsustainable manner.
- Population growth or growth in the rates of consumption of resources cannot be sustained.
- The usage of renewable resources ought to be less than or equal to the rate of natural replenishment, for achieving sustainability.
- To be sustainable, the use of non-renewable resources must proceed at a declining rate, and the rate of decline must be greater than or equal to the rate of depletion.
- The substances introduced into the environment from human activities need to be minimized and should be harmless to the environment.

Sustainability can be understood as both the process which generates well-being and social health now and in the future and those social institutions that facilitate environmental and economical sustainability.

8.2 Sustainability of Natural Dyes in Traditional Textiles

Natural dyes are important as they are naturally available in abundance, recyclable and decomposable in nature and are considered to be better than artificial dyes as natural dyes do not require toxic mordants like chromium, tin, copper, etc. They are also said to have a low impact on the environment. To go in line with sustainability in wet processing, natural dyes were widely used. Wet processing of fabric involves pre-treatment of the fabric before applying colour, dyeing and the post-processing or finishing processes. The usage of natural dyes in well known ancient dyeing and printing technologies van be one of the best options in sustainable wet processing. Due to the increasing awareness of sustainability, natural dye is widely used in modern days.

The husk and wastewater emitted from the production of Batik during natural colouring are biodegradable and provide nutrients to the environment. The animals and plants benefit from the waste of natural dyes. The natural dyes used are less harmful. The added benefit is that the wax that is being produced can be reused, therefore the usage of these helps in approaching the green vision of sustainable development [20].

8.3 Introducing Sustainability in Modern Traditional Textiles Through Natural Dyes

Sustainability can be introduced by using new technologies that mainly concentrate on minimizing energy consumption, wastewater elimination, and chemicals that are considered a threat to the environment. As all the textile manufacturing processes use a lot of energy and water and cause major impacts, opting for the conventional methods will lead to a reduction in the impact but no process provides zero impact. Hence some development in technology can aid in reducing the environmental load and thereby taking the path towards sustainability.

Many new technologies have been incorporated in the textile industry for the sustainable production of products, and these new technologies include supercritical dyeing technology which uses carbon dioxide as the dyeing medium and an anhydrous process that offers ecological advantages. Other techniques used involve foam coating technology, air dye, plasma treatment, digital printing technology and natural dyeing [35]. Among these, natural dyeing is considered to be the sustainable wet processing that was incorporated in colouring from the ancient period, as the awareness about sustainability and understanding of its importance have led people to look into its advantages and increase its usage. It has become popular in the modern-day industries, replacing synthetic dyes with some additional agents like mordants. Some natural dyes like indigo, madder and weld are considered to have become more popular and classic, and these dyes have the ability to impart any colour except fluorescent colour and electric blue. In today's context, the usage of natural colourants for dyeing fabrics tends to increase the competitive nature of the current textile industries, thereby decreasing the environmental load of dyeing with chemical or synthetic dyes.

Incorporating one-bath dyeing in modern textiles, that is, using both mordants and dye together in a single bath, enables the alteration of colour depth and shade variation by adding the required ratio of a mixture of mordants. This will help in increasing the reproducibility of dyeing compared to the other two-bath dyeing methods, and it also reduces the chemical load that is released into the environment and thus could also pave the way to ecological developments. A step towards sustainability can be taken by reducing carbon footprints in the textile industry. It can be done by incorporating one-step one-bath dyeing, pad dry and pad dry steam processes along with the cationization of cotton salt free dyeing of direct dyes. Also, the wastewaster from these processes can be treated with physical, biological and activated carbon systems [30]. All these methods will help to take a step towards reducing the carbon footprints and paving the way for a sustainable approach.

9 Conclusion

From this chapter, we can conclude that natural dyes have been a characteristic feature of traditional textiles. Developed by ancient methods, these textiles have used natural colourants extracted from a wide range of sources and have also employed various extraction and application methods of natural dyeing. The main properties of natural dyes, their application methods and their uses are discussed, through which their significance is understood. Natural dyes have many advantages like their sustainable nature. These days, various fashion brands are focusing on the incorporation of natural dyes due to the recent environmental awareness among people. The comparatively harmless nature of natural dyes has turned people's attention towards it because of the current increase in environmental pollution caused due to the usage of synthetic dyes in dyeing industries. The methods of application of natural dyes on traditional textiles can be taken as an inspiration by modern textile industries through which sustainability can be achieved.

References

1. Adeel, S., Kiran, S., Rafi, S., Ayesha, T., Fazal-Ur-Rehman, T. G., & Zuber, M. (2017). Radiation pretreatment: A potential novel technology to improve fastness properties of plant-derived natural dyes. In *Plant-based natural products: Derivatives and applications* (pp. 65–87). John Wiley & Sons, Inc., Scrivener Publishing LLC.
2. Ado, A., Yahaya, H., Kwalli, A. A., & Abdulkadir, R. S. (2014). Dyeing of textiles with eco-friendly natural dyes: A review. *International Journal of Environmental Monitoring and Protection, 1*(5), 76–81.
3. Ali, A., Ali, S., Saleem, H., & Hussain, T. (2010). Effect of tannic acid and metallic mordants on the dyeing properties of natural dye extracted from Acacia nilotica bark. *Asian Journal of Chemistry, 22*(9), 7065.
4. Ayele, M., Tesfaye, T., Alemu, D., Limeneh, M., & Sithole, B. (2020). Natural dyeing of cotton fabric with extracts from mango tree: A step towards sustainable dyeing. *Sustainable Chemistry and Pharmacy, 17*, 100293.
5. Bhatia, R., & Desai, D. (2014). Sustaining the traditional paithani saree weaving of Maharashtra. *ITMAR, 1*, 529–536.
6. Bowonder, B., & Sailesh, J. V. (2005). ICT for the renewal of a traditional industry: A case study of Kancheepuram silk sarees. *International Journal of Services Technology and Management, 6*(3–5), 342–355.
7. Cardon, D. (2010). Natural dyes, our global heritage of colors. *Textile Society of America Symposium Proceedings*, Paper 12. Available on http://digitalcommons.unl.edu/tsaconf/12
8. Chungkrang, L., Bhuyan, S., & Phukan, A. R. (2020). Natural dye sources and its applications in textiles: a brief review. *International Journal of Current Microbiology and Applied Sciences, 9*(10), 261–269.
9. Dave, H., Ledwani, L., Chandwani, N., Chauhan, N., & Nema, S. K. (2014). The removal of impurities from gray cotton fabric by atmospheric pressure plasma treatment and its characterization using ATR-FTIR spectroscopy. *The Journal of The Textile Institute, 105*(6), 586–596.
10. dos Santos Silva, P. M., Fiaschitello, T. R., de Queiroz, R. S., Freeman, H. S., da Costa, S. A., Leo, P., Montemor, A. F., & da Costa, S. M. (2020). Natural dye from Croton urucurana Baill. bark: Extraction, physicochemical characterization, textile dyeing and color fastness properties. *Dyes and Pigments, 173*, 107953.
11. Edwards, E. M. (2016). Ajrakh: From caste dress to catwalk. *Textile History, 47*(2), 146–170.
12. Fidalgo, A., Ciriminna, R., Carnaroglio, D., Tamburino, A., Cravotto, G., Grillo, G., Ilharco, L. M., & Pagliaro, M. (2016). Eco-friendly extraction of pectin and essential oils from orange and lemon peels. *ACS Sustainable Chemistry & Engineering, 4*(4), 2243–2251.
13. Ford, L., Rayner, C. M., & Blackburn, R. S. (2018). Comparative sorption isotherms for colorants present in Dyers' madder (Rubia tinctorum L.) provide new insights into historical dyeing. *Coloration Technology, 134*(1), 3–12.
14. Geetha, B., & Sumathy, V. J. H. (2013). Extraction of natural dyes from plants. *International Journal of Chemistry and Pharmaceutical Sciences, 1*(8), 502–509.
15. Gowri, D. P., & Ramachander, A. (2018). Case study: An overview of the growth of Kanchipuram silk industry. *Srusti Management Review, 11*(2), 71–73.

16. Gupta, D., Khare, S. K., & Laha, A. (2004). Antimicrobial properties of natural dyes against Gram-negative bacteria. *Coloration Technology, 120*(4), 167–171.
17. Haji, A., & Naebe, M. (2020). Cleaner dyeing of textiles using plasma treatment and natural dyes: A review. *Journal of Cleaner Production, 265*, 121866.
18. Heinberg, R., & Lerch, D. (2010). *The post carbon reader: Managing the 21st century's sustainability crises* (p. 523). Watershed Media.
19. Hou, X., Chen, X., Cheng, Y., Xu, H., Chen, L., & Yang, Y. (2013). Dyeing and UV-protection properties of water extracts from orange peel. *Journal of Cleaner Production, 52*, 410–419.
20. Hussin, N. S. M., Ismail, A. R., Hasbullah, S. W., & Kadir, N. A. (2020, August). A review on sustainable development and heritage preservation and its conceal detrimental in batik dyeing. *IOP Conference Series: Earth and Environmental Science, 549*(1), 012081. IOP Publishing.
21. Jothi, D. (2008). Extraction of natural dyes from African marigold flower (Tagetes erecta L) for textile coloration. *Autex Research Journal, 8*(2), 49–53.
22. Kale, S., Naik, S., & Devdhar, S. (2009). Dyeing paithani silk saree with natural colours. *Indian Silk, 48*, 24.
23. Kanchana, R., Fernandes, A., Bhat, B., Budkule, S., Dessai, S., & Mohan, R. (2013). Dyeing of textiles with natural dyes-an eco-friendly approach. *International Journal of ChemTech Research, 5*(5), 2102–2109.
24. Kaplan, P. (2016). *Silk stories*. Doctoral dissertation, The George Washington University.
25. Karolia, A., & Ladia, B. (2012). Traditional textiles and costumes of Karbi and Biate tribes of Meghalaya. *NISCAIR-CSIR, 11*(2), 309–316.
26. Kasinathan, K. (2011). A sustainable synergy of traditional and CAD based silk saree designing/production. In *Fashion & luxury: Between heritage & innovation*. Institut Français de la Mode (IFM). Retrieved from http://www.iffti.com/downloads/past_conferences/IFFTI%20 ABSTRACTS%20BROCHURE,202011
27. Kasiri, M. B., & Safapour, S. (2014). Natural dyes and antimicrobials for green treatment of textiles. *Environmental Chemistry Letters, 12*(1), 1–13.
28. Kavyashree, M. (2020). Printing of textiles using natural dyes: A global sustainable approach. In *Chemistry and technology of natural and synthetic dyes and pigments*. IntechOpen.
29. Křížová, H. (2015). Natural dyes: Their past, present, future and sustainability. In *Recent developments in fibrous material science* (pp. 59–71). Kanina Publishing.
30. Manickam, P., & Duraisamy, G. (2019). 3Rs and circular economy. In *Circular economy in textiles and apparel* (pp. 77–93). Woodhead Publishing.
31. Mansour, R. (2018). Natural dyes and pigments: Extraction and applications. In *Handbook of renewable materials for coloration and finishing* (Vol. 9, pp. 75–102). Scrivener Publishing LLC.
32. McDonald, F. (2012). *Textiles: A history*. Casemate Publishers.
33. Merdan, N., Eyupoglu, S., & Duman, M. N. (2017). Ecological and sustainable natural dyes. In *Textiles and clothing sustainability* (pp. 1–41). Springer.
34. Padfield, T., & Landi, S. (1966). The light-fastness of the natural dyes. *Studies in Conservation, 11*(4), 181–196.
35. Palamutcu, S. (2017). Sustainable textile technologies. In *Textiles and clothing sustainability* (pp. 1–22). Springer.
36. Panneerselvam, R. G., Sriramulu, V., Ramalingam, P., & Prakash, C. (2021). A study on the revival of jamdani weaving technique and Kalamkari drawing—Dyeing processes of traditional Kodali Karuppur sarees. *Journal of Testing and Evaluation, 50*(2). https://doi.org/10.1520/JTE20210337
37. Patel, B. H. (2011). Natural dyes. In *Handbook of textile and industrial dyeing* (pp. 395–424). Woodhead Publishing.
38. Peters, A. (2007). *The history of tribal textiles* [Online]. Available from https://www.fibre2fashion.com/industry-article/2928/the-history-of-tribal-textiles. Accessed on 17 Jan 2023.
39. Prabhu, K. H., & Bhute, A. S. (2012). Plant based natural dyes and mordants: A review. *Journal of Natural Product and Plant Resources, 2*(6), 649–664.

40. Pradhan, S., & Khandual, A. (2020). Community, local practices and cultural sustainability: A case study of Sambalpuri Ikat handloom. In *Sustainability in the textile and apparel industries* (pp. 121–139). Springer.
41. Samanta, A. K., & Agarwal, P. (2009). Application of natural dyes on textiles. *Indian Journal of Fibre & Textile Research, 34*(4), 384–399.
42. Savvidis, G., Karanikas, E., Nikolaidis, N., Eleftheriadis, I., & Tsatsaroni, E. (2014). Ink-jet printing of cotton with natural dyes. *Coloration Technology, 130*(3), 200–204.
43. Saxena, S., & Raja, A. S. M. (2014). Natural dyes: Sources, chemistry, application and sustainability issues. In *Roadmap to sustainable textiles and clothing* (pp. 37–80). Springer.
44. Sequin-Frey, M. (1981). The chemistry of plant and animal dyes. *Journal of Chemical Education, 58*(4), 301.
45. Singh, R., Jain, A., Panwar, S., Gupta, D., & Khare, S. K. (2005). Antimicrobial activity of some natural dyes. *Dyes and Pigments, 66*(2), 99–102.
46. Singh, M., Vajpayee, M., & Ledwani, L. (2021). Eco-friendly surface modification of natural fibres to improve dye uptake using natural dyes and application of natural dyes in fabric finishing: A review. *Materials Today: Proceedings, 43*, 2868–2871.
47. Sinnur, H. D., Samanta, A. K., Verma, D. K., & Kaware, R. (2018). Studies on coloration and UV protective action of anar peel (pomegranate rind) as an effective natural colorant for cotton khadi fabric. *Journal of The Institution of Engineers (India): Series E, 99*(1), 9–26.
48. Siva, R. (2007). Status of natural dyes and dye-yielding plants in India. *Current Science, 92*, 916–925.
49. Sudarmin, S., Sumarni, W., Azizah, S.N., Yusof, M. H. H., & Listiaji, P. (2020, June). Scientific reconstruction of indigenous knowledge of batik natural dyes using ethno-STEM approach. *Journal of Physics: Conference Series, 1567*(4), 042046. IOP Publishing.
50. Tayade, P. B., & Adivarekar, R. V. (2014). Extraction of Indigo dye from Couroupita guianensis and its application on cotton fabric. *Fashion and Textiles, 1*, 1–16.
51. Umbreen, S., Ali, S., Hussain, T., & Nawaz, R. (2008). Dyeing properties of naturaldyes extracted from turmeric and their comparison with reactive dyeing. *Research Journal of Textile and Apparel, 12*(4), 1–11.
52. Werede, E., Jabasingh, S. A., Demsash, H. D., Jaya, N., & Gebrehiwot, G. (2023). Eco-friendly cotton fabric dyeing using a green, sustainable natural dye from Gunda Gundo (Citrus sinensis) orange peels. *Biomass Conversion and Biorefinery, 13*(6), 5219–5234.
53. Yıldırım, F. F., Yavas, A., & Avinc, O. (2020). Printing with sustainable natural dyes and pigments. In *Sustainability in the textile and apparel industries: Production process sustainability* (pp. 1–35). Springer International Publishing.

Sustainable Extractions and Applications of Natural Dyestuffs

Elçin Emekdar Karaman and Umut Kıvanç Şahin

Abstract Each day, the importance of sustainability increases as natural sources decrease because of various human-based or nature-based causes. Using fewer chemicals, recycling, and reusing byproducts and end products are important strategies for being sustainable. From this point of view, natural dyestuffs are good choices for replacing synthetic dyestuffs that harm the environment. Natural dyestuffs are advantageous for being nonhazardous, relatively cheaper, and easy to extract from natural sources; they have various characteristics and colors; and their production methods can be selected according to the needs of customers and sustainability. In this chapter, preferred natural dyestuffs, their sources and properties, their production methods, and their various levels of sustainability are examined, and new approaches are evaluated according to the results of the latest studies. In addition, a new sustainable dyestuff extraction study is prepared and reported for dyeing cotton fabrics.

Keywords Sustainability · Natural dyestuff · Anthocyanin · Extraction · Recycle · Reuse

1 Introduction

Human activities, natural disasters, pollution, global warming, etc. destroy Earth's resources. Natural resources decrease with the usage of raw materials in industries, especially for food and textiles. Water sources are contaminated by the byproducts and wastes of industry, and freshwater is being depleted because of human activities and industrial processes. Every day, 1041 billion liters of surface water is consumed

E. Emekdar Karaman (✉) · U. K. Şahin
Istanbul Technical University, Textile Technologies and Design Faculty, Gumussuyu, Beyoglu, Istanbul, Turkey
e-mail: emekdar@itu.edu.tr

S. S. Muthu (ed.), *Natural Dyes and Sustainability*, Sustainable Textiles: Production, Processing, Manufacturing & Chemistry,
https://doi.org/10.1007/978-3-031-47471-2_2

by humans [1]. Also, the automation of the modern world causes high levels of carbon emissions to pollute the air, and soil pollution, caused by improper waste-disposal strategies, increases day by day.

With increasing environmental awareness comes the increasing importance of environmental sustainability. The definition of sustainability is given by United Nations Brundtland Commission as meeting "the needs and aspirations of the present without compromising the ability to meet those of the future" [2]. To support sustainability, human beings must focus on using less raw material and water, improving refining and reuse strategies, recycling pre- and postconsumer end products, reducing carbon emissions, and both reducing waste disposal and increasing waste-disposal regulations.

The reuse of raw materials in wastes and end products or byproducts is one of the most important strategies to limit chemical usage. A material that reaches the end of its life cycle but still has valuable ingredients is recycled or reused in another process as raw material. With this strategy, both raw materials are saved; waste material is repurposed; and waste disposal and waste treatment costs are eliminated. By following suitable methods for natural raw material usage, depending on the materials, more environmentally friendly or biodegradable ingredients can be produced without causing a chemical load. Natural dyestuffs are good examples of the types of products that can be sustainably extracted.

According to the results of several studies, this chapter evaluates the types of natural dyestuffs, their possible sources and properties, their production methods, and their various levels of sustainability during and after extraction. Finally, a new sustainable dyestuff extraction study on dyeing cotton fabrics reports on the fastness values of extraction.

2 Sustainability Strategies for Circular Economy

Sustainability is a broad term; it is an ecofriendly principle for every sector. It prioritizes a circular economy over a linear economy. A linear economy takes raw materials, makes new products, and disposes wastes, whereas a circular economy follows a cycle as it makes products from recycled materials, uses the products, recycles them, and returns to the first step [3].

A circular economy basically follows the strategy of the 3Rs of sustainability:

1. *Reducing* the usage of raw material, energy, water, chemicals, etc.; controlling how the materials are used; and eliminating the unnecessary use that leads to waste are the first steps. For example, in a process, chemical minimization decreases the chemical load, shortens refining processes, and reduces the energy needs for making more-natural and more-ecofriendly products. Also, using one bath textile process decreases the amount of water and energy used.
2. *Recycling* waste turns it into another form, and the new form is used as raw material for other processes. For example, paper waste can be turned into paper pulp,

and recycled paper is produced from the paper pulp instead of from new wood pulp from new trees. It decreases the amount of waste disposal, energy, and costs while saving trees and other resources.

3. *Reusing* waste retains the product for it to be used again in its same form but in other processes. For example, washing water that has a low level of contamination can be used in the color recipes of dyeing processes. If the contamination in the water is high, refining processes can be applied because refined water is still in the same form, as water. Thanks to this approach, water usage, refining energy, and costs decrease.

Many studies have applied the 3R sustainability strategy [4, 5], and the number of these studies will only increase because the European Union prioritizes the principles of the Green Deal, which require following these strategies for environmental sustainability. Natural dyestuff usage plays a significant role here because it decreases synthetic dyestuff usage, the harmful contaminants in wastewater, and the difficulty of refining.

3 Natural Dyestuffs

Natural or organic dyestuffs are cost-effective alternatives to inorganic dyestuffs and can be used as dye sensitizers. They can be reaped from various parts of different colorful plants, such as the fruits, leaves, or flowers, by using simple extraction methods. They have many color options, and the color can change from red to purple. Natural dyestuffs have many better properties than inorganic ones: They are cheaper, more biodegradable, more widely available, purer, and more ecofriendly; they are also more amenable to simple extraction methods and use fewer noble metals. However, they have some disadvantages as well, such as low stability, low efficiency, and low scalability. The dyestuff's reactions to sunlight change the wavelengths of that sunlight. The electrical results of dyestuff from plants can be examined to identify the plants' pigmentations. In this case, the maximum absorption wavelength (λ_{max}) is the most effective property for determining the color that the dyestuff will confer [6].

The advantages and disadvantages of natural dyestuffs are given in Table 1. Natural dyestuffs are classified into four groups: flavonoids, carotenoids, chlorophylls, and betalains, as given in Fig. 1.

3.1 Flavonoids

Flavonoids make up a very common and broad subgroup for natural dyestuffs. Flavonoid is the general term for natural dyestuffs that have C6–C3–C6 carbon structures—or, more clearly, phenylbenzopyran functionality. They include 15

Table 1 Advantages and disadvantages of natural dyestuffs

ADVANTAGES	DISADVANTAGES
• Easily available	• Low efficiency
• Biodegradable	• Low scalability
• Low cost	• Low stability
• Highly pure	
• Environmentally friendly	
• Simple extraction method	
• Less noble metal	
• Low cost of synthesis	

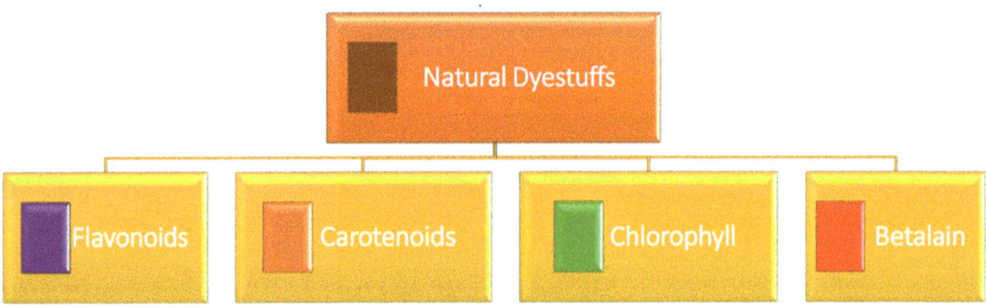

Fig. 1 Classification of natural dyestuffs

carbons with two phenyl rings joined by three carbon bridges, forming the third ring. Some flavonoids cannot absorb visible light. Lower levels of energy are needed to excite flavonoid electrons because they have loose electrons; thus, low levels of visible light can provide energy to these dyestuff molecules. They can be found on many fruits and plants that animals prefer to consume. There are many types of flavonoids; they are named according to their color; and this group is bonded at the C3 regions [6]. The following is a list of the types of flavonoids:

- Anthocyanin can be found in red and purple vegetables and fruits.
- Anthocyanidin can be found in red and purple vegetables and fruits.
- Proanthocyanidin's structure is similar to the polymer form of anthocyanidin, and it can be found in apples and cacao, among others.
- Flavanone has an unsaturated C-C bond and is mostly found in citrus fruits.
- Flavon can be found in celery, parsley, and oregano.
- Flavonol can be found in onions, ginger, broccoli, asparagus, and other leafy greens.
- Isoflavone (3-benzopyrans) can be found in soy-based foods, such as soy milk, tofu, tempeh, etc. [7].

- Neoflavanoids (4-bezopyrans) can be found in soy-based foods.
- Catechin can be found in fresh tea leaves, apricots, black grapes and strawberries.
- Chalcone can be found in citrus, apple, tomato and potato.

3.2 Carotenoids

The chloroplasts and chromoplasts of plants are the main sources of carotenoids. In plants, they protect chlorophyll from photodamage and absorb energy for photosynthesis. They produce red, orange, and yellow colors in and various aromas for flowers and fruits. They have eight isoprene molecules with 40 carbons in their structures—known as tetraterpenoids. The absorption wavelengths of carotenoids are generally between 400 and 550 nm, and these wavelengths refer to violet to green light. The many types of carotenoids can be divided according to their structural differences into two subclasses [6]:

- Carotenes are pure hydrocarbons.
- Xanthophylls contain oxygen.

3.3 Chlorophyll

Chlorophyll is found in green plants, cyanobacteria, and some algae. It is a large organic molecule with a metal ion in its center (mostly Mg in the center of a chlorine ring); ingredients such as hydrogen, carbon, oxygen, and nitrogen in its side chains; and a hydrocarbon trail, depending on the type of chlorophyll. Chlorophyll is responsible for absorbing energy during photosynthesis and for harvesting light. Thus, it is vital to life on Earth. Finally, 670 nm is its maximum absorption point. Chlorophylls come in six types, but the most popular ones are chlorophyll A and B:

- Chlorophyll A—the chemical formula of chlorophyll A is $C_{55}H_{72}O_5N_4Mg$. Its absorption peaks are 430 nm and 662 nm, and it produces a grass-green color.
- Chlorophyll B—The chemical formula of chlorophyll B is $C_{55}H_{70}O_6N_4Mg$. Its molecular structure features an aldehyde group instead of a methyl group (in chlorophyll A). Its absorption peaks are 453 nm and 642 nm, and it produces a blue-green color [6].

3.4 Betalain

Betalain is a water-soluble form of alkaloid pigments and includes nitrogen. Betalains can be found in the plant suborder Chenopodineae and some higher-order fungi. The stable pH range for betalains is between 3 and 7. The wavelength of

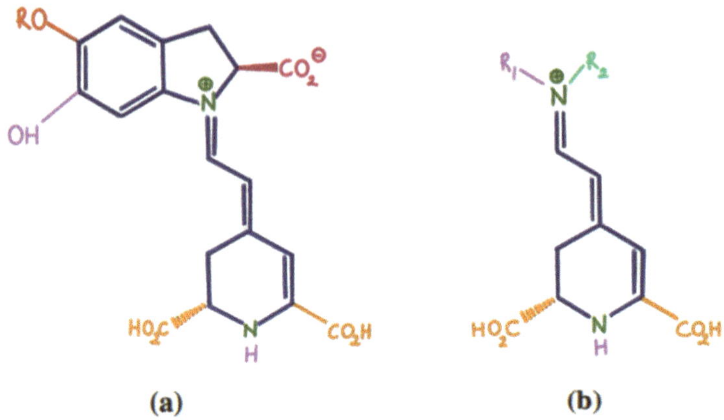

Fig. 2 General structures of (**a**) betacyanin (R = b-D-glucose) and (**b**) betaxanthin (R1 = H; R2 = amine or amino acid group)

maximum absorption decreases when pH falls below 3.5. Thus, the pH value should be checked when betalains are used as sensitizers. They can absorb visible light with wavelengths between 476 and 600 nm [8]. They can be classified according to their structures into two subclasses, as given in Fig. 2:

- Red-violet betacyanin has a wavelength of 535 nm as its absorption peak.
- Yellow-orange betaxanthin has a wavelength of 480 nm as its absorption peak [8].

A comparison of natural dyestuffs appears in Table 2.

4 Extraction Methods for Natural Dyestuffs

In ancient methods, natural fruits and vegetables were added directly to the dye bath, and those fruits and vegetables contained only small numbers of dyeing molecules. Thus, they were not stable, uniform, or especially useful. Also, these fruits and vegetables acted as obstacle in dye baths during the procedure, and they became waste products after the dyeing process was completed. For these reasons, finding and choosing the optimal extraction method is vital. Extraction methods are classified according to the extracting equipment and extraction media that they require, as shown in Figs. 3 and 4.

Table 2 Comparison of natural dyestuffs

Pigments	Molecular Formula	Chemical Structure	Solvency	Color Shade	Natural Sources
Chlorophyll	$C_{35}H_{30}MgN_4O_5$	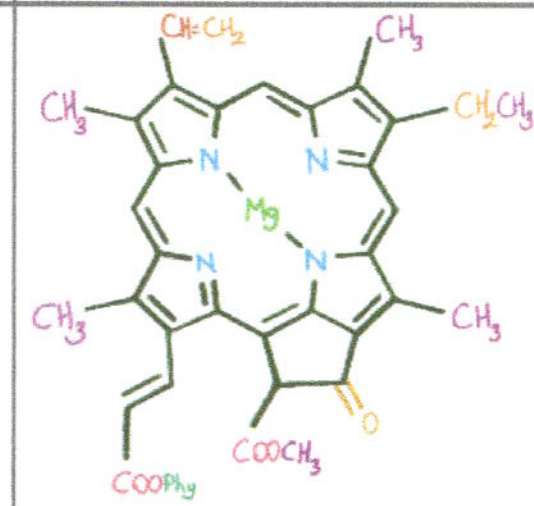	Insoluble in water Soluble in ethanol, ether, pyridine	Green	Spinach Mint Algae Green leaves
Anthocyanin	$C_{15}H_{11}O^+$	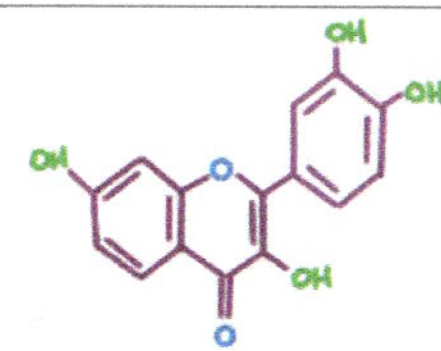	Soluble in water, ether, alcohol	Purple-red-blue	Blackberry Black carrot Red beet Eggplant
Carotenoid	$C_{40}H_{56}$	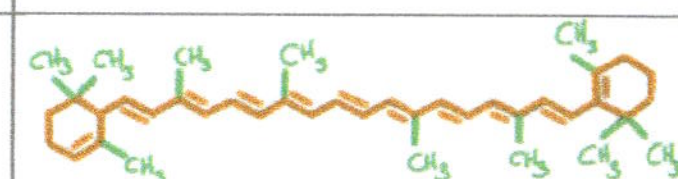	Insoluble in water Soluble in ethyl ether, acetone, alcohol, ethyl acetate	Yellow- orange	Carrot Plum Apricot Mango
Flavonoid	$C_{10}H_{12}O_2$	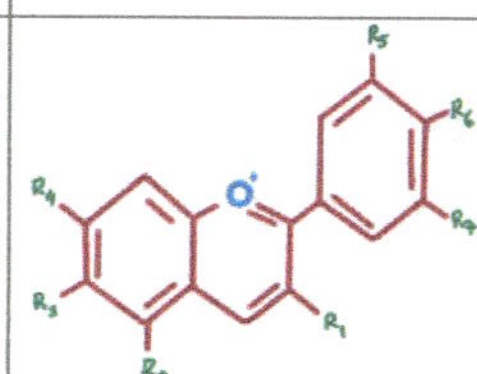	Soluble in water, ether, alcohol	Yellow Red (acidic) Blue (Alkaline)	Grapefruit Soybean Green tea Broccoli

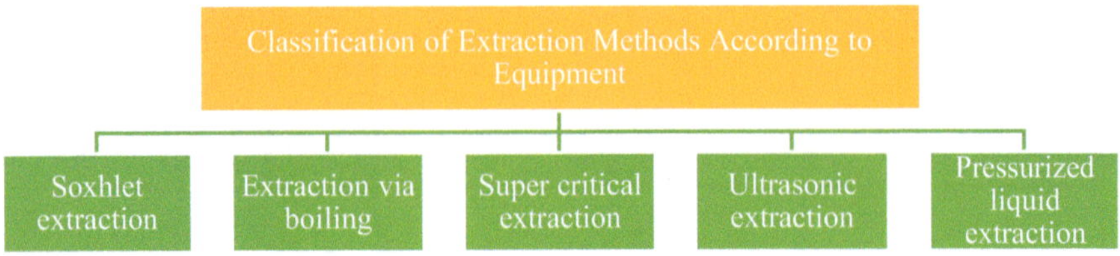

Fig. 3 Classification of extraction methods according to the extracting equipment

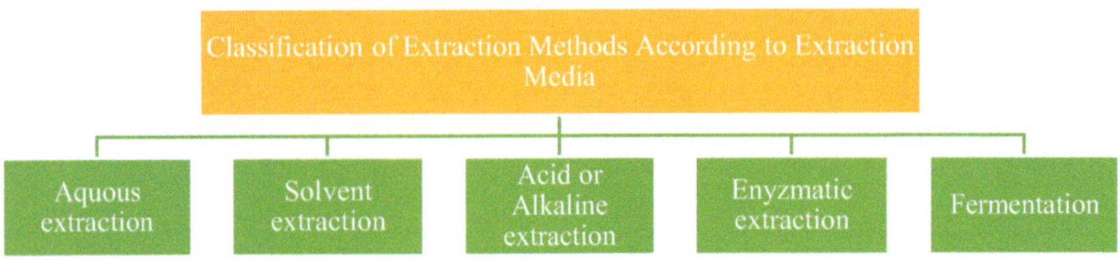

Fig. 4 Classification of extraction methods according to the extraction media

4.1 Classification of Extraction Methods According to Equipment

4.1.1 Soxhlet Extraction

Semivolatile extracts can be continuously taken from solid or semisolid materials by using this method. Various organic solvents can be used in this procedure. It has a system that includes a heater, a balloon flask, a Soxhlet extractor, and a reflux condenser, from bottom to top. All these glass units are bonded together to provide an inert extraction medium. The balloon flask carries the solvent, and the Soxhlet extractor includes a solid substrate in a thimble [9]. The Soxhlet extraction system is shown in Fig. 5.

4.1.2 Extraction via Boiling

A water bath in a beaker and a heater are used as the extraction equipment. A crushed extracting substrate is dissolved in water and boiled for at least two hours. Other solvents can be preferred.

4.1.3 Supercritical Fluid Extraction (SCFE)

The supercritical fluid extraction (SCFE) method needs a special chamber set to precise temperatures and pressures. The solvent used in this method is a gas that liquidizes after reaching a critical temperature and pressure point. Extraction is carried out with the liquidized gas, and the liquid moves inside the chamber in cycles.

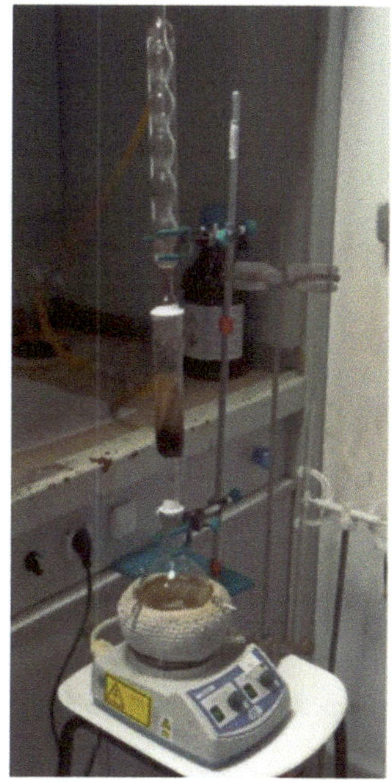

Fig. 5 Soxhlet extraction system [10]

The liquid turns into a gaseous form again after removing the temperature and pressure needed for extraction. Also, the solvent can be reclaimed after the extraction, so it is not wasted. This method is advantages if nonhazardous and cheap gases such as carbon dioxide are used. Energy consumption can be another advantage, but the requirement of using a special chamber can be costly.

4.1.4 Ultrasonic Extraction

The ultrasonic method uses sound energy to produce small bubbles, which are called cavitation in the bath. These bubbles hit the extracting samples like a microjet so that the extracting solvent can achieve deeper penetration and particles can come out more easily. The bath generates energy, so this method is especially suitable for use under limited energy conditions. Finally, it also works well in low temperatures.

4.1.5 Pressurized Liquid Extraction

Pressurized liquid extraction requires high temperatures and high pressures during its process. The solvent that it uses is sent to the substrate at a precise pressure and temperature setting. In this method, polarity can choose extracts. It is also an eco-friendly method, especially because it needs only a small amount of solvent [11].

Table 3 Classification of extraction methods according to their extracting equipment

Extracting Method	Advantage	Disadvantage
Soxhlet Extraction	Any solvent Same solvent, no waste Continuous Various cycles	Cold water waste in condenser
Extraction via Boiling	Basic instruments Easy	Takes time Evaporation of solvent
Super Critical Fluid Extraction	No waste of solvent Less energy consumption	Needs a special chamber
Ultrasonic Extraction	Extra energy introduced into the bath Works at low temperatures	Needs an ultrasound generator
Pressurized Liquid Extraction	Small volume of solvent Identification of extracted compounds Green technique	Needs a pressure generator Needs high temperature

A classification of dyestuff extraction methods according to their extracting equipment is given in Table 3.

4.2 Classification of Extraction Methods According to Extraction Media

4.2.1 Aqueous Extraction

This method is very basic and natural because it uses water as an extracting medium. It can also be used for extracting water-soluble ingredients. It does not require extra equipment for extraction, but additional equipment and optimal conditions can enhance extraction efficiency.

4.2.2 Solvent Extraction

This method needs two immiscible liquids, such as an organic solvent and water. The extracting substrate can be either liquid or solid. Compounds are separated from each other because they have different solubilities in extraction solvents.

4.2.3 Acidic or Alkaline Extraction

The pH value is important parameter for extraction because some natural ingredients are soluble in acidic or alkaline conditions. Using a purely acidic or alkaline structure can be harmful for extracting substrates, so they are mixed with water or other solvents to increase the extraction efficiency. Acidic extraction requires mixing extracting solvents with an acid, whereas alkaline extraction requires mixing extracting solvents with a base.

4.2.4 Enzymatic Extraction

Some enzymes, such as amylase, pectinase, and cellulase, can be used to extract coloring molecules from plant-based substrates that contain starch, pectin, and cellulose. These are particularly advantageous for hard substrates [12].

4.2.5 Fermentation Extraction

Microorganisms are used to produce colorants from plants and fungi. Colorants are produced in cultures, and colorants depend on which microorganisms were used. These variables mean that the products will not be homogenous [13]. Sharmila et al. examined the use of microalgae types as substrates to identify which one is the best. They used the fungus *Aspergillus niger* as a fermentation microorganism. Carotene, chlorophyll A, and chlorophyll B were produced from various substrates [13]. In addition, Santis et al. showed that red dyestuff can be produced from the genus *Monascus* [6].

A classification of dyestuff extraction methods according to extraction media is given in Table 4.

Table 4 Classification of extraction methods according to extraction media

Extracting method	Advantage	Disadvantage
Aqueous Extraction	Basic instruments Natural Natural pH of around 7	Limited source Valid for water-soluble contents Relatively low efficiency
Solvent Extraction	Fast and deeper production with suitable substrates	Harmful solvent options Evaporates volatile solvents Valid for solvent-soluble contents
Acidic or Alkaline Extraction	Fast and deeper production with suitable substrates	Can harm the substrate Can be dangerous for workers Valid for special substrates
Enzyme Extraction	Fast and deeper production with suitable substrates Focuses on only the needed parts	Works only at suitable temperatures and pH values Suitable for specific procedures
Fermentation Extraction	Cheaper culture media with fast and easy growth Independence from weather conditions and colors of different shades	Seasonal and geographic variability in the production Heterogenous Depends on microorganisms, so needs suitable temperatures and pH values

5 Sustainability in Dyestuff Extraction and Dyeing

Natural dyestuff extraction is preferable when the extraction process is sustainable. To carry out sustainable dyestuff extraction and dyeing, three main options are available: using wastes as raw materials, using less water and fewer chemicals in the extraction solvents, and opting for bath dyeing with natural mordants.

First, raw materials for extraction can be selected from the waste of any industry, such as the food, the agriculture industry, the textile industry, etc. Raw materials can be any natural materials. However, using a new plant that could be a food source, for example, is not a good choice for extracting dyestuff, because that plant would be more valuable as food. But if the dyestuff can be extracted from a waste product, such as the roots, shells, stems, etc., then the process can recycle this waste instead of disposing of it. Also, fallen leaves are also good sources for natural dyestuff because they contain chlorophyll but have become wastes and been discarded by their trees.

Naturally extracted red dyestuff from alkanet root, which can dye nylon and polyester, contain anchusin as a coloring unit, and alcohol is used to extract the phenolic chemical first. This structure can be used to dye natural fibers because there is some alcohol in the media for dispersal, whereas synthetic dyes disperse dyestuff in the presence of only water [14].

Vankar et al. showed that supercritical fluid extraction is better for eucalyptus bark [15]. Vankar and Dixit used *Illicium griffithii* leaves to extract dye liquor by using the extraction solvent method, and color evaluations of the dyed fabric showed that using this method is more advantageous than using conventional methods [16].

Second, some chemicals can be added to the extraction solvent to improve extraction efficiency and the quality of the dyestuff, but it also increases the chemical load in waste extraction solvent. Therefore, it is better to use as few chemicals as possible or to use less-harmful chemicals and their natural alternatives. Natural dyes are good alternatives because high-quality dyestuffs can still be extracted with fewer and less-harmful chemicals. Also, enzymes can be added to the process to break down the hard plant parts, such as seeds, stems, etc., and to acquire more and better dyestuff instead of adding harmful chemicals. In addition, water usage must be minimized or reused as much as possible.

Vankar and Srivastava showed that using ethanolic extraction on the *Canna indica* flower produced superior dyeability [17]. Srivastava et al. examined the acidic extraction of anthocyanin dyestuff by using methanol and citric acid as solvents and hibiscus as a substrate. They proved that lower pH values work better on hibiscus [18]. Ali et al. indicated that better extraction efficiency, dyeing, and fastness can be achieved by using alkaline extraction on henna leaves [19]. Tiwari et al. extracted dyestuff from pomegranate skin by using a 2% concentration of a 1:2 cellulase–pectinase mixture. Substrates were soaked in extraction liquor overnight; the enzymatic part was washed; and a pH of 10 was set to dye the fabrics. Tiwari et al. also combined ultrasound with enzyme extraction, and the best result came from the enzyme and ultrasound combination [20].

Third, dyestuff extraction media can be used directly as dyeing media if the extraction method and textile material are suitable. After the extraction has finished in the ultrasonic chamber, the fabric can be inserted into the chamber. In this case, some mordants, which must be ecofriendly, can be applied before or after the dyeing operation. If natural mordants are applied in this stage, the dyeing process will be more ecofriendly and sustainable. Also, dyeing more fabric in the same ultrasonic bath can improve the dyeing efficiency and decrease the amount of mordant needed.

Sivakumar et al. proved that an ultrasound additive can enhance extraction efficiency by at least 13% [21]. Mansour examined the effects of using both solvents and ultrasound in extraction methods. This study showed that using a 9:1 water–acetone solution performed better than using only water with ultrasound [22]. According to a study by Shukla and Vankar, black carrots can dye the silk fibers that were mordanted before the dyeing procedure, and they produced high spectrophotometer values. They grated and froze the black carrot and extracted the dyestuff by using a 1:20 black carrot—water solution [23]. Sivakumar et al. showed that the ultrasonic extraction method can decrease the energy needs and increase the extraction efficiency of beetroot [24]. According to Vankar et al. the anthocyanin in *Cayratia* fruits can be extracted by first crushing the substrate—the skin of the fruits—and then boiling the substrate in distilled water for 3 hours [25]. Finally, Summoogum-Utchanah et al. discovered a shorter method: It uses a 1:1 water–ethanol mixture and adds a microwave oven to the system [26].

6 Sustainability After Dyestuff Extraction

After dyestuff extractions, their natural raw materials are still useful and valuable. Thus, they can be used after the extraction processes have been completed, or extra processes can be conducted after the extraction to use the remaining raw material. This reuse strategy can decrease the amount of waste, decrease the energy needed to dispose of the waste, and serve as a more environmentally friendly approach. The waste of dyestuff extraction has great potential to be reused in different forms, just like other agricultural wastes. These forms can be subsumed under five classes: biochar, filler, biocompost, new chemical synthesis, and feed for animals.

6.1 Biochar

Bio-based substances, especially agricultural plant wastes, are used to form biochar through pyrolysis: the application of very high temperatures—i.e., above 350 °C—without oxygen. At the end of the pyrolysis operation, bio-based charcoal, biochar for short, is produced, along with a gas that can be turned into bio-oil. Thus, the pyrolysis operation can generate two recycled end products, namely biochar and bio-oil [27].

There are slow, intermediate, and fast options for pyrolysis, but the slow one is simple and more suitable for agricultural wastes, such as the residual raw material of crops used for dyestuff extraction [28]. The characteristics of biochar can be changed by using activation methods that add functional groups to the biochars, but generally, a biochar has a very high surface area and a porous structure, so it has high potential to be used as an adsorbent. The properties and uses of biochar are detailed in the chapter by Emekdar and Şahin [27].

6.2 Filler

Raw materials become waste materials after dyestuff extraction. They lose a bit of their color, but if hazardous chemicals that can damage the cellulose structure and enzymes that can break cellulose into smaller molecules (an enzyme such as cellulase) are not used during extraction, they will still be valuable sources of cellulose. Their forms are small and their qualities are too low to dye most fibers, but they can be used as very small fillers. The remaining waster materials are dried, and further processes can be added, depending on the type of cellulose form that is needed. Waste materials from dyestuff extraction can be advantageous because they can act as pretreated, washed, and color-eliminated materials for new processes.

El Achaby et al. used waste algae, after producing agar, as sources of cellulose nanocrystals (CNCs) by using alkaline, bleaching, and acidic hydrolysis processes. Next, the extracted CNCs were added to Polyvinyl alcohol (PVA) to check the increase in the tensile strength, and as a result, a 215% Young's modulus and a 150% tensile strength increase was produced thanks to the addition of 8% CNC. Also, the addition did not affect the transparency [29].

Marcos dos Santos et al. studied CNC extraction from pineapple leaf wastes by using acidic hydrolysis, and the best CNCs were extracted after a 30-minute treatment. Their results proved that thermally stable and high-crystalline CNCs can be extracted from agricultural wastes and can be used as fillers in composite structures [30].

Bahloul et al. noted the potential of using cellulose microfibers (CMFs) and CNCs that are extracted from eggplant wastes as fillers for composites. Their process included an alkaline treatment, bleaching, and acidic hydrolysis. In their study, 13.6 μm diameter CMFs and CNCs with crystallinity rates higher than 71% were extracted [31].

Mahmud et al. added waste cellulose fiber as a filler and epoxidized soybean oil as plasticizer into PLA composites and investigated mechanical properties of composites. According to the test results, the impact strength, bending strain, and elongation at break values increased, so cellulose-based filler addition increased the mechanical properties [32].

In a review paper, Sydow et al. detailed the use and positive effects of using plant-based wastes, such as cellulosic agricultural wastes, as fillers in different composites [33].

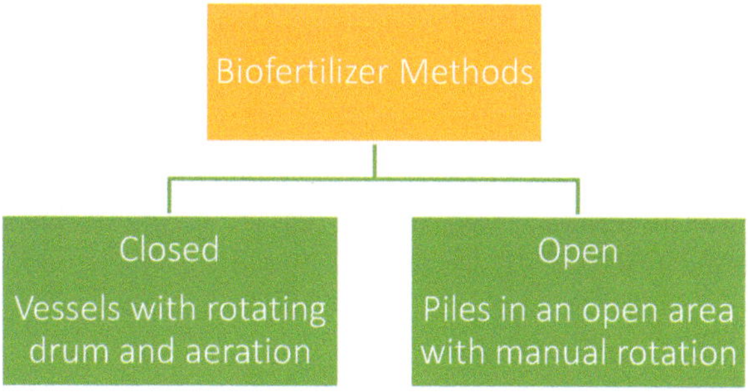

Fig. 6 Biocompost production methods

6.3 *Biocompost*

Yielding biocompost from organic waste by using aerobic microorganisms is one of the ancient methods for waste management and the process is known as composting. Composting needs the waste, microorganisms, oxygen, carbon/nitrogen (C/N) amount, and moisture. C and N are important because microorganisms use them for energy and metabolic activities. Microorganisms require moisture for the process, but a moisture level higher than 75% would decrease the oxygen flow by clogging the pores of the process media [34]. Biofertilizer production methods are shown in Fig. 6.

In a review paper, Shu et al. demonstrated in detail the use of agricultural waste, which is a source of cellulose, in reuse strategies, especially composting [35].

6.4 *New Chemical Synthesis*

The cellulosic waste of dyestuff extraction can be modified and turned into different forms of cellulose, depending on the need. The most commonly used cellulose modification turns cellulose into carboxymethylcellulose (CMC), which can act as a thickening and sizing agent. Also, other functional groups can be added, depending on the needs of desired chemical.

Biswas et al. produced CMC by using rice hulls and straw from barley and wheat [36]. In addition, by using the waste of raw materials after dyestuff extraction, other chemicals, such as glucose, pectin, fatty acids, phenolic compounds, and minerals, can be extracted too [35].

6.5 *Feed for Animals*

Another ancient reuse strategy uses organic and still-valuable wastes as feed for animals. Herbivorous animals eat cellulosic plants, grass, and other crops, and the cellulosic wastes from natural dyestuff extraction are still-valuable sources of cellulose for them. Therefore, if the dyestuff extraction takes place under ideal conditions (without chemicals), then the cellulosic waste can be used as animal feed, after other necessary processes have been carried out.

Food wastes from human sources can also be used as feed for animals, but in this case, a human disease may be passed on to an animal [37]. However, using natural dyestuff extraction on cellulosic waste that is free from harmful chemicals produces safer feed for animals in that it is free from the pathogens that infect humans.

7 A Study on Alternatives for Sustainable Natural Dyestuff Extraction

7.1 Extraction

The Soxhlet extraction method was preferred to obtain natural dyestuff from black carrots. There are three main parts in the experiment. The volumetric flask at the bottom is heated at the maximum temperature in the flask heater during the extraction. The chemicals used in this flask are purified water and sodium carbonate. The solution, which evaporates with the help of the pouch heater, condenses thanks to the mechanism connected to the water circulation at the top and drips into the filter paper bag in the middle part, which contains grated black carrots. When a sufficient amount of solution is reached, the device siphons off, thus completing a cycle. Three cycles were completed for each extraction. Then, the obtained dyed solution is taken into petri dishes to be heated and thus extract is obtained in the powder form. Approximately 2 g of dyestuff extract is obtained from 55 g of black carrot after this heating process and they are scraped and taken into sample boxes in powder form.

7.2 Dyeing

For the dyeing process, 5 g of cotton fabrics and 1 g dyestuff were prepared by using a Gyrowash machine. For the dyeing process, the machine was fixed at 60 °C, and dyeing was carried out for 60 minutes for cotton dyeing. A 1:25 liquor ratio was used.

7.3 Postmordanting

A 1:20 liquor ratio was used, and 6% (on weight of fiber, or owf) tannic acid was added to the distilled water. Postmordanting was performed on dyed fabrics for 30 minutes in a Gyrowash machine at 70 °C. After postmordanting, the fabrics were rinsed and left at room temperature to dry.

Table 5 Sample details

Sample Name	Grated Black Carrot	Water (%)	Sodium Carbonate (%)	Tannic Acid After Treatment
WD [10]	✓	100	0	–
WDT [10]	✓	100	0	✓
5SD	✓	95	5	–
5SDT	✓	95	5	✓

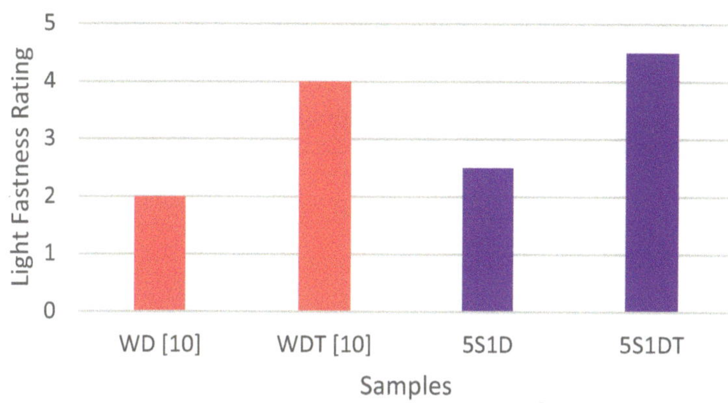

Fig. 7 Lightfastness results of samples

7.4 Results

The sample details are given in Table 5. All the samples were tested by using TS EN ISO 105-B02 for lightfastness. Figure 7 presents the lightfastness results for the samples. The lightfastness results were poor without postmordanting in that most of the natural dyestuffs were ineffective, but adding sodium carbonate to the dyestuff extraction process increased the lightfastness value from 2 to 2–3.

Postmordanting enhanced the lightfastness results for both samples, and those enhanced values are within the acceptable range, which sets a minimum value at 4. However, the sample to which sodium carbonate was added had higher lightfastness values, between 4 and 5. These results indicate that sodium carbonate is useful for dyestuff extraction and that lightfastness values are more efficient when sodium carbonate is added. Also, the dyestuff extracted with the addition of sodium carbonate is compatible with using mordant.

8 Conclusion

Earth's resources decrease every day, and sustainability continues to become more popular because it saves natural sources. Important strategies to ensure or improve sustainability, such as reducing, recycling, and reusing, can be used.

Natural dyestuff selection is an option both during dye production and after dye production thanks to its superior properties: Natural dyestuff is easy to produce; it uses nonhazardous materials; and the materials come from inexpensive sources. Also, many methods can yield the specified dyestuffs in sustainable ways. Natural dyestuffs can be sustainable when extracted from wastes, and they require less solvent and less mordant for dyeing. In addition, the sustainability of natural dyestuffs can be maintained after extraction by turning the waste into biochar, fillers, biocompost, new chemicals, and/or feed for animals. Finally in this chapter, a study for anthocyanin extraction by using alkaline conditions as a sustainable form is conducted and it is presented that similar dyeing characteristics can be yielded within the study.

Acknowledgments The results presented in this study are part of a PhD dissertation prepared by Elçin Emekdar under the supervision of Umut Kıvanç Şahin.

References

1. White, S., et al. (2010). *Water markets of the united states and the worls: A strategic analysis for the Milwaukee Water Council* (Vol. Final). U.S. Economic Development Administration.
2. United Nations. (1987). *1987: Brundtland Report*. Accessed: 25 Oct 2022. [Online]. Available: https://www.are.admin.ch/are/en/home/media/publications/sustainable-development/brundtland-report.html
3. Rezvani Ghomi, E., Khosravi, F., Tahavori, M. A., & Ramakrishna, S. (2021). Circular economy: A comparison between the case of Singapore and France. *Materials Circular Economy, 3*(1). https://doi.org/10.1007/S42824-020-00016-W
4. Mohammed, M., Shafiq, N., Abdallah, N. A. W., Ayoub, M., & Haruna, A. A review on achieving sustainable construction waste management through application of 3R (reduction, reuse, recycling): A lifecycle approach. *IOP Conference Series: Earth and Environmental Science, 476*, 12010. https://doi.org/10.1088/1755-1315/476/1/012010
5. Dahiya, D., Sharma, H., Rai, A. K., & Nigam, P. S. (2022). Application of biological systems and processes employing microbes and algae to Reduce, Recycle, Reuse (3Rs) for the sustainability of circular bioeconomy. *AIMS Microbiology, 8*(1), 83–102. https://doi.org/10.3934/microbiol.2022008
6. Khir, H., et al. (2022). Recent advancements and challenges in flexible low temperature dye sensitised solar cells. *Sustainable Energy Technologies and Assessments, 53*. https://doi.org/10.1016/J.SETA.2022.102745
7. Samantha. Flavonoids in your foods: Here's where to get them. *American Institute for Cancer Research* [Online]. Available: https://www.aicr.org/resources/blog/flavonoids-in-your-foods-where-to-get-them/
8. Isah, K. U., et al. (2015). Betalain pigments as natural photosensitizers for dye-sensitized solar cells: The effect of dye pH on the photoelectric parameters. *Materials for Renewable and Sustainable Energy, 4*(1). https://doi.org/10.1007/s40243-014-0039-0
9. Shukla, D., & Vankar, P. S. (2017). Innovative dye extraction methods. In *Natural dyes for textiles: Sources, chemistry and applications* (pp. 191–203). https://doi.org/10.1016/B978-0-08-101274-1.00005-7
10. Emekdar, E., Şahin, U. K., Ötev, B., Özdem, B. Ş., & Masır, Y. (2022). Natural dyestuff extraction from black carrot and its textile application. In *Proceedings of 3rd international congress*

of innovative textiles, ISBN: 978-605-4265-68-8, (pp. 201–208), Namık Kemal University, Tekirdag, Turkiye.
11. Mustafa, A., & Turner, C. (2011). Pressurized liquid extraction as a green approach in food and herbal plants extraction: A review. *Analytica Chimica Acta, 703*(1), 8–18. https://doi.org/10.1016/j.aca.2011.07.018
12. Helmy, H. (2020). Extraction approaches of natural dyes for textile coloration. *Journal of Textiles, Coloration and Polymer Science, 0*(0), 0–0. https://doi.org/10.21608/jtcps.2020.30990.1037
13. Sharmila, S., et al. (2017). Production of natural dye by solid state fermentation. *International Journal of ChemTech Research, 10*(4), 219–225.
14. Tiwari, V., & Vankar, P. S. (2001). Unconventional natural dyeing using microwave and sonicator with alkanet root bark. *Colourage, 48*(5), 25–28.
15. Vankar, P. S., Tiwari, V., & Ghorpade, B. (2002). Supercritical fluid extraction: Of natural dye from eucalyptus bark used for cotton dyeing in microwave and sonicator. *Canadian Textile Journal, 119*(5), 32–35.
16. Vankar, P. S., & Dixit, S. (2011). Natural dyeing of cotton, wool and silk with the stem and leaves extract of Illicium griffithii. *Research Journal of Textile and Apparel, 15*(2), 77–83. https://doi.org/10.1108/RJTA-15-02-2011-B009
17. Vankar, P. S., & Srivastava, J. (2010). Ultrasound-assisted extraction in different solvents for phytochemical study of Canna indica. *International Journal of Food Engineering, 6*(3). https://doi.org/10.2202/1556-3758.1599
18. Srivastava, J., Seth, R. P., Shanker, R., & Vankar, P. S. (2008). Solubilisation of red pigments from Canna indica flower in different media and cotton fabric dyeing. *International Dyer, 193*(1), 31–36.
19. Ali, S., Hussain, T., & Nawaz, R. (2009). Optimization of alkaline extraction of natural dye from henna leaves and its dyeing on cotton by exhaust method. *Journal of Cleaner Production, 17*(1), 61–66. https://doi.org/10.1016/j.jclepro.2008.03.002
20. Tiwari, H. C., Singh, P., Mishra, P. K., & Srivastava, P. (2010). Evaluation of various techniques for extraction of natural colorants from pomegranate rind- ultrasound and enzyme assisted extraction. *Indian Journal of Fibre & Textile Research, 35*(3), 272–276.
21. Sivakumar, V., Vijaeeswarri, J., & Anna, J. L. (2011). Effective natural dye extraction from different plant materials using ultrasound. *Industrial Crops and Products, 33*(1), 116–122. https://doi.org/10.1016/j.indcrop.2010.09.007
22. Mansour, H. F. (2010). Environment and energy efficient dyeing of woollen fabric with sticta coronata. *Clean Technologies and Environmental Policy, 12*(5), 571–578. https://doi.org/10.1007/s10098-009-0267-7
23. Shukla, D., & Vankar, P. S. (2013). Natural dyeing with black carrot: New source for newer shades on silk. *Journal of Natural Fibers, 10*(3), 207–218. https://doi.org/10.1080/15440478.2012.757031
24. Sivakumar, V., Anna, J. L., Vijayeeswarri, J., & Swaminathan, G. (2009). Ultrasound assisted enhancement in natural dye extraction from beetroot for industrial applications and natural dyeing of leather. *Ultrasonics Sonochemistry, 16*(6), 782–789. https://doi.org/10.1016/j.ultsonch.2009.03.009
25. Vankar, P. S., & Shanker, R. (2006). Dyeing cotton, silk and wool with Cayratia carnosaGagn or Vitis trifolia. *Asian Textile Journal, 15*(10), 38–45.
26. Summoogum-Utchanah, J., Sindra, L., & Hashita. (2015). An investigation on the potential of extracting natural dyes from beetroot and turmeric. *International Journal of Research in Engineering and Technology, 04*(02), 401–416. https://doi.org/10.15623/ijret.2015.0402054
27. Emekdar, E., & Şahin, U. K. (2021). Dyestuff adsorbing natural composites for wastewater treatments. In S. S. Muthu (Ed.), *Advances in textile waste water treatments* (pp. 103–133). Springer. https://doi.org/10.1007/978-981-16-0065-4_4
28. Manyà, J. J., Azuara, M., & Manso, J. A. (2018). Biochar production through slow pyrolysis of different biomass materials: Seeking the best operating conditions. *Biomass and Bioenergy, 117*, 115–123. https://doi.org/10.1016/J.BIOMBIOE.2018.07.019

29. El Achaby, M., Kassab, Z., Aboulkas, A., Gaillard, C., & Barakat, A. (2018). Reuse of red algae waste for the production of cellulose nanocrystals and its application in polymer nanocomposites. *International Journal of Biological Macromolecules, 106*, 681–691. https://doi.org/10.1016/J.IJBIOMAC.2017.08.067
30. dos Santos, R. M., Flauzino Neto, W. P., Silvério, H. A., Martins, D. F., Dantas, N. O., & Pasquini, D. (2013). Cellulose nanocrystals from pineapple leaf, a new approach for the reuse of this agro-waste. *Industrial Crops and Products, 50*, 707–714. https://doi.org/10.1016/J.INDCROP.2013.08.049
31. Bahloul, A., et al. (2021). Micro- and nano-structures of cellulose from eggplant plant (Solanum melongena L) agricultural residue. *Carbohydrate Polymers, 253*(October 2020), 117311. https://doi.org/10.1016/j.carbpol.2020.117311
32. Mahmud, S., Long, Y., Wang, J., Dai, J., Zhang, R., & Zhu, J. (2020). Waste cellulose fibers reinforced polylactide toughened by direct blending of epoxidized soybean oil. *Fibers and Polymers, 21*(12), 2949–2961. https://doi.org/10.1007/s12221-020-0111-6
33. Sydow, Z., Sydow, M., & Wojciechowski, Ł. (2021). Tribological performance of composites reinforced with the agricultural, industrial and post-consumer wastes: A review. *Materials, 14*(8), 1863.
34. El-Mashad, H. M., Van Loon, W. K. P., Zeeman, G., Bot, G. P. A., & Lettinga, G. (2003). Reuse potential of agricultural wastes in semi-arid regions: Egypt as a case study. *Reviews in Environmental Science and Biotechnology, 2*(1), 53–66. https://doi.org/10.1023/B:RESB.0000022933.77648.09
35. Shu, H., Zhang, P., Chang, C., Wang, R., & Zhang, S. (2015). Agricultural waste. *Water Environment Research, 87*(10), 1256–1285. https://doi.org/10.2175/106143015X14338845155660
36. Biswas, A., Kim, S., Selling, G. W., & Cheng, H. N. (2014). Conversion of agricultural residues to carboxymethylcellulose and carboxymethylcellulose acetate. *Industrial Crops and Products, 60*, 259–265. https://doi.org/10.1016/j.indcrop.2014.06.004
37. Sinha, S., & Tripathi, P. (2021). Trends and challenges in valorisation of food waste in developing economies: A case study of India. *Case Studies in Chemical and Environmental Engineering, 4*. https://doi.org/10.1016/J.CSCEE.2021.100162

Dyes and Pigments from Agricultural Wastes in the Coloration of Textiles

Noureddine Baaka, Shahid Adeel, Fozia Anjum, Aamir Ali, and Rehab Ahmed Abdelghaffar

Abstract Agriculture and the food industry create many organic waste streams that still include pigments and dyes. Their removal should no longer be an issue, thanks to their sustainable use in textile coloring. To provide direction for future research, this chapter reports on the use of agricultural and industrial wastes; presents the history of natural dyes; analyzes the advantages and disadvantages of these dyes; lists the essential food industry byproducts used as sources of natural dyes for textile use; and details those byproducts' dyeing applications on textile fibers. The composition and production specifics of a few wastes employed in textile dyeing, provided by different authors, are emphasized, and their utilization is systematically discussed. The documented investigations are organized into case studies to demonstrate the authors' various explanations for why these wastes are effective as sources of textile dyes. Most experiments have shown that textiles dyed using these wastes are strong and have good dyeability. There have also been claims that some scraps confer antibacterial and light-protection qualities. According to the literature, particular attention should be paid in this chapter to the history of using natural dyes,

N. Baaka (✉)
Faculty of Sciences of Monastir, Research Laboratory – Environmental Chemistry and Clean Processes (LR21ES04), University of Monastir, Monastir, Tunisia

Department of Textile and Fashion Management, Higher Institute of Fashion of Monastir, Monastir, Tunisia

S. Adeel
Department of Applied Chemistry, Government College University Faisalabad, Faisalabad, Pakistan

F. Anjum
Department of Chemistry, Government College University Faisalabad, Faisalabad, Pakistan

A. Ali
Department of Botany, Government College University Faisalabad, Faisalabad, Pakistan

R. A. Abdelghaffar
Dyeing, Printing, and Textile Auxiliaries Department, Textile Industries Research Division, National Research Center, Cairo, Egypt

S. S. Muthu (ed.), *Natural Dyes and Sustainability*, Sustainable Textiles: Production, Processing, Manufacturing & Chemistry,
https://doi.org/10.1007/978-3-031-47471-2_3

the advantages and limitations of using natural dyes, and the extraction and mordanting methods. Furthermore, some case studios for specific natural dyes extracted from wastes are presented.

Keywords Food industry · Byproduct · Natural dye · Waste valorization · Textile application

1 Introduction

The foundations of sustainable development, a significant issue for the future of humanity and the planet, are environmental preservation and energy management. Climate change and ecological deterioration will impact society as a result. The textile industry is a water-consuming sector and, above all, one that uses color pollutants. In addition to their toxic character, synthetic dyes are produced from petrochemical resources through dangerous chemical processes. Leaking colored liquid effluents into the ecosystem represents a danger to the environment and human health because of their toxicity and bioaccumulation in nature.

The growing environmental concern about synthetic dyes and additives has led to considering natural dyes as alternatives for textiles, cosmetics, and food coloring.

Natural dyes are renewable and biodegradable and have a low environmental impact [3]. However, they still cause problems, such as the lack of a standardized procedure for extraction and application [50], low-to-moderate color fastness, the nonreproducibility of dyes [51], pollution produced by the use of metallic mordants [2], and high energy depletion throughout extraction and exhaustion dyeing [17]. Several efforts have been made around the world to fill the gaps in using natural dyes by finding suitable alternative sources because of their enormous environmental benefits.

A substantial amount of the dye source is needed to color small amounts of textile material because the dye component content of most natural dyes is only a tiny portion of their overall solid weight. Therefore, dyeing textiles in this way would require enormous amounts of the dye source, which could result in the overexploitation of natural resources, mainly if they are derived from plants or animals. Additionally, it would jeopardize several threatened plant and animal species [52]. In this regard, using inexpensive or readily available waste materials for textile dyeing should reduce the cost of natural dyeing while enhancing its sustainability.

Nature offers us flora and fauna rich in colored compounds, some of which have been exploited in dyeing since ancient times because all civilizations have used the knowledge required to reproduce recipes, improve them, or revamp them. This allowed dyeing to build up a vast field of experience. Today, vegetable dyes are part of the current interest in alternatives to chemical dyes. Natural dyes are becoming popular again across the globe as people have tended to return to natural and eco-friendly products.

The plant world is not lacking in sources of dyes, the exploitation of which could constitute a new economic sector and a factor for future rural development in many regions of the world. This new financial opportunity comes with the challenge of sustainably developing a natural dye industry by responsibly managing natural resources.

In several countries, the development of research into new sources of dyes from plants has shown interest in various valuable parts of plants exploited for other uses. Researchers are increasingly turning to forestry waste, agricultural and food sector waste, and byproducts as sources of natural dyes to color textile fibers. Tree fruits might be the most sustainably produced in this regard, where fresh items are being grown and processed each year for their added value, while the fruit's inedible components are thrown away as trash. Agricultural wastes like orange peels [11], pomegranate peels [18, 59], onion peels [47], almond shells [46], eucalyptus leaves [45], kermes oak [15, 16], fennel leaves [29, 30], red cabbage [31], jackfruit wood [27], harmala seeds [1], etc. are abundant in pigments and can be used to create natural dyes. According to the literature, particular attention should be paid in this chapter to the history of using natural dyes, the advantages and limitations of using natural dyes, and the extraction and mordanting methods. Finally, some case studies on the specific natural dyes extracted from wastes are presented.

2 The Usage of Natural Colors Throughout History

Until the end of the nineteenth century, colors could be obtained only from natural elements. The first known dyes were those used in Lascaux (France) or Altamira (Spain), dating back to the Magdalenian period. These dyes were mineral pigments: iron oxides for yellows; ochers; reds; and manganese oxides for browns. As early as 1500 BCE, the Egyptians used saffron (yellow), wood (blue), and madder (red) as dyes. Since then, most of the pigments that have been used were of plant origin, extracted from plants, trees, or lichens, or of animal origin, taken from insects such as kermes or mollusks.

Yellow dyes, the most numerous, are generally of mediocre quality in terms of both dyeing power and lightfastness compared to red dyes and blue dyes. As a whole, they are obtained from vegetable sources. The most widely used, especially in the Middle Ages, is weld, which, unlike most other yellow dyes, is based on a flavone-type chromogenic group, relatively resistant to atmospheric oxidation. Red dyes include kermes and cochineal, obtained from insects, while the most important, madder, is a plant extract. They are all hydroxyl derivatives of anthraquinone. The famous Turkish red (madder), used with a metallic mordant in a mixture of calcium and aluminum salts, leads to the favorite Turkish red for its brightness and lightfastness. A dye for purple (extracted from murex), which is similar to that of indigo, was discovered in 1909 by Friedlander.

3 Natural Colors Have Advantages and Disadvantages

Some natural dyes are extracted from renewable resources and are therefore inexpensive and cost-effective. They are of natural origin, and although that does not mean that they are completely safe, they are less toxic than their synthetic counterparts [5, 54]. Many are permitted as food additives and have cosmetic and pharmacological effects.

Natural dyes tend to be earthy and subtle for the most part. Artificial dyes frequently result in garish, harsh, or muddy hues. It takes hours to achieve a natural color and requires combining several mordants, adjusting the pH of dye baths, and mixing different dyes to achieve brightness. Preparing natural dyes is easy, and it barely or rarely involves any chemical reactions; if necessary, it may require simple and light chemical reactions. However, these dyes are susceptible to physicochemical factors and have properties in their native environments that are often altered by extraction processes.

Natural dyes are biodegradable and do not cause disposal problems; their effluents or degradation products are not hazardous and can be fully degraded under natural conditions.

Thus, the major problem with natural dyes is their low stability.

Studies on stability enhancement have been the subject of much research. Several avenues are being pursued, such as the following:

- The use of additives
- Encapsulation with cyclodextrins [42]
- Complementation with amino acids/proteins/polymers [7, 49]

4 Extraction and Mordanting Methods

4.1 Extraction Techniques Throughout the Ages

Since ancient times, humankind has used colorings, perfumes, aromas, and extracts of natural products through the following techniques:

- Pressing exerts pressure on a product to obtain its juice or extract its aromas.
- In the ancient method of decoction, the root or bark of a plant is placed in cold water and brought to a boil, where the constituents dissolve in the water.
- During enfleurage, fragile flowers are placed on frames coated with pure, odorless animal fat, which absorbs the perfume of the flowers on contact; after drying, the odorous substances are extracted by using alcohol.
- Infusion requires pouring boiling water over finely chopped leaves or flowers, and they are left to soak so that their active ingredients dissolve.
- For maceration, while a substance is cold, it is left in an organic solvent, which extracts the soluble constituents.

- Steam distillation (hydrodistillation), a technique that dates back to ancient Egypt, uses steam to carry away the perfumes of the plant (perfumed or essential oils).
- Solvent extraction is a more recent process (starting in nineteenth century) that allows the extraction of compounds that are not soluble in water.

All these techniques have been known for a long time, have been perfected over time, and are still used today.

4.2 *New Extraction Technologies*

4.2.1 Conventional Solid–Liquid Solvent Extraction

Maceration or homogenization, followed by centrifugation or sedimentation, is necessary for solid–liquid extractions. The extraction effectiveness of the solvent and the intended use of the extracted compounds must be considered when choosing a solvent to extract phenolic compounds. If the extracted components are to be used in the food industry, closer monitoring is required to prevent the solvent from contaminating the finished product. Hydroalcoholic mixtures are generally used to extract phenolic compounds because of their high selectivity [33].

The phenols in olive pomace have been extracted by using an 80:20 mixture of methanol and water that has been acidified (pH 2). Compared with methanol and other solvents, ethanol may be less hazardous, making it a safer choice [8].

4.2.2 Hydrolysis

Lipophilic components or cell wall polysaccharides that are linked to the matrix by ester or glycosidic linkages cannot be fully recovered by using extraction procedures that use ethanol–water or methanol–water mixtures. Primary acids and enzymatic hydrolysis are typically used to disrupt these bonds and increase recovery rates [26]. In basic hydrolysis, the sample is treated with a NaOH or KOH solution to dissolve the ester bonds; in acid hydrolysis, the sample is treated with an HCl solution to dissolve the ester and glycosidic bonds. Glycosidases and esterases are necessary for enzymatic hydrolysis [61].

4.2.3 Ultrasound-Assisted Extraction

This technique produces high-frequency waves, modifies the medium's physicochemical properties, and allows cavitation bubbles to form. These cause cell membranes to burst, facilitating interaction between the solvent and the matrix.

Ultrasound-assisted extraction is typically carried out with an ultrasonic bath or probe [37].

These conditions make it a suitable process for the extraction of phenolic compounds. Some authors have recently used ultrasonic devices to recover phenolic compounds from olive industry waste [33]. This technique has been used to extract carotenoids from paprika; in the presence of organic solvents, acceptable levels have been obtained [21].

4.2.4 Extraction by Supercritical Fluid (CO_2)

A procedure known as supercritical fluid extraction raises a fluid's critical temperature and pressure to enhance its solvating capabilities. It transforms into a supercritical fluid with greater diffusion into the sample matrix, resulting in better extraction rates [36]. But because phenolic chemicals are polar, supercritical fluid extraction requires adding a polar modifier, like ethanol or methanol [40]. Compared with other extraction methods, extracts obtained via supercritical fluid extraction have the lowest content of phenolic compounds [58].

4.2.5 Extraction via a Pressurized Liquid

This extraction technique uses organic solvents, which are monitored electronically at high temperatures and pressures. The system usually consists of a stainless-steel chamber or cell, an oven, a pump for solvent distribution, and valves [39]. Some authors have obtained higher levels of phenolic compounds from olive waste through liquid pressure extraction than through other processes.

However, previous studies have shown that the obtained extracts had lower contents of phenolic compounds than those obtained from solid–liquid extraction with a methanol–water solvent mixture [48].

4.2.6 Microwave Extraction

In microwave extraction, microwave energy is used to treat an organic solvent close to the dry or wet sample. The composition and the temperature of the solvent affect how the analytes are distributed throughout the matrix and extractant [21]. The solvent absorbs all the microwave energy, which then heats up to the solvent's boiling point, diffuses into the sample matrix, and solubilizes the analytes. In this instance, there are no significant differences between the mechanism of traditional solid–liquid extraction and that of microwave-assisted extraction. However, microwaves offer a perfectly adjustable energy source that can be instantly controlled. This technique is increasingly used for natural product extraction [32, 41].

4.3 Mordanting Methods

Most natural dyes are solid, so a mordant is required to fix the cloth and stop the color from fading when it is washed or exposed to light. These substances connect textiles and natural dyes. An ingredient known as a mordant helps the dye and fiber's chemical reaction to absorb the dye.

The most popular mordants are blue vitriol, alum, and tin, which are typically used with cream of tartar and which promote uniformity while dulling or darkening colors and bringing out shades of green. The container is also frequently used with cream of tartar and brings out colors, especially reds, oranges, and yellows. Iron (or copper) also serves as a common mordant (which dulls colors and brings out shades of green).

The mordant has been added at different stages of the dyeing process. There are three methods of mordanting:

Premordanting has been the most commonly used method in European industrial dyeing since the Middle Ages. It allows the dye bath to remain unaltered and several fabric samples to be prepared in advance.

Simultaneous mordanting is particularly fast. It yields satisfactory results with wool, especially if light colors are required. The dye is first extracted via decoction if the pigment is not available in powder form. The mordant is added, and the fiber is added after it has completely dissolved.

Postmordanting allows the dye to be intensified, fixed even more firmly, and shaded. It is often applied along with a prestain mordant.

5 Waste Sources for Natural Dyes

5.1 Wastes from Agriculture

Agricultural wastes are byproducts of various farm operations. Most of this waste is disposed of via incineration and by dumping it into or piling it onto landfills. On the one hand, this wastes resources; on the other hand, waste disposal poses a severe environmental threat. The efficient use of agricultural waste could also provide additional income to the farmers or processing industries that produce the waste [23].

5.2 Wastes from Forestry

Because of deforestation and the clearing of plants that can be used for textile dyeing, forestry operations can produce significant amounts of wood chips, bark, fallen branches, and leaves. The primary uses for waste products from the wood industry are as sources of renewable energy and for the production of biofuels.

Because these residues contain natural pigments and dyes, they can be used to dye textiles. Many investigations have been undertaken to dye textiles by using forestry wastes. Eucalyptus byproducts were used to extract natural dyes. Moreover, cotton has been dyed with a dye extracted from eucalyptus bark [4]. Eucalyptus leaves have been used to dye silk and wool fabrics [45]. Finally, extracts of kermes oak (Quercus coccifera) have been applied to cotton and wool fabrics [9, 14].

5.3 Wastes from Industry

Most food industries produce significant amounts of solid and liquid wastes. These wastes represent potential pollutants that come with disposal problems. These wastes are either used as animal feed, composted, or disposed of.

Agrifood industries generate massive amounts of colored plant wastes. Some studies on the use of plant wastes from the food and beverage industry have shown that such wastes can be suitable as sources for the natural dyeing of natural and synthetic fibers, such as in the winemaking industry [12, 14–16], the olive oil industry [24, 44, 43], the tomato-processing industry [13], the red pepper industry [25], and the date paste–processing industry [55, 56].

6 Case Studies on Specific Natural Dyes from Wastes

6.1 Olive

The cultivation of olive trees is part of the Mediterranean tradition. It is a symbol of civilization for the Mediterranean people and, at the same time, a common delicacy on their tables. There are about 840 million olive trees in the Mediterranean and 90 million in the rest of the world. Olive groves cover about 10 million hectares, having an average density of 80 olive trees per hectare. Olive growing plays a significant role in the national economies of many countries, particularly Spain, Italy, Greece, Turkey, Tunisia, and Morocco. Paradoxically to the production of olive oil, these countries discharge wastewaters, which currently pose environmental risks for the Mediterranean region. Indeed, more than 30 million m^3 of wastewater are produced annually in the Mediterranean basin, compared to a world volume of less than 40 million m^3/year.

The olive oil extraction processes generate residues of different solid and liquid natures [22]:

- The vegetation water is a brown aqueous liquid residue that is separated from the oil via centrifugation or decantation after pressing.
- Leaves collected at the oil mill are not the residues of the pruning but rather the leaves obtained after washing and cleaning the olives at the entrance to the oil mill. In Greece, the quantity of these leaves is about 5% of the weight of the olives.
- Crude pomace is the residue from the first oil extraction, which is carried out by pressing the whole olive. Its water content (24%) and oil content (9%) are relatively high, thus favoring its rapid alteration in the open air.
- Depleted pomace is the residue obtained after de-oiling the crude pomace with a solvent, generally hexane.
- Partially destoned pomace results from partially separating the core from the pulp by sieving it or via ventilation. It is "fatty" if it has not been de-oiled and is "defatted or exhausted" if it has been de-oiled by a solvent.

Case Study 1: The Valorization of Solid Olive Byproducts in the Production of Dyes for Textile Use [24]

Elksibi et al. [24] developed an environmentally friendly nonconventional process for the aqueous extraction of dyes from olive pomace. The modeling and optimization of the extraction process allowed for determining the optimal operating conditions for the highest extraction yields. The textile application of this extraction resulted in good dyeing yields, and good to very good fastnesses was obtained.

The application of the extracted dyes on textile supports resulted in dyes with excellent fastnesses and high biological activity, which was comparable to or better than the commonly used standards. The evaluation of the effects of these products on the environment was quite encouraging given their low toxicity and moderate pollution characteristics.

Case Study 2: Dyeing Synthetic Textile Fibers (Polyamide and Acrylic) with Dyes Prepared from the Olive Tree [29–31]

Haddar et al. [29, 30] showed that acrylic fiber, a synthetic fiber, does not have a very high affinity for the dyeing substances in olive mill wastewater. Hence, an ionization pretreatment is needed to improve the dyeing yield of this natural dye. According to a comparative study featuring comparative studies of other quaternary ammonium ionizing agents, Croscolor DRT is the best ionizing agent for this purpose. By evaluating the effects of the experimental parameters on the obtained dyeing quality, the authors were able to model and optimize the dyeing strength (K/S) by using the design of the experimental method with the MINITAB 15 software. Indeed, the optimal conditions obtained with the help of the complete factorial design are as follows: a pH of 3, a temperature of 100 °C, and a treatment time of 60 min.

Haddar et al. [31] studied the effects of ultrasonic energy on the dyeing quality of polyamide fiber with olive mill wastewater. However, the ultrasound-assisted dyeing resulted in significant improvements in the studied samples' respective dyeing strengths (K/S) and improvements in their fastness scores.

Case Study 3: Dyeing of Natural Textile Fibers (Wool and Cotton) with Olive Mill Wastewater [43, 44]

The study carried out by Meksi et al. [43] revealed that the presented method of dyeing wool with oil mill wastewater was attractive because it solves the environmental problems of using oil mill wastewater. Furthermore, a study on the influences of various parameters on the suitability of wool dyeing with oil mill wastewater showed that the temperature and pH of the dyeing bath and the dyeing time have important effects on the color strength. The best results were obtained at pH 2, at 100 °C, and for 90 min.

Meksi et al. [44] investigated the valorization of oil mill wastewater on the dyeing of cationized cotton fabrics. They found that cotton cationized with Croscolor DRT generally imparted a higher color yield than cotton modified with tannic acid did. The authors also studied the environmental impact of the residual dye bath. The results confirmed that dyeing with this aqueous waste considerably decreases organic pollution (the reduction in chemical oxygen demand (COD) and biological oxygen demand (BOD5)).

6.2 *Grape*

Grapes (*Vitis vinifera*) are among the world's most famous and widely grown fruits [60]. Viticulture is one of the leading agricultural activities in the world. Like most fermentation systems, the winemaking cycle results in byproducts. Grape byproducts can be classified into three categories: pomace and its products (pulp and pips), concentrated grape juice, and leaves and branches.

Grape marc is the main byproduct of winemaking. It consists of the solid residues (pips, skins, and stalks) remaining after the critical stage of making white or red wine [20]. Grape marc constitutes about 20% of the weight of the grapes used for winemaking, amounting to about nine million tons per year [53].

Case Study 4: Extraction of Dyes from Grape Pomace and Dyeing Wool (Baraka et al. 2015)

Baaka et al. [10] developed processes for extracting coloring matter from grape pomace. They studied the experimental parameters acting on the extraction process. Next, they determined the optimal extraction conditions. Baaka et al. also studied the process of dyeing wool with the aqueous extracts of grape marc. They conducted a comparative study on the two dyeing processes: conventional wool dyeing and ultrasonic wool dyeing.

Case Study 5: Study on Dyeing Cotton and Acrylic Fibers with the Aqueous Extracts of Grape Marc

Despite its high affinity for protein fibers, the coloring matter extracted from grape marc does not provide a high yield when the extracted grape marc does not allow for a high yield, such as in the cases of cotton dyeing and acrylic dyeing. Moreover, pretreating these fibers seems an indispensable approach to improving the quality of dyeing.

6.3 *Tomato*

The tomato is the second-most-consumed vegetable in the world (grown on more than 4.8 million hectares worldwide). The tomato is also one of the most widespread horticultural species in the world. Tomato can be used in many forms, from its consumption in its raw state to its use as a manufactured derivative (sauces, juices, ketchup, etc.).

The industrial processing chain for tomatoes produces enormous amounts of residue, generally called tomato pomace or tomato byproducts. These quantities can represent up to 10% of the weight of fresh tomatoes. Tomato byproducts include a rich variety of biologically active substances, mainly lycopene, whose antioxidant, hypolipidemic, and anticarcinogenic activities have been demonstrated by in vitro and in vivo studies.

Case Study 6: Carotenoids from Tomato-Processing Wastes: Application on Textile Dyeing and Assessment of Its Antioxidant Activity [11]

The objectives of case study 6 were to investigate the yield of carotenoids extracted from tomato waste by using various solvents and solvent mixes and to improve extraction conditions. A central composite design featuring five levels and five factors was used to optimize the extraction conditions. Under the following ideal circumstances, 80.7 μg/g of carotenoids were produced:

(i) Solvent solid ratio—90:1 (v/w)
(ii) Percentage of hexane in the solvent mixture—60%
(iii) Extraction time—50 min
(iv) Number of extractions—four
(v) Extraction temperature—35 °C

The authors demonstrated the use of natural pigments from tomato pomace to dye wool, silk, and polyamide materials with reddish-yellow shades and excellent fastnesses.

6.4 Red Pepper

Pepper, of the genus *Capsicum*, contains a high level of phytochemicals that may contribute to its antioxidant activity [19]. It represents a rich source of phenolic compounds, capsaicinoids, carotenoids, chlorophylls [6], and vitamins (A, B1, B2, B3, C, and E).

Case Study 7: Extraction of Colored Compounds from the Byproducts of the Red Pepper–Processing Industry [25]
The study by Elksibi et al. revealed that the recovery of colored phenolic compounds from red pepper waste depended on the solvent system used. The mixing plane methodology determined the optimal composition of the most efficient solvent mixture for extracting polyphenols and flavonoids, namely 65% water and 35% acetone. In addition, the quality assessment of the dyed fabrics showed higher fastness values and stabler antioxidant power.

Elksibi et al. also studied the antibacterial activities of red pepper dyes. The results showed that the extracted dye could be a promising antibacterial agent for developing bioactive textile materials.

6.5 Date Palm

The date palm is a hardy tree that can adapt to the aridest regions in the world. It is a monocotyledonous tree, of the family Palmaceae or Phoenicicaceae, of the subfamily Coryphineae, and of the genus and species *Phoenix dactylifera* L. It is the primary source of life for the Saharan population.

In addition to dates, date palms offer various agricultural byproducts that have several applications, from basketry crafts to timber and livestock feed. Enormous quantities of fibrous materials, in the form of stipe and palm, poor-quality dates, and seeds, are discarded into the environment. This waste has massive potential to be used as a raw material in many areas, such as construction, making furniture, energy production and storage, food, pharmaceuticals, and biological research.

Case Study 8: Extraction of Colored Compounds from the Date Paste–Processing Industry [55, 56]
First, Souissi et al. [55] optimized the extraction process of colorant from date palm seeds by using a Taguchi experimental design. The optimal extraction conditions were as follows:

- An extraction temperature of 80 °C
- An extraction solvent composed of 10% acetone and 90% water
- An extraction duration of 90 min

Second, Souissi et al. [56] studied the dyeability of cotton fabrics with colorant extract from date palm seeds. They investigated using a Box-Behnken experimental design to optimize the dyeing process. Next, several metallic and natural mordants were applied to improve fastness.

6.6 *Fennel*

Fennel (*Foeniculum vulgare*) is a shrub that grows to a height of 80 to 150 cm; has a strong fragrance; belongs to the Apiaceae (Umbelliferae) family. It was thought to be indigenous to the Mediterranean regions, yet it has become extensively naturalized elsewhere.

It is used extensively in traditional medicine and cuisine. Its extracts contain a variety of useful properties, including analgesic, anti-inflammatory, antispasmodic, carminative, diuretic, expectorant, and laxative properties, and these extracts are useful for treating neurological disorders.

Case Study 9: Extraction of Natural Colorant from Fennel [31]
Maximizing the value of the aqueous fennel leaf extract on the dyeing of cationized cotton fibers was the aim of case study 9 [31]. The most effective dyeing conditions were found by implementing a response surface design. The results showed that a pH of 8.22, a temperature of 99.83 °C, and a dyeing time of approximately 59.38 min were used to achieve the best dyeing quality.

6.7 *Orange*

One of the most significant fruits in the world is the orange. Numerous byproducts, like pulp, seeds, and peels, are produced during their industrial processing and consumption and are frequently dumped into nature [28, 38]. Typically, molasses made from peels is used to make fuel and animal feed [35]. These peels are sources of physiologically active complexes and are abundant in phenolic compounds like flavonoids, according to recent studies [34, 57].

Case Study 10: Cotton Dyeing with Orange Peel Waste [11]
In case study 10, the scientists showed how natural dyes can be produced from waste from the orange industry and can be viable substitutes for synthetic colors in underdeveloped nations.

Carotenoid recovery relies on the solvent system, as demonstrated by Baaka et al.

7 Future Perspective

Sustainable development requires transitioning industry to meet present human needs without compromising the environment. This concept includes balancing ecology, economics, social impacts, politics, cultures, and human health. The use of natural dyes in their current state and their availability raises major concerns about the sustainability of dyeing textiles. The nontoxic, noncarcinogenic, biodegradable, and ecofriendly properties of natural dyes have reached the hearts of consumers who want to live healthy lifestyles.

Using natural dyes in textile substrates will bring about new business ventures. In a number of applicable industries, including the leather, textile and apparel, cosmetics, culinary, pharmaceutical, and paint industries, among others, the natural dye market has significant potential and a promising future.

In addition, many of the plants reviewed in this chapter have antimicrobial, antibacterial, antioxidant, anti-inflammatory, and many other medicinal properties that can be administered through skin absorption.

8 Conclusion

Around the world, ecological considerations are becoming important factors in the selection of consumer goods. Consumers are looking for quality products that do not harm the environment. In this context, the use of byproducts from various sources for textile dyeing can be beneficial for two reasons. First, they constitute a durable alternative to waste-disposal management. Second, they are cheaper sources for textile dyeing.

While a number of studies have been developed to find new sources of natural dyes, most notably from by products, much remains to be conducted in a systemic and scientific manner. A systematic method for collecting wastes at their sources and sustainable options for transforming wastes into suitable concentrates and dye formulations to make reproducible dyes still need to be investigated.

References

1. Adeel, S., Rehman, F., Zuber, M., Batool, F., Habib, N., Hussaan, M., & Amin, N. (2021). Environmental friendly application of ultrasonic rays for extraction of natural colorant from Harmal (P. Harmala) for dyeing of bio-mordanted silk. *Journal of Engineered Fibers and Fabrics, 16*, 1–8.
2. Ajmal, M., Adeel, S., Azeem, M., Zuber, M., Akhtar, N., & Iqbal, N. (2014). Modulation of pomegranate peel colourant characteristics for textile dyeing using high energy radiation. *Industrial Crops and Products, 58*, 188–193.
3. Alam, M. M., Rahman, M. L., & Haque, M. Z. (2007). Extraction of henna leaf dye and its dyeing effect on textile fibre. *Bangladesh Journal of Scientific and Industrial Research, 42*(2), 217–222.

4. Ali, S., Nisar, N., & Hussain, T. (2007). Dyeing properties of natural dyes extracted from eucalyptus. *The Journal of The Textile Institute, 98*(6), 559–562.
5. Ali, N. F., EL-Mohamedy, R. S. R., & El-Khatibl, E. M. (2011). Antimicrobial activity of wool fabric dyed with natural dyes. *Research Journal of Textile and Appeal, 15*, 1–10.
6. Alvarez Parilla, E., De La Rosa, L. A., Amarowicz, R., & Shahidi, F. (2011). Antioxidant activity of fresh and processed Jalapenõ and serrano peppers. *Journal of Agricultural and Food Chemistry, 59*, 163–173.
7. Apanasenko, I. E., Selyutina, O. Y., Polyakov, N. E., Suntsova, L. P., Meteleva, E. S., Dushkin, A. V., Vachali, P., & Bernstein, P. S. (2015). Solubilization and stabilization of macular carotenoids by water soluble oligosaccharides and polysaccharides. *Archives of Biochemistry and Biophysics, 572*, 58–65.
8. Araújo, M., Pimentel, F. B., Alves, R. C., Beatriz, M., & Oliveira, P. P. (2015). Review phenolic compounds from olive mill wastes: Health effects, analytical approach and application as food antioxidants. *Trends in Food Science & Technology, 45*, 200–211.
9. Baaka, N. (2022). Sustainable dyeing of wool fabric using kermes oak (*Quercus Coccifera* L) as source of natural colorant. *Journal of Natural Fibers, 19*(1), 37–45.
10. Baaka, N., Ben Ticha, M., Haddar, W., Hammami, S., & Mhenni, M. F. (2015). Extraction of natural dye from waste wine industry: Optimization survey based on a central composite design method. *Fibers and Polymers, 16*(1), 38–45.
11. Baaka, N., Mahfoudhi, A., & Mhenni, M. F. (2017a). Orange peels waste as a low cost cotton natural dye. *Moroccan Journal of Chemistry, 5*(2).
12. Baaka, N., Ben Ticha, M., Haddar, W., Amorim, M. T. P., & Mhenni, M. F. (2017b). Sustainability issues of ultrasonic wool dyeing with grape pomace colourant. *Natural Product Research, 31*(14), 1655–1662.
13. Baaka, N., Elksibi, I., & Mhenni, M. F. (2017c). Optimisation of the recovery of carotenoids from tomato processing wastes: Application on textile dyeing and assessment of its antioxidant activity. *Natural Product Research, 31*(2), 196–203.
14. Baaka, N., Ben Ticha, M., Haddar, W., & Mhenni, M. F. (2018). Upgrading of UV protection properties of several textile fabrics by their dyeing with grape pomace colorants. *Fibers and Polymers, 19*(2), 307–312.
15. Baaka, N., Mahfoudhi, A., & Mhenni, M. F. (2019a). Tannin-rich natural dye extracted from kermes oak (Quercus coccifera L.): Process optimization using response surface methodology (RSM). *Journal of Natural Fibers, 16*(8), 1209–1220.
16. Baaka, N., Ben Ticha, M., Haddar, W., & Mhenni, M. F. A. (2019b). Challenging approach to improve the low affinity of acrylic fibres to be successfully dyed with a bio-colorant extracted from grape Marc. *Journal of Renewable Materials, 7*(3), 289–300.
17. Bhargava, D., & Shahnaz, J. (2013). Microbial dyes: A new dimension to natural dyes. *Colourage, 60*(7), 42–43.
18. Bouaziz, A., Dridi, D., Gargoubi, S., Chelbi, S., Boudokhane, C., Kenani, A., & Aroui, S. (2021). Analysis of the Coloring and antibacterial effects of natural dye: Pomegranate Peel. *Coatings, 11*, 1277.
19. Chen, L., & Kang, Y.-H. (2013). Anti-inflammatory and antioxidant activities of red pepper (Capsicum annuum L.) stalk extracts: Comparison of pericarp and placenta extracts. *Journal of FunctionalFoods, 5*, 1724–1731.
20. Crespo, J. G., & Brazinha, C. (2010). Membrane processing: Natural antioxidants from winemaking by-products. *Filtration & Separation, 47*, 32–35.
21. Csiktusnádi Kiss, G. A., Forgács, E., Cserháti, T., Mota, T., Morais, H., & Ramos, A. (2000). Optimisation of microwave-assisted extraction of pigments from paprika (*Capsicum annum* L.) powders. *Journal of Chromatography, 889*, 41–49.
22. Dermeche, S., Nadour, M., Larroche, C., Moulti-Mati, F., & Michaud, P. (2013). Review olive mill wastes: Biochemical characterizations and valorization strategies. *Process Biochemistry, 48*, 1532–1552.
23. Doty, K., Haar, S., & Kim, J. (2016). Black walnut, Osage orange and eastern redcedar sawmill waste as natural dyes: Effect of aluminum mordant on color parameters. *Fashion & Textiles, 3*, 22–38.

24. Elksibi, I., Haddar, W., Ben Ticha, M., Gharbi, R., & Mhenni, M. F. (2014). Development and optimisation of a non conventional extraction process of natural dye from olive solid waste using response surface methodology (RSM). *Food Chemistry, 161*, 345–352.
25. Elksibi, I., Ben Slama, R., Faidi, K., Ben Ticha, M., & Mhenni, M. F. (2015). Mixture approach for optimizing the recovery of colored phenolics from red pepper (Capsicum annum L.) by-products as potential source of natural dye and assessment of its antimicrobial activity. *Industrial Crops and Products, 70*, 34–40.
26. Fazary, A. E., & Ju, Y. H. (2007). Feruloyl esterases as biotechnological tools: Current and future perspectives. *Acta Biochimica et Biophysica Sinica, 39*, 811–828.
27. Gala, S., Sumarno, S., & Mahfud, M. (2021). Optimization of microwave-assisted extraction of natural dyes from jackfruit wood (Artocarpus heterophyllus Lamk) by response surface methodology. *Engineering and Applied Science Research, 49*(1), 29–35.
28. Gorinstein, S., Martín-Belloso, O., Park, Y. S., Haruenkit, R., Lojek, A., Ĉíž, M., Caspi, A., Libman, I., & Trakhtenberg, S. (2001). Comparison of some biochemical characteristics of different citrus fruits. *Food Chemistry, 74*, 309–315.
29. Haddar, W., Elksibi, I., Meksi, N., & Mhenni, M. F. (2014a). Valorization of the leaves of fennel (Foeniculum vulgare) as natural dyes fixed on modified cotton: A dyeing process optimization based on a response surface methodology. *Industrial Crops and Products, 52*, 588–596.
30. Haddar, W., Baaka, N., Meksi, N., Elksibi, I., & Mhenni, M. F. (2014b). Optimization of an ecofriendly dyeing process using the wastewater of the olive oil industry as natural dyes for acrylic fibres. *Journal of Cleaner Production, 66*, 546–554.
31. Haddar, W., Baaka, N., Meksi, N., Ben Ticha, M., Guesmi, A., & Mhenni, M. F. (2015). Use of ultrasonic energy for enhancing the dyeing performances of polyamide Fibers with olive vegetable water. *Fibers and Polymers, 16*(7), 1506–1511.
32. Islam, S., Shahid, M., & Mohammad, F. (2013). Review perspectives for natural product based agents derived from industrial plants in textile applications. *Journal of Cleaner Production, 57*, 2–18.
33. Klen, J. T., & Vodopivec, B. M. (2011). Ultrasonic extraction of phenols from olive mill wastewater: Comparison with conventional methods. *Journal of Agricultural and Food Chemistry, 59*, 12725–12731.
34. Koolaji, N., Shammugasamy, B., Schindeler, A., Dong, Q., Dehghani, F., & Valtvhev, P. (2020). Citrus Peel flavonoids as potential cancer prevention agents. *Current Developments in Nutrition, 4*, 1.
35. Kumar, K. A., Nayarani, M., Subanthini, A., & Jayakumar, M. (2011). Antimicrobial activity and phytochemical analysis of citrus fruit peels -utilization of fruit waste. International journal of engineering. *Science and Technology, 3*(6), 5414–5421.
36. Lang, Q., & Wai, C. M. (2001). Supercritical fluid extraction in herbal and natural product studies e a practical review. *Talanta, 53*, 771–782.
37. Leonelli, C., & Mason, T. (2010). Microwave and ultrasonic processing: Now a realistic option for industry. *Chemical Engineering and Processing, 49*, 885–900.
38. Li, B. B., Smith, B., & Hossain, M. (2006). Extraction of phenolics from citrus peels I. Solvent extraction method. *Separation and Purification Technology, 48*, 182–188.
39. Lozano-Sanchez, J., Castro-Puyana, M., Mendiola, J. A., Segura-Carretero, A., Cifuentes, A., & Ibaez, E. (2014). Recovering bioactive compounds from olive oil filter cake by advanced extraction techniques. *International Journal of Molecular Sciences, 15*, 16270–16283.
40. Machado, B. A. S., Pereira, C. G., Nunes, S. B., Padilha, F. F., & Umsza-Guez, M. A. (2013). Supercritical fluid extraction using CO_2: Main applications and future perspectives. *Separation Science and Technology, 48*, 2741–2760.
41. Majetich, G., & Hicks, R. (1995). The use of microwave heating to promote organic reactions. *The Journal of Microwave Power and Electromagnetic Energy, 30*, 27–30.
42. Mangolim, C. S., Moriwaki, C., Nogueira, A. C., Sato, F., Baesso, M. L., Neto, A. M., & Matioli, G. (2014). Curcumin–b-cyclodextrin inclusion complex: Stability, solubility, characterisation by FT-IR, FT-Raman, X-ray diffraction and photoacoustic spectroscopy and food application. *Food Chemistry, 153*, 361–370.

43. Meksi, N., Haddar, W., Hammami, S., & Mhenni, M. F. (2012). Olive mill wastewater: A potential source of natural dyes for textile dyeing. *Industrial Crops and Products, 40*, 103–109.
44. Meksi, N., Haddar, W., Baaka, N., Mejri, M., & Mhenni, M. F. (2018). Eco-friendly dyeing of modified cotton fabrics using olive dyes extracted from olive mill wastewater: Focus on dyeing qualities and environmental performances. *International Journal Of Applied Research On Textile, 6*(1), 46–64.
45. Mongkholrattanasit, R., Kryštfek, J., & Wiener, J. (2010). Dyeing and fastness properties of natural dyes extracted from eucalyptus leaves using padding techniques. *Fibers and Polymers, 11*(3), 346–350.
46. Moussa, I., Baaka, N., Khiari, R., Moussa, A., Mortha, G., & Mhenni, M. F. (2018). Application of Prunus amygdalus by-products in eco-friendly dyeing of textile fabrics. *Journal of Renewable Materials, 6*(1), 55–67.
47. Nguyen, H. L., & Bechtold, T. (2021). Thermal stability of natural dye lakes from Canadian goldenrod and onion peel as sustainable pigments. *Journal of Cleaner Production, 315*.
48. Perez-Serradilla, J. A., Japon-Lujan, R., & Luque de Castro, M. D. (2008). Static-dynamic sequential superheated liquid extraction of phenols and fatty acids from alperujo. *Analytical and Bioanalytical Chemistry, 392*, 1241–1248.
49. Polyakov, N. E., Leshina, T. V., Meteleva, E. S., Dushkin, A. V., Konovalova, T. A., & Kispert, L. D. (2009). Enhancement of the photocatalytic activity of TiO_2 nanoparticles by water-soluble complexes of carotenoids. *The Journal of Physical Chemistry, 113*, 275–282.
50. Saima, N., Bhatit Ijaz, A., & Shahid, A. (2011). Dyeing properties of cotton fabric using un-irradiated and gamma irradiate extracts of Eucalyptus camaldulensis bark powder. *Indian Journal of Fibre & Textile Research, 36*(June), 132–136.
51. Samanta, A. K., Agarwal, P., & Dutta, S. (2007). Dyeing of jute and cotton fabric using jack-fruit wood extract. Part 1: Effects of mordanting in dyeing process variables on color yield and color fastness properties. Indian journal of Fiber and textile. *Research, 32*(4), 466–476.
52. Saxena, S., & Raja, A. S. M. (2014). Roadmap to sustainable textiles and clothing. In *Natural dyes: Sources, chemistry, application and sustainability issues*. Springer.
53. Schieber, A., Stintzing, F., & Carle, R. (2001). By-products of plant food processing as a source of functional compounds – Recent developments. *Trends in Food Science and Technology, 12*, 401–413.
54. Siva, R. (2007). Status of natural dyes and dye-yielding plants in India. *Current Science, 92*, 916–925.
55. Souissi, M., Guesmi, A., & Moussa, A. (2018a). Valorization of natural dye extracted from date palm pits (Phoenix dactylifera) for dyeing of cotton fabric. Part 1: Optimization of extraction process using Taguchi design. *Journal of Cleaner Production, 202*, 1045–1055.
56. Souissi, M., Guesmi, A., & Moussa, A. (2018b). Valorization of natural dye extracted from date palm pits (Phoenix dactylifera) for dyeing of cotton fabric. Part 2: Optimization of dyeing process and improvement of colorfastness with biological mordants. *Journal of Cleaner Production, 204*, 1143–1153.
57. Suleria, H. A., Barrow, C. J., & Dunshea, F. R. (2020). Screening and characterization of phenolic compounds and their antioxidant capacity in different fruit peels. *Food, 9*, 1206.
58. Taamalli, A., Arraez-Roman, D., Barrajon-Catalan, E., Ruiz-Torres, V., Perez-Sanchez, A., & Herrero, M. (2012). Use of advanced techniques for the extraction of phenolic compounds from Tunisian olive leaves: Phenolic composition and cytotoxicity against human breast cancer cells. *Food and Chemical Toxicology, 50*, 1817–1825.
59. Tian, Y., Liu, Q., Zhang, Y., Hou, X., & Zhang, Y. (2022). Obtaining colored patterns on polyamide fabric with laser pretreatment and pomegranate peel dyeing. *Textile Research Journal, 92*(12), 2080–2089.
60. Xu, C., Zhang, Y., Cao, L., & Lu, J. (2010). Phenolic compounds and antioxidant properties of different grape cultivars grown in China. *Food Chemistry, 119*, 1557–1565.
61. Zheng, H., Hwang, I.-W., & Chung, S.-K. (2009). Enhancing polyphenol extraction from unripe apples by carbohydrate-hydrolyzing enzymes. *Journal of Zhejiang University Science B, 10*, 912–919.

Recent Approaches and Advancements in Natural Dyes

Silpinwita Das and Sankar Roy Maulik

Abstract Natural dyes have played significant roles in shaping and rewriting history worldwide. Natural colorants, mainly from plant and animal origins, have been used for dyeing various things. They bear a crucial historical testimony that widens the horizon of looking at the glorious past and promising future. Recently, the detrimental environmental pollution caused by synthetic dyes has created room to reconsider natural dyes as alternative sources. The newly discovered natural dye sources and nontoxic and sustainable processes for their application have broadened the spectrum of natural dyes that can be employed in traditional and newer interventions. Today, advanced, optimized, and unconventional techniques, such as ultrasonic and microwave-assisted extraction, have been applied contrary to conventional extraction to obtain the maximum color yield with minimum environmental pollution. Statistically designed programs are now commonly exploited for optimization and process modeling. In recent years, relatively novel and scarcely investigated microbial dyes were being utilized to achieve high production yields, bright colors, and unique functional properties. These natural dye sources can also positively affect the textile industry and the environment by reducing water and contamination, energy consumption, etc. Besides textiles, eco-leather is now becoming an option for environmentally friendly, nontoxic leather production by overcoming the limitation of utilizing natural dyes. These dyes set the trend for the textile industry from being fashionable to functional for their innocuous, antiallergenic, and biodegradable qualities. More recently, certain artificial color additives have demonstrated that they can lead to negative health issues following their consumption. On the other hand, natural colors have been proven safe in these areas. Therefore, research into newer sources of pigments and their occurrence is coming to the front to find natural alternatives for synthetic dyes. Natural dyes in food, beverages, pharmaceuticals, etc. enhance their functional and aesthetic impressions. Moreover, the intelligent food packaging industry is trying to adopt natural dyes to monitor the

S. Das · S. Roy Maulik (✉)
Department of Silpa-Sadana, Visva-Bharati (A Central University), Sriniketan, India
e-mail: s_r_moulik@yahoo.co.in

S. S. Muthu (ed.), *Natural Dyes and Sustainability*, Sustainable Textiles: Production, Processing, Manufacturing & Chemistry,
https://doi.org/10.1007/978-3-031-47471-2_4

freshness of foodstuffs, although color stability in different circumstances of these dyes still needs to be worked on. In the past few decades, natural dyes have been considered the feasible alternatives to the toxic, expensive, and rare inorganic sensitizers in solar cells. Natural sensitizers containing plant pigments responsible for chemical reactions absorb sunlight's photons and convert them into electrical energy.

This review provides comprehensive breakthroughs in using newer ways of applying natural dyes, with technological developments in the functional finishing of textiles, food coloration, food packaging, cosmetics, leather, dye-sensitized solar cells, microencapsulation, etc.

Keywords Biodegradable · Coloration · Cosmetic · Food · Natural dye · Packaging · Solar cell

1 Introduction

One of the aspects of nature that adds beauty and fascination to daily life is color. Since ancient times, plants, insects/animals, and minerals have been used to color textiles, foods, and cosmetics. Some dyes were later found to be used to color paper, leather, shoe polish, candles, and wood [6, 41]. Among these dyes, vegetable dyes are obtained from different plant parts, such as roots, bark, leaves, flowers, and fruits. In contrast, animal dyes such as lac, cochineal, tyrian purple, etc. are obtained from the animal kingdom. Mineral dyes are derived from various inorganic metal salts [42, 108]. These dyes have also been used as traditional medicines since antiquity. Many well-preserved textiles discovered at archaeological sites systematically have allowed us to understand the wide range of natural dyes used in ancient periods [101]. With the discovery and arrival of coal tar dyes (now synthetic dyes), natural dyes gradually diminished in use [53, 104]. Technologies for using natural dyes on textiles still need to advance through stages of sophistication and refinement to meet newer demands, address known disadvantages, improve processing efficiency, and increase the end product's constancy and comfortability to make them more acceptable and appealing [116]. Recently, research and development on natural dyes have been carried out because of increasing environmental concerns and the growing demand for a natural manner of living [118]. The recent COVID-19 pandemic situation created new room for nontoxic, natural, or herbal products [68]. Various natural sources can be used in functional textiles, food coloration, cosmetics, smart food packaging, dye-sensitized solar cells, and several other applications [22].

This chapter focuses mainly on recent approaches and advancements in using natural dyes. Effective and nontoxic extraction methods, how extraction and dyeing techniques can be optimized, and various characterization methodologies are discussed. The chapter also highlights research on edible natural pigments for a sustainable way of coloring food, beverages, and pharmaceutical preparations and using natural pigments in various cosmetics production for their different pharmacological effects. Natural dyes with promising potentials as nondestructive monitoring materials in intelligent food packaging and as sensitizers for dye-sensitized solar cells are also reported here. The use of artificial neural networks to predict dyeing behavior and measure color strength is also discussed in the chapter. Studies on using natural dyes as counters to synthetic dyes are still in progress.

2 Approaches and Advancements

2.1 Coloration and Functional Finishing

Natural dyes can be categorized as substantive and adjective, depending on their affinity for the substrate [87]. A substantive dye have a positive affinity for the substrate. In contrast, an adjective dye can be applied only in the presence of metal salts, tannins, etc., known as mordants [3]. The commonly used mordants are alum, cupric sulfate, ferrous sulfate, potassium dichromate, stannous chloride, and stannic chloride [40, 93]. In various mordants, many natural dyes appear to produce multiple colors on natural fibers and fabrics. Other than natural fabrics, polyester fabric can also be dyed with natural indigo dye by using sodium dithionite as a reducing agent in an alkaline bath [14]. *Bixa orrellana* dye was also applied in polyester fabric dyeing, showing poor lightfastness [23].

The mordant and dye concentrations influence the shades of the dyed substrates. The percentage of these metal mordants for textiles should not exceed established eco-parameters, because the disposal of dye baths generates severe environmental pollution, and it can be more harmful if the mordants consist of heavy metals [84]. Most metal mordants are considered toxic, and bio-mordants with high tannin contents can replace them. Wool fibers dyed with the aqueous extraction of madder and weld dyed by using walnut husks and myrobalan as bio-mordants showed good color strength and lightfastness [49]. The unmordanted wool fabric dyed with fallen pomegranate leaves showed good washing and lightfastness. The response surface method was adopted to optimize variables to produce the highest color intensity of the dyed fabric [43]. Cotton fabrics, treated with a bio-mordant, namely aloe vera, and dyed with turmeric showed good light, washing, rubbing, and perspiration fastness over the fabrics mordanted with copper sulfate [5]. Various shades of natural colors were imparted to wool fabric from plant sources such as Kalanchoe pinnata, papaya, peepal, and banyan in the presence of plant-based (harda, amla, pomegranate, and orange) and mineral mordants (alum, copper, and ferrous sulfate). All four natural extracts, along with natural mordants, produced good colorfastness to washing and light, opening up a path toward the sustainable coloration of wool [80].

Moreover, the quality of the water plays an important role in the dyeing process. Silk fabrics dyed with *Ceriops decandra* and *Diospyros malabarica* showed effective surface color strength and improved overall colorfastness. This was because using the river water, which contains several kinds of heavy metals and other metals, as a dye liquor medium elevated the dyeing quality of the textile substrates [65]. Thus, a controlled dyeing process is the key to optimizing results. Given the health hazards and environmental pollution created by synthetic dyes, scientific interventions into using natural dyes are encouraged for time- and energy-saving, cost-effective, and ecofriendly alternatives [29].

Natural dyes can be obtained in different forms. The efficient extraction of active and functioning pigments from one complex dye source is essential for standardization and optimization. Organic solvents help retain the efficient coloring components, which may otherwise be degraded in conventional water extraction at high temperatures [84].

Microwave-assisted extraction (MAE) is nontoxic when compared with conventional approaches because the former avoids thermal degradation and oxidation and reduces the extraction time. This process intensifies the extraction yield by using less of each solvent. However, the associated maintenance costs have been found to be higher than the traditional extraction process [73, 81]. The wool fabrics dyed with microwave-extracted cochineal show good colorfastness properties [36]. Using aqueous organic solvents as part of an ultrasound technique can improve the extraction efficiency of phenolic and flavonoid compounds. Ziziphus bark is a natural dye source containing many anthraquinones, phenolic compounds, saponins, and alkaloid compounds, which impart functional properties to dyed wool [64].

Other than the MAE, ultrasound-assisted extraction (UAE), gamma-assisted extraction (GAE), and infrared-assisted extraction (IAE) are becoming popular nowadays [32] for the effective extraction of colorants. The UAE technique efficiently extracted the flavonoids of Hawthorn fruits. Nylon fabric dyed with this extraction method showed high colorfastness and antimicrobial properties [86]. Sonochemical mordanting using soy protein produced higher color strength on cotton fabrics by enhancing the dye's affinity for *Artemisiae argyi* as a natural dye source. The UV-protection and antibacterial properties of ultrasonic processed soy-mordanted fabrics were significantly improved [120].

Some natural dyes could be used as effective blocking agents against harmful UV radiation. Such dyes have also been used to produce photoprotective clothing. *Punica granatum* rind extract imparted excellent ultraviolet protection factor (UPF) values to dyed cotton fabric. In this case, zinc oxide enhanced sunscreen's coverage and increased the UPF of the dyed cotton fabric [74]. Alum-mordanted silk fabric dyed with *Artemisia herba-alba* leaves provided different colors and shades and a good flame-retardant property when exposed to fire by improving the molecular structure of silk fabric [61]. The ammonium salt of phosphorylated safflower yellow colorant was used to provide a flame-retardant finish to the silk fabric. The char residue and the limiting oxygen index were increased more in the flame-retardant silk sample than in the control sample, also conferring excellent UV-resistance properties as well [117].

A few newly discovered natural sources opened further investigations into and possibilities for dyeing. Cotton fabric dyed with *Crotalaria globosa*, a tropical weed, was imparted with various functional properties because of the flavonoid compounds in the leaves. Multiple mordants and different types of mordanting techniques have been employed to optimize dyeing conditions. The pad-dye technique improved fastness properties more than the conventional dyeing process did [25]. Mango leaf extraction produces beautiful shades of yellow in chocolate in an optimized dyeing condition on Microwave (MW)-irradiated and unirradiated cotton fabric using ecofriendly mordants [50]. The anthocyanin compounds in *Hibiscus sabdariffa* L. (rosella) extract imparted a protective properties against pathogenic microorganisms and UV radiation to the dyed cotton fabric [82].

A statistically designed program is now the focus of multiobjective optimization experiments. The values and levels in this program help to determine the best conditions [106, 107, 112]. Such a design had been adopted to predict suitable combinations for cationized cotton dyeing with *Juglans regia* bark extract. A phytochemical study confirmed the antioxidant property imparted by polyphenols, flavonoids, and tannins [99]. An optimized condition has also been explored for acrylic fabric dyeing that uses microwaves and the same dye source [103]. An orthogonal design of experiments optimized the dyeing conditions of polylactic acid yarn with natural indigo to obtain darker shades. The colorfastness of dyed yarn has been found to be higher than the gray-scale rating of 4 [114].

Research has emerged in the past two decades on using natural bacterial pigments for textile applications for yielding bright colors and unique functional properties. These secondary metabolites generally have prodigiosin, violacein, and melanin as their main coloring materials. The use of metal mordants can improve the optimal dyeing conditions and colorfastness properties of a dyed substrate [54].

2.2 *Application in Food and Pharmaceutical*

Developing foods with attractive appearances is one of the crucial goals of the food and beverage industry. Because specific negative health impacts have been increased after consuming artificial color additives, food producers are turning to natural food colors to meet consumer preferences for healthy food habits. Researchers prefer to select food colorants that do not affect the taste or aroma of the food itself. Several accredited organizations around the world have advocated for using safe dosages of natural colorants in food, drugs, and cosmetics products [30, 90]. Rosella flowers, kale leaves, pandan leaves, strawberry fruit, dragon fruit peels, annatto seeds, turmeric, saffron, beetroot, etc. can produce food colors [100]. Recent research has offered a more natural way of coloring foods and providing a clean label. Natural dyes come in several pigments: flavonoids, tannins, alkaloids, indigoids, quinones, chlorophyll, and carotenoids. Carotenoids, chlorophylls, anthocyanins, and betalains are the standard pigments used for food additives. Canthaxanthin, a carotenoid, is commercially produced from the algae *Haematococcus lacustris* and the orange-yellow edible mushroom *Cantharellus cibarius*, and it is used as a poultry feed additive to produce a red color in skins, yolks, fruit products, beverages, etc. [9]. Carminic acid, the cochineal's primary pigment, is accepted as a food colorant in many countries and is used mainly for foods, pharmaceuticals, cosmetics, textiles, and plastics [20]. Recent studies have revealed that *Morus* (Mulberry) extract can be used in the pharmaceutical industry and food industry for its functional properties [51]. The primary constituent of beetroot red dye, betanin, has been approved for the food and pharmaceutical industries as a natural coloring agent. Its anti-inflammatory, anti-carcinogenic, and hepatoprotective effects point to its potential uses in medicines as well [35, 94].

Several natural colorants, such as red rosella anthocyanin, are pH sensitive [110]. Thus, modern food processing aims to improve the stability of the coloring extract under various conditions [109]. The intense yellow colorant of food and beverages can also be sourced from annatto seeds (*Bixa orellana* L.), which are rich in norbixin, a carotenoid with antioxidant activity. The microencapsulation of norbixin, added to isotonic tangerine soft drinks, favorably affects beverage stability under accelerated storage [105]. Recently, several researchers have opened up new arenas for using encapsulating natural dyes as the core materials used as colorants in foods and beverages [83].

Microbial pigments are preferable because of their scalability, ease of use, lower production costs, and various medicinal advantages [90]. The scarcity of readily available blue pigments in nature induced the researchers to find a new source of blue colorant that is safe for food and beverages. Jagua blue, produced from the fruit of *Genipa americana* L., has markedly high storage stability. It could be a promising alternative for potential food applications [21]. A genipin-extracted blue colorant increased the cold storage stability of milk samples without modifying functional amino acid groups [15, 102]. These fruit extracts have shown excellent antioxidant and antiproliferative activities due to the presence of genipin, genipin 1-β-gentiobioside, and geniposide as their main glycoside compounds [69].

Adding color to tablet and syrup preparations imparts an attractive hue, increases attractiveness to consumers, and helps patients differentiate between drugs. Natural dyes, produced from either vegetable or animal sources, provide different appearances to pharmaceutical preparations [8]. Carminic acid, obtained from the cochineal insect, is used primarily in pharmaceuticals [18]. Recently, researchers applied natural dyes extracted from blue pea flowers and dragon fruit peels to tablets and paracetamol syrups [85]. A color extract from *Hibiscus sabdarrifa* calyces has been found to be a suitable coloring agent for several pediatric syrups [37].

2.3 Intelligent Food Packaging

The food industry is seeking ecofriendly packaging solutions for petroleum-derived plastics, where food quality will be monitored for any environmental changes by using sensory colorants from natural dyes. The change in color of intelligent packaging sensors depends on the pH, time and temperature, microorganisms, oxidizing agents, etc. It helps consumers to identify their quality and freshness. Many fruits, vegetables, and flowers, which contain anthocyanins, curcumin, betalains, carotenoids, chlorophyll, brazilin, quercetin, etc. as natural colorants, can be used as natural pH indicators [13, 16, 44, 59, 62]. Curcumin and anthocyanins can be used in pH-sensitive films to monitor the freshness and spoilage of packaged bighead carp (*Hypophthalmichthys nobilis*). The colorimetric film, using the same colorants, can monitor fish products' freshness [27, 28]. Edible gelatin, gellan gum, and the pH-sensitive film from red radish anthocyanins extract can act as gas sensors, showing visible color changes as consequences of milk and fish spoilage [119]. The purple sweet potato was also used as a pH indicator, exhibiting a noticeable color difference (from red to yellow) as a result of the changing pH, indicating superior colorimetric performance [27, 28]. Tea polyphenols demonstrated a high UV blocking capacity, radical scavenging activities, and enhanced antioxidant and antibacterial performance in cellulose-based bio-hybrid films [115]. One study developed antimicrobial food packaging films from composite nanofibers by using prodigiosin as a microbial pigment, isolated from *Serratia plymuthica* [11]. Future studies on smart packaging films need to focus more on commercially viable aspects [7].

2.4 Cosmetics

The worldwide interest in a safe lifestyle and healthy living has created a demand for nontoxic cosmetics for human health. Various shades of color, textures, and lusters are widely added to lipstick to enhance the beauty of lips. The filtered and evaporated methanolic extraction of *Bixa orenella* can be used as a pigment in lipstick [26]. Extracts from flower petals of *Papaver rhoeas*, *Clitoria ternatea*, and *Carthamus tinctorius* exhibited the most substantial coloring ability and storage stability, making them potential dye sources for cosmetics products. These natural coloring sources have antioxidant and anti-inflammatory qualities, inhibitory effects, and UV-protection effects without exhibiting cytotoxic activity on skin cells. Furthermore, the plants positively impacted skin hydration, reduced transepidermal water loss, and slowed aging processes [24]. The red carotenoid lycopene acts as an anticancer therapeutic agent that is used in sunscreen lotions and moisturizers [17, 97]. *Ipomoea batatas* L. extract has been used for producing several green cosmetics samples without using any heavy metals [71]. The anti-oxidative, anti-inflammatory, anti-itching, and stability qualities of the purple anthocyanin content in *Ipomoea batatas* L. has made it a suitable bio-antioxidant body scrub cream [111]. As a natural edible pigment, purple sweet potato anthocyanins are safe, non-toxic, odorless, and bright in color with high yields, making them potential choices for industrial applications [58]. Recently, microencapsulation and nanocarriers have been introduced for advanced hair colorants obtained from natural sources [31].

2.5 Dye-Sensitized Solar Cells

Environmental consciousness has led scientists to search for a green chemistry route for solar cells. Dye-sensitized solar cells (DSSCs) consist of transparent oxide-conducting glass, a nanocrystalline semiconductor anode (TiO_2), a cathode, iodine electrolytes, and dye as a photosensitizer [52, 91]. Several organic and inorganic dyes are used here. Natural dyes have been found to be the best alternatives to inorganic metal complexes used in DSSCs due to their cost-effectiveness, biocompatibility, accessible arrangement, and wide availability [4]. Recently, organic solvent–extracted plant-based pigments used as sensitizers for DSSCs have been in progress because the current stability, interaction, and efficiency levels are poor compared with those of synthetic dyes. The commonly used anthocyanins are pH and light sensitive, highly unstable, and highly susceptible to degradation [19, 46]. The efficiency of anthocyanin from red cabbage has been found to be low [77]. Rosella anthocyanin extraction under solar light and at a high pH suffered more rapid degradation than those kept in the dark. One of the main reasons for this decadence is the formation of hydroxyl radicals in the presence of solar light, which is responsible for the demineralization process [12, 63]. The efficiency of that extract can be improved in an ethanol solvent [10]. Rutin in mangosteen pericarp is considered the most productive component of the DSSC sensitizer. An ethanol solvent enhanced the photo-to-electric conversion efficiency of the mangosteen pericarp extract [121]. The absorption coefficiency of anthocyanin from blue peas has been shown to be 15 times higher than that of N719 ruthenium dye [38]. Red turnip betalain recorded the highest efficiency, at 1.70% [2]. Anthocyanins present in ivy gourd fruits were found to be more efficient than beta-carotenes of red frangipani [10, 77]. Red cabbage anthocyanin showed the highest absorption rate, best quantum effects, and best efficiency, higher than corresponding values for mangosteen peels and dragon fruit anthocyanin [78]. Malabar spinach seed extraction showed exemplary behavior and potential as a photosensitive dye source [33]. The black rice extract exhibited better photosensitized activity than the *Erythrina variegata* flower, *Rosa xanthina* flower, and capsicum [45]. Black currant crush and mixed berry extraction showed higher efficiency than Indian jamun and plum. The high concentrations of anthocyanins increased light absorption and enhanced photocurrent densities, resulting in better dye loading [89]. Fresh *Coffea arabica* husk extract can also be a potential dye source for anthocyanin compounds [47]. Anthocyanins and carotenoids had been isolated from *Tropaeolum majus* flower petals via Fourier Transform Infrared (FT-IR) spectroscopy. The light-to-electricity conversion efficiency of orange nasturtium flower extraction has also been reported in literatures [95, 96]. Sweet pomegranate extract outperformed the sour one in anthocyanin-based solar cells, because of its higher intensity light absorption and higher interaction quality between the extract and the TiO_2 anode [48]. *Schinus terebinthifolius* leaves performed better for the DSSC than 14 other selected sources of natural dyes (flowers: bougainvillea, mint, basil, and royal poinciana; leaves: black tea, green tea, henna, and fig; plant peel: passion, onion, and eggplant; and other natural sources: carob fruit, clove grain, and eggplant pulp) [1]. Purified bixin-sensitized TiO_2 solar cells exhibited better efficiencies than the norbixin-sensitized cells did. These coloring components were isolated from *Bixa orellana* L. seeds [39].

2.6 Few Other Usages

Researchers recently found at least 16 natural dyes by analyzing about 400 dyed textile samples from archaeological research dating from the seventeenth century BCE to the tenth century CE in Northwest China [60]. The ultrasonic-assisted ethanol–water extraction of beetroot dye has been shown to be an acceptable alternative for leather and paper dyeing [98]. An environmentally friendly, nontoxic leather product can be produced by using gallnut, sappan wood, and iron mordant ($FeSO_4.7H_2O$) to manufacture a black dye. An anionic fat liquor agent evaluated the leather's softness and color change. The simultaneous mordanting produced a superior result compared to the pre- and post-mordanting techniques, resulting in excellent rubbing fastness [57]. Several surface modification techniques have been adopted to improve the dye uptake capacity of leather. One such ecofriendly method is the atmospheric pressure dielectric barrier discharge (DBD) process, which improves the rubbing fastness of the treated samples [34]. Eco-printed leather using mangrove bark extract exhibited better dry and wet rubbing and washing fastness and excellent perspiration fastness in the presence of ferrous sulfate. It also improved properties such as tensile strength, elasticity, and cracking resistance, but the elongation of eco-printed leather has decreased [75].

Recently, research has been carried out to make an alternative indicator of red litmus with rosella flower material made from a macerated solution of an acid-activated and isolated compound. The color change in the rosella indicator paper was found to be the same as that in the comparison indicator [72].

3 Conclusion

Natural dyes were once the main sources of color for textiles, before the advent of petrochemical-based dyes. The advent of synthetic dyes also pushed aside the art and knowledge of natural dyeing [66]. However, growing environmental concerns have prompted us to reconsider using natural dyes as alternative sources. The present generation's consciousness of ecological preservation and reducing pollution has resulted in more scientific investigations into using natural dyes in various fields, such as textiles [55]. Recent advancements in unconventional dye extraction processes, such as ultrasound-assisted, microwave-assisted, and gamma-assisted processes, are now considered better options for standardizing dyeing performance, resulting in improved color yields, functional properties, and fastness qualities while minimizing environmental pollution [76]. The complex types of colorant compounds in a single dye source necessitate efficient extraction so that the actual color components do not decompose during the extraction process. Natural dyes, such as those in food, beverages, pharmaceuticals, etc., enhance product appearances. Artificial color additives have adverse health effects; therefore, producers have turned to natural food colors as ecofriendly alternatives. The microencapsulation of natural dyes as core materials is fulfilling the nontoxic food additives and pharmaceutical demands in recent years. In the past few decades, research has been conducted on natural bacterial pigments or microbial dyes for dyeing textiles, due to their high production yields, bright colors, and unique functional properties. Besides textiles, eco-leather is now becoming an option for environmentally friendly, nontoxic leather production because it has overcome the limitations of using natural dyes. Furthermore, natural colorants are being used in the smart food packaging industry to detect the freshness of foodstuffs. The global market for herbal cosmetics is expanding. Nontoxic natural hair dyes have adequate color strength, good washing, and high lightfastness properties. Growing environmental consciousness has led to thinking about plant pigments as sustainable solutions to the inorganic sensitizers in solar cells [92].

References

1. Abdel-Latif, M., Abuiriban, M., Dahoudi, N., Al-Kahlout, A., Taya, S., El-Agez, T., & El-Ghamri, H. (2015). Dye-sensitized solar cells using fifteen natural dyes as sensitisers of nanocrystalline TiO2. *Science, Technology and Development, 34*(3), 135–139.
2. Adedokun, O., Titilope, K., & Awodugba, A. O. (2016). Review on natural dye-sensitized solar cells (DSSCs). *International Journal of Engineering Technologies, 2*(2), 34–41.
3. Aggarwal, S. (2021). Indian dye yielding plants: Efforts and opportunities. *Natural Resources Forum, 45*(1), 63–86. Blackwell Publishing Ltd.
4. Ahmad, M. S., Pandey, A. K., & Abd Rahim, N. (2017). Advancements in the development of TiO2 photoanodes and its fabrication methods for dye-sensitized solar cell (DSSC) applications. A review. *Renewable and Sustainable Energy Reviews, 77*, 89–108.
5. Alam, S. S., Ghosh, J., & Das, D. (2022). The coloration of cotton fabric with natural dye extracted from turmeric powder. *Journal of Textile Engineering and Fashion Technology, 8*(4), 134–138.
6. Ali, N. F. (2018). Green strategy for the production of antimicrobial textiles. In *Fashion and textiles: Breakthroughs in research and practice* (pp. 442–461). IGI Global.
7. Alizadeh-Sani, M., Mohammadian, E., Rhim, J. W., & Jafari, S. (2020). pH-sensitive (halochromic) smart packaging films based on natural food colourants for the monitoring of food quality and safety. *Trends in Food Science and Technology, 105*, 93–144.
8. Allam, K., & Kumar, G. (2011). Colourants the cosmetics for the pharmaceutical dosage forms. *International Journal of Pharmacy and Pharmaceutical Sciences, 3*(3), 13–21.
9. Amaya, E., & Nickell, D. (2015). 11 – Using feed to enhance the colour quality of fish and crustaceans. In I. D. A. Davis (Ed.), *Feed and feeding practices in aquaculture* (pp. 269–298). Woodhead Publishing.
10. Amogne, N., Ayele, D., & Tsigie, Y. (2020). Recent advances in anthocyanin dyes extracted from plants for dye. *Materials for Renewable and Sustainable Energy, 9*(23), 1–16. https://doi.org/10.1007/s40243-020-00183-5

11. Amorim, L., Mouro, C., Riool, M., & Gouveia, I. (2022). Antimicrobial food packaging based on prodigiosin-incorporated double-layered bacterial cellulose and chitosan composites. *Polymers, 14*(2). https://doi.org/10.3390/polym14020315
12. Askar, K., Alsawad, Z., & Khalaf, M. (2015). Evaluation of the pH and thermal stabilities of rosella anthocyanin extracts under solar light. *Beni-Suef University Journal of Basic and Applied Sciences, 4*(3), 262–268.
13. Azman, N., Khairul, W., & Sarbon, N. (2022). A comprehensive review on biocompatible film sensor containing natural extract: Active/intelligent food packaging. *Food Control, 141*, 109189. https://doi.org/10.1016/j.foodcont.2022.109189
14. Baig, G. A. (2011). Dyeing of polyester fabrics with indigo. *Research Journal of Textile and Apparel, 15*(4), 19–25.
15. Bentes, A., de Souza, A. L., Amaya-Farfan, H., Lopes, J., & Faria, A. (2015). Influence of the composition of unripe genipap (*Genipa americana* L.) fruit on the formation of blue pigment. *Journal of Food Science and Technology, 52*(6), 3919–3924.
16. Bhargava, N., Sharanagat, V., Mor, R., & Kumar, K. (2020). Active and intelligent biodegradable packaging films using food and food waste-derived bioactive compounds: A review. *Trends in Food Science & Technology, 105*, 385–401.
17. Bin-Jumah, M., Alwakeel, S. S., Moga, M., Buvnariu, L., Bigiu, N., & Zia-Ul-Haq, M. (2021). Application of carotenoids in cosmetics. *Carotenoids: Structure and Function in the Human Body*, 747–756.
18. Biswal, P., Mishar, M., Bhadouriya, A., & Yadav, V. (2015). An updated review on colorants as pharmaceutical excipients. *International Journal of Pharmaceutical, Chemical and Biological Sciences, 5*(4), 1004–1017.
19. Boranbayeva, T., Karadeniz, F., & Yılmaz, E. (2014). Effect of storage on anthocyanin degradation in black mulberry juice and concentrates. *Food and Bioprocess Technology, 7*, 1894–1902.
20. Borges, M., Tejera, R., Díaz, L., Esparza, P., & Ibáñez, E. (2012). Natural dyes extraction from cochineal (Dactylopius coccus). New extraction methods. *Food Chemistry, 132*(4), 1855–1860.
21. Brauch, J., Zapata-Porras, S., Buchweitz, M., Aschoff, J., & Carle, R. (2016). Jagua blue derived from Genipa americana L. fruit: A natural alternative to commonly used blue food colorants? *Food Research International, 89*(1), 391–398.
22. Brudzyńska, P., Sionkowska, A., & Grisel, M. (2021). Plant-derived colorants for food, cosmetic and textile industries: A review. *Materials, 14*(13), 3484.
23. Bueno, A., Hoffmann, T., Souza, C., Carvalho, L., Bertoli, S., Barcellos, I., & Gonçalves, M. (2022). Optimal process conditions for recycled polyester dyeing using natural annatto dye. *Journal of Cleaner Production, 370*.
24. Bujak, T., Zagórska-Dziok, M., Ziemlewska, A., Nizioł-Łukaszewska, Z., Lal, K., Wasilewski, T., & Hordyjewicz-Baran, Z. (2022). Flower extracts as multifunctional dyes in the cosmetics industry. *Molecules, 27*(3), 922.
25. Chandru, P., & Praveena, R. (2021). Organic hybrid pigments from Crotalaria globosa for cotton dyeing – A sustainable approach. *Materials Today: Proceedings, 47*(9), 1773–1776.
26. Chaudhari, N., Chaudhari, N., Chaudhari, H., Premchandani, L., Dhankani, A., & Pawar, S. (2018). A review on herbal lipstick from different natural colouring pigment. *Indian Journal of Drugs, 6*(3), 174–179.
27. Chen, H.-Z., Zhang, M., Bhandari, B., & Yang, C.-H. (2020a). Novel pH-sensitive films containing curcumin and anthocyanins to monitor fish freshness. *Food Hydrocolloids, 100*. https://doi.org/10.1016/j.foodhyd.2019.105438
28. Chen, S., Wu, M., Lu, P., Gao, L., Yan, S., & Wang, S. (2020b). Development of pH indicator and antimicrobial cellulose nanofibre packaging film based on purple sweet potato anthocyanin and oregano essential oil. *International Journal of Biological Macromolecules, 149*, 271–280.

29. Chequer, F. D., De Oliveira, G. R., Ferraz, E. A., Cardoso, J. C., Zanoni, M. B., & De Oliveira, D. P. (2013). Textile dyes: Dyeing process and environmental impact. *Eco-friendly Textile Dyeing and Finishing, 6*(6), 151–176.
30. Coultate, T., & Blackburn, R. S. (2018). Food colorants: Their past, present and future. *Coloration Technology, 134*(3), 165–186.
31. Cui, H., Xie, W., Hua, Z., Cao, L., Xiong, Z., Tang, Y., & Yuan, Z. (2022). Recent advancements in natural plant colorants used for hair dye applications: A review. *Molecules, 27*(22), 8062.
32. Cuong, D. X., Hoan, N. X., Dong, D. H., Thuy, L. T. M., Van Thanh, N., Ha, H. T., et al. (2019). Tannins: Extraction from plants. *Tannins-Structural Properties, Biological Properties, and Current Knowledge*, 57–76.
33. Das, S. K., Ganguli, S., Kabir, H., Khandaker, J., & Ahmed, F. (2020). Performance of natural dyes in dye-sensitized solar cells as photosensitizer. *Transactions on Electrical and Electronic Materials, 21*, 105–116.
34. Dave, H., Ledwani, L., & Nema, S. (2016). Surface modification by atmospheric pressure air plasma treatment to improve dyeing with natural dyes is an environment-friendly approach for leather processing. *Plasma Chemistry and Plasma Processing, 36*, 599–613.
35. Devadiga, D., & Ahipa, T. (2020). Betanin: A red-violet pigment – Chemistry and applications. In A. K. Samanta, N. S. Awwad, & H. Algarni (Eds.), *Chemistry and technology of natural and synthetic dyes and pigments*. https://doi.org/10.5772/intechopen.88939
36. Elshemy, N. (2011). Unconventional natural dyeing using microwave heating with cochineal as natural dyes. *Research Journal of Textile and Apparel, 15*(4), 26–36.
37. Frimpong, G., Adotey, J., Ofori-Kwakye, K., Kipo, S., & Dwomo-Fokuo, Y. (2014). Investigating the suitability of *Hibiscus sabdariffa* calyx extract as a colouring agent for paediatric syrups. *Journal of Applied Pharmaceutical Science, 4*(12). https://doi.org/10.7324/JAPS.2014.41201
38. Gokilamani, N., Muthukumarasamy, N., Thambidurai, M., Ranjitha, A., & Velauthapillai, D. (2013). Utilization of natural anthocyanin pigments as photosensitizers for dye-sensitized solar cells. *Journal of Sol-Gel Science & Technology, 66*, 212–219.
39. Gómez-Ortíz, N., Vázquez-Maldonado, I., Pérez-Espadas, A., Mena-Rejón, G., Azamar-Barrios, J., & Oskam, G. (2010). Dye-sensitized solar cells with natural dyes extracted from achiote seeds. *Solar Energy Materials and Solar Cells, 94*(1), 40–44. https://doi.org/10.1016/j.solmat.2009.05.013
40. Gulrajani, M. (1999). Natural dyes. Part I: Present status of natural dye. *Colourage, 46*(7), 19–28.
41. Gulrajani, M. (2001). Present status of natural dyes. *Indian Journal of Fibre & Textile Research, 26*(1–2), 191–201.
42. Gulrajani, M., & Gupta, D. (1992). *Natural dyes and their application to textiles*. IIT New Delhi.
43. Haji, A., Ghaheh, F., & Indrie, L. (2022). Pomegranate fallen leaves as a source of natural dye for the mordant-free dyeing of wool. *Coloration Technology*. https://doi.org/10.1111/cote.12651
44. Halonen, N., Pálvölgyi, P., Bassani, A., Fiorentini, C., Nair, R., Spigno, G., & Kordas, K. (2020). Bio-based smart materials for food packaging and sensors – A review. *Frontiers in Materials*. https://doi.org/10.3389/fmats.2020.00082
45. Hao, S., Wu, J., Huang, Y., & Lin, J. (2006). Natural dyes as photosensitizers for dye-sensitized solar cell. In X. Mathew (Ed.), *Solar energy* (pp. 209–214). Elsevier.
46. Haq, Z. U., Riaz, M., Saad, M., Riaz, B., Zia-Ul-Haq, M., & Saad, B. (2016). Biosynthesis and stability of anthocyanins. *Springer International Publishing, 1*(1), 71–86.
47. Hasby, N., Pamungkas, G., & Akbar, S. A. (2020). Dye-sensitized solar cells properties from natural dye as light-reaping materials extracted from Gayo arabica coffee husks. *Rasayan Journal of Chemistry, 13*(1), 38–43.

48. Hosseinnezhad, M., Moradian, S., & Gharanjig, K. (2015). Fruit extract dyes as photosensitizers in solar cells. *Current Science, 109*(5), 953–956.
49. Hosseinnezhad, M., Gharanjig, K., Rouhani, S., Imani, H., & Razani, N. (2022). Environmentally dyeing using dried walnut husk as bio-mordant: Investigation of creating new red and yellow shades on wool. *Journal of Natural Fibers, 19*(15), 10953–10963.
50. Jabar, J., Owokotomo, I., & Ogunsade, A. (2022). Sustainable dyeing of cotton fabric with mangiferin: Roles of microwave-rays and bio-mordants on fabric colorimetric and fastness properties. *Sustainable Chemistry and Pharmacy, 29*, 100822.
51. Jan, B., Parveen, R., Zahiruddin, S., Khan, M., Mohapatra, S., & Ahmad, S. (2021). Nutritional constituents of mulberry and their potential applications in food and pharmaceuticals: A review. *Saudi Journal of Biological Sciences, 28*(7), 3909–3921.
52. Kabir, F., Bhuiyan, M. M. H., Manir, M. S., Rahaman, M. S., Khan, M. A., & Ikegami, T. (2019). Development of dye-sensitized solar cell based on the combination of natural dyes extracted from Malabar spinach and red spinach. *Results in Physics, 14*, 102474.
53. Kassinger, R. (2003). *Dyes: From sea snails to synthetics*. Twenty-First Century Books.
54. Kramar, A., & Kostic, M. (2022). Bacterial secondary metabolites as biopigments for textile dyeing. *Text, 2*(2), 252–264.
55. Křížová, H. (2015). Natural dyes: Their past, present, future and sustainability. In *Recent developments in fibrous material science* (pp. 59–71). Kosmas Publishing.
56. Latos-Brozio, M., & Masek, A. (2020). The application of natural food colorants as indicator substances in intelligent biodegradable packaging materials. *Food and Chemical Toxicology, 135*.
57. Lee, S., Shin, E., Kim, W., & Park, S. (2012). Dyeing process for improving properties of black color using natural dyes and mordant. *Journal of the American Leather Chemists Association, 107*(2), 33–39.
58. Li, A., Xiao, R., He, S., An, X., He, Y., Wang, C., et al. (2019). Research advances of purple sweet potato anthocyanins: Extraction, identification, stability, bioactivity, application, and biotransformation. *Molecules, 24*(21). https://doi.org/10.3390/molecules24213816
59. Li, X., Liu, D., Pu, Y., & Zhong, Y. (2023). Recent advances of intelligent packaging aided by artificial intelligence for monitoring food freshness. *Food, 12*(15), 2976. https://doi.org/10.3390/foods12152976
60. Liu, J., Li, W., Kang, X., Zhao, F., He, M., She, Y., & Zhou, Y. (2021). Profiling by HPLC-DAD-MSD reveals a 2500-year history of the use of natural dyes in Northwest China. *Dyes and Pigments, 187*. https://doi.org/10.1016/j.dyepig.2021.109143
61. Mahmoud, R., & Adelzaher, N. (2022). Investigation of thermal and structure properties of silk dyed with a natural dye. *Letters in Applied NanoBioScience, 12*(2).
62. Martins, V., Santos, L., Romani, V., & Fernandes, S. (2022). Bio-based sensing: Role of natural dyes in food freshness indicators. In A. Shukla (Ed.), *Bio- and nano-sensing technologies for food processing and packaging* (pp. 37–62). The Royal Society of Chemistry.
63. Meshram, S., Limaye, R., Ghodke, S., Nigam, S., Sonawane, S., & Chikate, R. (2011). Continuous flow photocatalytic reactor using ZnO–bentonite nanocomposite for degradation of phenol. *Chemical Engineering Journal, 172*(2–3), 1008–1015.
64. Mirnezhad, S., Sadeghi-Kiakhani, M., & Hashemi, E. (2022). Wool dyeing using Ziziphus bark extract as a natural dye: Studies on the dyeing, antibacterial and antioxidant characteristics. *Environmental Science and Pollution Research*. https://doi.org/10.21203/rs.3.rs-1894001/v1
65. Mridha, A., Mukhopadhyay, A., Das, D., Sarkar, B., & Roy Maulik, S. (2021). Sustainable dyeing of silk with natural dyes using natural sources of water. *Journal of The Institution of Engineers (India): Series E, 102*(1), 61–74.
66. Mussak, R. A., & Bechtold, T. (2009). Natural colorants in textile dyeing. In *Handbook of natural colorants* (pp. 315–335).
67. Narayan, M. R. (2012). Dye sensitized solar cells based on natural photosensitizers. *Renewable & Sustainable Energy Reviews, 16*(1), 208–215.

68. Natarajan, T., & Jayadevan, G. R. (2022). COVID-19 pandemic and the consumption behaviour of branded functional beverages in India: A conceptual framework. *Nutrition & Food Science, 52*(3), 423–444.
69. Neri-Numa, I., Angolini, C., Bicas, J., GoisRuiz, A., & Pastore, G. (2018). Iridoid blue-based pigments of Genipa americana L. (Rubiaceae) extract: Influence of pH and temperature on color stability and antioxidant capacity during in vitro simulated digestion. *Food Chemistry, 263*, 300–306.
70. Neri-Numa, I., Pessôa, M., Arruda, H., Pereira, G., Paulino, B., Angolini, R. C., & Pastore, G. (2020). Genipap (Genipa americana L.) fruit extract as a source of antioxidant and antiproliferative iridoids. *Food Research International, 134.* https://doi.org/10.1016/j.foodres.2020.109252
71. Nguyen, T., & Nguyen, H. (2022). Green cosmetic from natural colorant of anthocyanin extracted from Ipomoea batatas L. *Materials Science Forum*, 41–50.
72. Nuryanti, S., Puspitasari, D., & Supriadi. (2019). Rosella (Hibiscus sabdariffa) flowers as alternative indicators of blue and red litmus. *Oriental Journal of Chemistry, 35*(1), 476–480.
73. Ondruschka, B., & Asghari, J. (2006). Microwave-assisted extraction–a state-of-the-art overview of varieties. *Chimia, 60*(6), 321–321.
74. Otaviano, B., Sannomiya, M., Lima, F., Tangerina, M., Tamayose, C., Ferreira, M., et al. (2022). Pomegranate peel extract and zinc oxide as a source of natural dye and functional material for textile fibers aiming for photoprotective properties. *Materials Chemistry and Physics, 293.*
75. Pancapalaga, W., Ishartati, E., & Ambarwati, T. (2022). The Colorfastness and quality of eco-printed leather with different types of mordant in natural dyes from mangrove extract (Rhizophora mucronata). *Tropical Animal Science Journal, 45*(3).
76. Pizzicato, B., Pacifico, S., Cayuela, D., Mijas, G., & Riba-Moliner, M. (2023). Advancements in sustainable natural dyes for textile applications: A review. *Molecules, 28*(16), 5954.
77. Prabavathy, N., Shalini, S., Balasundaraprabhu, R., Prasanna, S., Prasanna, S., & Muthukumarasamy, N. (2017). Enhancement in the photostability of natural dyes for dye-sensitized solar cell (DSSC) applications: A review. *International Journal of Energy Research, 41*(10), 1372–1396.
78. Pratiwi, D., Nurosyid, F., Supriyanto, A., & Suryana, R. (2016). Optical properties of natural dyes on the dyesensitized solar cells (DSSC) performance. In *Journal of physics: Conference series*. IOP Publishing. https://doi.org/10.1088/1742-6596/776/1/012007
79. Puri Sikka, M., Sarkar, A., & Garg, S. (2022). Artificial intelligence (AI) in textile industry operational modernization. *Research Journal of Textile and Apparel*. https://doi.org/10.1108/RJTA-04-2021-0046
80. Rani, N., & Jajpura, L. (2021). Sustainable dyeing of wool by natural dyes in conjunction with natural mordants. In *Recent trends in traditional and technical textiles* (pp. 127–144). Springer.
81. Reddy, A., Moniruzzaman, M., Madhavi, V., & Jaafar, J. (2020). Chapter 8 – Recent improvements in the extraction, cleanup, and quantification of bioactive flavonoids. In Atta-ur-Rahman (Ed.), *Studies in natural products chemistry* (Vol. 66, pp. 197–223). Elsevier.
82. Rehana, M., Ibrahim, G., Mashaly, H., Hasanin, M., GamalRashad, H., & Mowafi, S. (2022). Simultaneous dyeing and multifunctional finishing of natural fabrics with hibiscus flowers extract. *Journal of Cleaner Production, 374.*
83. Ribeiro, J., & Veloso, C. (2021). Microencapsulation of natural dyes with biopolymers for application in food: A review. *Food Hydrocolloids, 112.* https://doi.org/10.1016/j.foodhyd.2020.106374
84. Roy Maulik, S. (2015). Application and characterization of natural dye. *Asian Dyer, 12*(3), 30–37.
85. Saati, E., Mulandari, R., Wachid, M., & Winarsih, S. (2018). The utilization of Telang flower as healthy-natural food coloring on dawet drink. In *SolarPACES 2017: International*

conference on concentrating solar power and chemical energy systems. AIP Publishing. https://doi.org/10.1063/1.5064356

86. Sadeghi-Kiakhani, M., Tehrani-Bagha, A. R., Safapour, S., Eshaghloo-Galugahi, S., & Etezad, S. M. (2021). Ultrasound-assisted extraction of natural dyes from hawthorn fruits for dyeing polyamide fabric and study its fastness, antimicrobial, and antioxidant properties. *Environment, Development and Sustainability, 23*, 9163–9180.
87. Samanta, A. K., & Konar, A. (2011). Dyeing of textiles with natural dyes. *Natural Dyes (Ed. E. Perrin Akçakoca Kumbasar), 3*, 30–56.
88. Saranraj, P., & Sivasakthi, S. (2014). *Spirulina platensis* – Food for future: A review. *Asian Journal of Pharmaceutical Science & Technology, 4*(1), 26–33.
89. Sawhney, N., & Satapathi, S. (2016). Utilization of naturally occurring dyes as sensitizers in dye sensitized solar cells. *IEEE Journal of Photovoltaics, 7*(2), 539–544.
90. Sen, T., Barrow, C., & Desh, S. (2019). Microbial pigments in the food industry – Challenges and the way forward. *Frontiers in Nutrition, 6*. https://doi.org/10.3389/fnut.2019.00007
91. Shakunthala, N. (2021). Natural dyes as a photosensitizer in dye sensitized solar cells – Review. *Journal of Scientific Research, 65*(8). https://doi.org/10.37398/JSR.2021.650813
92. Shalini, S., Prasanna, S., Mallick, T. K., & Senthilarasu, S. (2015). Review on natural dye sensitized solar cells: Operation, materials, and methods. *Renewable & Sustainable Energy Reviews, 51*, 1306–1325.
93. Shukla, D., & Vankar, P. S. (2017). Dyeing of cotton by different new natural dyeing sources. *Natural Dyes for Textiles: Sources, Chemistry, and Applications*, 111.
94. Singh, H., & Bharati, K. (2015). *Handbook of natural dyes and pigments*. Woodhead Publishing India.
95. Singh, M., Vajpayee, M., & Ledwani, L. (2021a). Eco-friendly surface modification of natural fibres to improve dye uptake using natural dyes and application of natural dyes in fabric finishing: A review. *Materials Today: Proceedings, 43*, 2868–2871.
96. Singh, S., Maurya, I., Sharma, S., Kushwaha, S., Srivastava, P., & Bahadur, L. (2021b). Application of new natural dyes extracted from Nasturtium flowers (*Tropaeolum majus*) as photosensitizer in dye-sensitized solar cells. *Optik, 243*. https://doi.org/10.1016/j.ijleo.2021.167331
97. Singh, D., Kumari, K., & Ahmed, S. (2022). Chapter 17 – Natural herbal products for cancer therapy. In B. Jain & S. Pandey (Eds.), *Understanding cancer* (pp. 257–268). Academic.
98. Sivakumar, V., Lakshmi Anna, J., Vijayeeswarri, J., & Swaminathan, G. (2009). Ultrasound-assisted enhancement in natural dye extraction from beetroot for industrial applications and natural dyeing of leather. *Ultrasonics Sonochemistry, 16*(6), 782–789.
99. Slama, N., Ticha, M., Skhiri, W., Boudokhane, C., & Dhaouadi, H. (2022). Development and optimization of a sustainable process assisted by microwave energy to dye cellulosic fabrics by *Juglans regia* bark residues extract. *Sustainability, 14*(13), 7534.
100. Solymosi, K., Latruffe, N., Morant-Manceau, A., & Schoefs, B. (2015). *Food colour additives of natural origin* (pp. 3–34). Woodhead Publishing.
101. Stevens, C. V. (2023). *Handbook of natural colorants*. Wiley.
102. Strieder, M., LandimNeves, M., Silva, E., & Meireles, M. (2020). Low-frequency and high-power ultrasound-assisted production of natural blue colorant from the milk and unripe Genipa americana L. *Ultrasonics Sonochemistry*. https://doi.org/10.1016/j.ultsonch.2020.105068
103. Ticha, M., Slama, N., Dhouibi, N., Boudokhane, C., & Dhaouadi, H. (2022). Valorization of *Juglans regia*. L bark residues as a natural colorant based on response surface methodology: A challenging approach to a sustainable dyeing process for acrylic fabrics. *Sustainability, 14*(7), 4134.
104. Trotman, E. (1994). *Dyeing and chemical technology of textile fibres* (1st ed.). B I Publications Pvt. Ltd.
105. Tupuna-Yerovi, D., Paese, K., Flo^res, S., Guterres, S., & Rios, A. (2020). Addition of norbixin microcapsules obtained by spray drying in an isotonic tangerine soft drink as a natural dye. *Journal of Food Science and Technology, 57*(3), 1021–1031.

106. Vadood, M., & Haji, A. (2022a). A hybrid artificial intelligence model to predict the color coordinates of polyester fabric dyed with madder natural dye. *Expert Systems with Applications, 193*. https://doi.org/10.1016/j.eswa.2022.116514
107. Vadood, M., & Haji, A. (2022b). Application of ANN weighted by optimization algorithms to predict the color coordinates of cellulosic fabric in dyeing with binary mix of natural dyes. *Coatings, 12*. https://doi.org/10.3390/coatings12101519
108. Verma, S., & Gupta, G. (2017). Natural dyes and its applications: A brief review. *International Journal of Research & Analytical Reviews, 4*(4), 57–60.
109. Vinha, A., Rodrigues, F., Nunes, M., & Oliveira, M. (2018). 11 – Natural pigments and colorants in foods and beverages. In C. M. Galanakis (Ed.), *Polyphenols: Properties, recovery, and applications* (pp. 393–427). Woodhead Publishing.
110. Winarti, S., & Firdaus, A. (2010). Stability of red color rosella extract for food and beverage colorant. *Jurnal Teknologi Pertanian, 11*(2), 87–93.
111. Wirasuta, I., Triastuti, N., Deviyanthi, K., Sartika, D., & Utari, P. (2018). Formulation of the body scrub cream from purple sweet potato (Ipomoea batatas L.). *Indonesian Journal of Pharmaceutical Science and Technology, 5*(1), 26–30.
112. Witek-Krowiak, A., Chojnacka, K., Podstawczyk, D., Dawiec, A., & Pokomeda, K. (2014). Application of response surface methodology and artificial neural network methods in modelling and optimization of biosorption process. *Bioresource Technology, 160*, 150–160.
113. Wongcharee, K., Meeyoo, V., & Chavadej, S. (2007). Dye-sensitized solar cells using natural dyes extracted from rosella and blue pea flowers. *Solar Energy Materials and Solar Cells, 91*(7), 566–571.
114. Wu, M., Chena, Y., Penga, P., Zuo, D., Wang, Q., Liu, L., et al. (2022). Optimization of dark dyeing poly(lactic acid) yarn with indigo dyes applied to denim fabric. *The Journal of The Textile Institute*. https://doi.org/10.1080/00405000.2022.2047320
115. Xia, G., Ji, X., Xu, Z., & Ji, X. (2022). Transparent cellulose-based bio-hybrid films with enhanced anti-ultraviolet, antioxidant and antibacterial performance. *Carbohydrate Polymers, 298*. https://doi.org/10.1016/j.carbpol.2022.120118
116. Yadav, S., Tiwari, K. S., Gupta, C., Tiwari, M. K., Khan, A., & Sonkar, S. P. (2022). A brief review on natural dyes, pigments: Recent advances and future perspectives. *Results in Chemistry, 5*, 100733.
117. Yu, D., Liu, Y., Ren, Y., Liu, X., & Qu, H. (2022). One-pot strategy to simultaneously prepare dyed, flame retardant and UV-resistant silk fabric based on a safflower yellow derivative. *Royal Society of Chemistry, 12*, 26147–26159.
118. Yusuf, M., Shabbir, M., & Mohammad, F. (2017). Natural colorants: Historical, processing, and sustainable prospects. *Natural Products & Bioprospecting, 7*, 123–145.
119. Zhai, X., Li, Z., Zhang, J., Shi, J., Zou, X., Huang, X., et al. (2018). Natural biomaterial-based edible and pH-sensitive films combined with electrochemical writing for intelligent food packaging. *Journal of Agricultural & Food Chemistry, 66*(48), 12836–12846.
120. Zhang, Y., Zhou, Q., Xia, W., Rather, L., & Li, Q. (2022). Sonochemical mordanting as a green and effective approach in enhancing cotton bio-natural dye affinity through soy surface modification. *Journal of Cleaner Production, 336*.
121. Zhou, H., Wu, L., Gao, Y., & Ma, T. (2011). Dye-sensitized solar cells using 20 natural dyes as sensitizers. *Journal of Photochemistry and Photobiology A: Chemistry, 219*(2–3), 188–194.

Mushroom-Based Natural Dyes for Sustainable Color in Textiles

Muhammad Zulqurnain Haider, Shahid Adeel, Naveed Sultan, Noman Habib, Muhammad Arslan Ashraf, and Mozhgan Hosseinnezhad

Abstract Colors are crucial on Earth, but their sourcing and production face sustainability challenges. Humans historically used a wide range of colors from nature, but overexploitation due to growing demands threatens the environment. Synthetic dyes gained popularity in the nineteenth century for their color options and commercial viability. However, they are carcinogenic and persist in the environment. Today, there is a dire need to move toward sustainable alternatives like biodegradable pigments sourced from plants, bacteria, and fungi, particularly mushrooms. Mushroom dyes offer vibrant colors present in the natural palette that decompose easily and can be produced on a large scale. Chemically, mushroom dyes are anthraquinones, xanthophylls, and carotenoids mostly in nature. This chapter explores mushroom-based pigments as a sustainable substitute for synthetic dyes in the textile industry. It discusses isolation methods and the current/future prospects of these eco-friendly pigments.

Keywords Sustainable textile industry · Mushroom-based biopigments · Natural dyes · Biodegradable colors · Eco-friendly color alternatives · Microbial dyes

1 Introduction

Colors play a vital role in the existence of life on Earth. Various species, including humans, are drawn to the vibrancy and diversity of colors, which permeate their everyday lives through food, clothing, and other daily essentials. Through

M. Z. Haider (✉) · N. Sultan · N. Habib · M. A. Ashraf
Department of Botany, Government College University Faisalabad, Faisalabad, Pakistan

S. Adeel
Department of Applied Chemistry, Government College University Faisalabad, Faisalabad, Pakistan

M. Hosseinnezhad
Department of Organic Colorants, Institute for Color Science and Technology, Tehran, Iran

S. S. Muthu (ed.), *Natural Dyes and Sustainability*, Sustainable Textiles: Production, Processing, Manufacturing & Chemistry,
https://doi.org/10.1007/978-3-031-47471-2_5

archaeological studies, it has become evident that humans have utilized different coloring agents since ancient times [46]. Further research on human culture and art has revealed that early humans primarily relied on four pigments: yellow, red, black, and white [34, 38]. These natural colors, composed of carotene, canthaxanthin, and riboflavin, have been utilized since ancient times.

Moreover, with the exponential growth of the global population, the demand for biopigments in the textile industry has led to the depletion of natural sources such as algae, fungi, and plants. This depletion coincided with the introduction of synthetic dyes in the nineteenth century, which gradually replaced natural dyes [1]. Synthetic dyes, unlike natural ones, are man-made and offer advantages such as easy availability and a wide range of vibrant color options [34]. Furthermore, synthetic dyes have emerged as a cost-effective alternative, offering brighter and more consistent colors compared to natural dyes. Consequently, the textile industry shifted toward their use [1].

However, the use of synthetic dyes in the textile industry has had detrimental effects on human health. These dyes, which are toxic and carcinogenic during preparation, necessitate the use of hazardous chemicals during fabric treatment (dyeing, printing, and finishing). Additionally, synthetic dyes exhibit fiber-specific properties, binding only to certain types of fibers [30]. Not only do inorganic dyes (synthetic dyes) pose significant health risks, but they also contribute to pollution in aquatic ecosystems due to their carcinogenic nature, persistence, and harmful impact on the environment. Consequently, there has been growing global interest in natural and eco-friendly dyes. These natural pigments find application as colorants in various industries, including pharmaceuticals, textiles, foods, paints, and cosmetics [2]. They are derived from diverse sources, including plants, algae, bacteria, and fungi, with fungi exhibiting significant potential for pigment production. Fungi possess the ability to produce multiple biopigments with enhanced stability under light and heat. Furthermore, fungi can be cultivated on a large scale using cost-effective growth media and demonstrate resilience in harsh conditions, providing them with a competitive edge over other organisms [40, 64].

The chapter examines the historical background of mushroom-based natural pigments, uncovering their origins and significance. It explores the wide range of biopigments produced by various mushroom species across different genera. These pigments are responsible for the diverse array of colors observed in mushrooms, representing the different fungal families and genera. Moreover, the chapter covers techniques utilized to isolate mycopigments, covering both traditional and modern approaches. These methods enable researchers to extract and study the broad spectrum of color palettes produced by mushrooms. Furthermore, the chapter discusses prospects for the future in this field, providing recommendations for further research and drawing conclusions. In summary, the chapter explores the extensive realm of mushroom-based natural pigments, examining their origins, diversity, isolation techniques, and color palette and providing insights into future prospects. It concludes with key recommendations and conclusions drawn from research findings.

2 Historical Background

Color and the art of coloring have been deeply intertwined with human traditions since ancient times. Both China and Europe have a rich history of using natural dyes obtained from various parts of plants, such as stem bark and leaves, as well as insects, dating back more than 5000 years. In the Indus Valley Civilization (2600–1900 BC), madder (*Coccus ilicis*) dyes were widely employed in the textile industry to color different types of fabric. The Romans sourced red dye from marine mollusks, while the Egyptians naturally obtained indigo from the *Isatis tinctoria* plant. Notably, the extraction of Tyrian purple from murex shellfish marked the first globally traded color during the Bronze Age (1700 BC) [6, 7, 54]. Some of the earliest known examples of dyeing with mushroom pigments date back to the Neolithic period, around 10,000 BCE. During the Middle Ages in northern Europe, the fly agaric (*Amanita muscaria*) mushroom gained renown for its bright red dye. The successful use of mushrooms as dyes resulted in a range of colors known as "earth hues," encompassing yellow, gold, orange, burnt sienna, brown, and everything in between. The *Phaeolus schweinitzii* mushroom played a significant role in achieving these colors. Another breakthrough occurred when a small brown mushroom (*Cortinarius* family) yielded rose, red, and burgundy hues, expanding the available color palette (namyco.org). The Zapotec people of Oaxaca, Mexico, have a rich history of using mushrooms to dye textiles. They utilize the caps of the parasol mushroom (*Macrolepiota procera*) to produce a deep, reddish-brown shade. This dyeing technique has been passed down through generations and holds immense cultural significance for the Zapotec community (namyco.org). In Finland and other parts of Scandinavian countries, the use of mushrooms as dyes can be traced back to the Viking era. The Finns, in particular, have a tradition of dyeing wool and other natural fibers using the dyer's polypore (*Phaeolus schweinitzii*). This technique produces a range of colors, including greens, yellows, and browns (namyco.org). Mushroom-dyed textiles are often incorporated into traditional Finnish dress and various cultural contexts. It is worth noting that Miriam C. Rice pioneered the technique of using mushrooms to dye fibers in 1972 [55].

3 Biopigment Palette of Mushrooms

Mushrooms, which are the visible structures of microscopic filamentous fungi, exhibit a remarkable array of colors. The *Agaricus* genus, for example, produces yellow and red-pink hues attributed to the presence of pigments called agaricone and quinone, respectively (Fig. 1). *Agaricus xanthodermus* specifically synthesizes the yellow agaricone pigment, while *A. bisporus* produces quinone pigments that result in red and pink colors. These pigments hold potential for utilization in the textile industry [67].

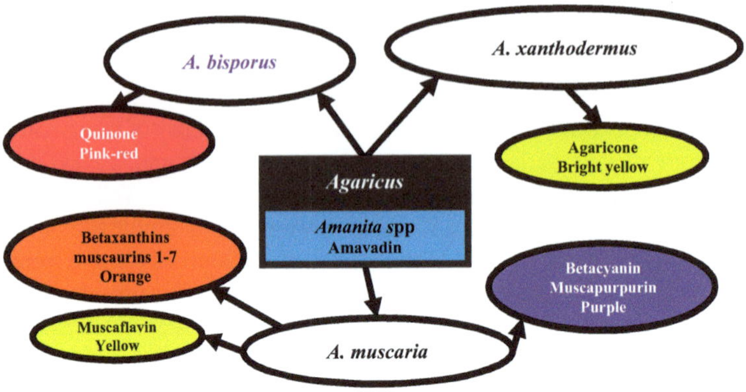

Fig. 1 *Agaricus* and *Amanita* genera biopigments production (Table 1, Serial # 1–4)

Similarly, various species within the *Amanita* genus exhibit a biological diversity of pigments (Fig. 1). Examples include betacyanin muscapurpurin, betaxanthins muscaurins 1–7, muscaflavin, and amavadin, which give rise to purple, orange, yellow, and pale blue colors, as observed in *A. muscaria* [68]. Another genus, *Hygrocybe*, synthesizes the yellow muscaflavin pigment. In particular, *Hygrocybe conica* has been reported to synthesize a yellow pigment that is stored in the cap region and can be extracted [66].

Furthermore, the *Cortinarius* genus is widely recognized as the largest genus of agarics, comprising thousands of species worldwide (Fig. 2). It is closely related to the *Dermocybe* and *Tricholoma* genera. Various species within the *Cortinarius* genus biologically synthesize anthraquinones of different colors. For instance, *C. cinnabarinus*, commonly known as the toadstool, produces fallacinol 6-O-methoxycitreorosein, which imparts a dark orange color. Similarly, *C. cinnamomeoluteus* produces the bright yellow flavomannin-6,6′-di-O-methyl ether pigment [22]. *C. citrinus* and *C. croceus* give rise to dihydroanthracenone and torosachrysone compounds, resulting in a green color. It is worth noting that macrofungal species found in the Australian region, such as *C. atrovirens* and *C. odoratus*, synthesize atrochrysone and torosachrysone-8-O-methyl ether, contributing to the green coloration [20].

Furthermore, *Cortinarius* species also biosynthesize xanthone and dermoxanthone metabolites that exhibit bright yellow and deep violet colors when exposed to UV light. For instance, *C. semisanguineus* synthesizes a bright yellow pigment, while *C. violaceus* produces a deep violet pigment [53]. In an example of the *Tricholoma* genus, Klostermeyer et al. [36] proposed that benzotropolone metabolites derived from aurantricholone, extracted from the cap of *Tricholoma aurantium*, impart a dark orange-red color (Fig. 2). These compounds are closely related to the compounds found in Boletales mushrooms. Aurantricholone derivative exhibits metal chelation properties. Additionally, *T. aurantium* produces aurantricholides A-B, compounds responsible for the yellow color. Another investigation reported that

Table 1 List of natural biopigments obtained from various fungal genera that produce mushroom fruiting bodies

Serial #	Fungal genera	Color	Pigments responsible	Fungal species	References
1	*Agaricus*	Yellow	Agaricone	*Agaricus xanthodermus*	Walton et al. [67]
2		Red-pink	Quinone	*A. bisporus*	Weijn et al. [68]
3	*Amanita*	Purple	Betacyanin muscapurpurin	*Amanita muscaria*	Velíšek and Cejpekn [66]
		Orange	Betaxanthins muscaurins 1–7		
		Yellow	Muscaflavin		
4		Pale blue	Amavadin	*Amanita species*, e.g., *A. muscaria*	Bayer and Kneifel [5]
5	*Hygrocybe*	Yellow	Muscaflavin	*Hygrocybe conica*	Velíšek and Cejpekn [66]
6	*Cortinarius*	Dark orange	Fallacinol	*Cortinarius* spp., e.g., *C. cinnabarinus*	Gill [20]
7		Bright yellow	Anthraquinone flavomannin-6,6′-di-O-methyl ether	*C. cinnamomeoluteus*	
8		Green	(3R)-atrochrysone	*C. atrovirens*	
9				*C. odoratus*	
10		Bright yellow florescence under UV	Xanthone dermoxanthone	*C. semisanguineus*	Gill [21]
11		Violet color	β-DOPA-$Fe3^{+}$	*C. violaceus*	Räisänen [53]
12	*Tricholoma*	Bright yellow	Anthraquinone flavomannin-6,6′-di-O-methyl ether	*Tricholoma equestre*	Gill [20]
13		Orange-red	Benzotropolone	*T. aurantium*	Klostermeyer et al. [36]
		Yellow	Aurantricholides A-B		
14	*Gymnopilus*	Yellow-brown	Styrylpyrone (bisnoryangonin and hispidin)	*Gymnopilus* spp.	Gill [20]
15	*Hypholoma*			*Hypholoma* spp.	
16	*Pholiota*			*Pholiota* spp.	
17	*Oudemansiella*	Orange-yellow	Dihydroxerulin	*Oudemansiella melanotricha*	Velíšek and Cejpekn [66]
			Xerulin		
			Xerulinic acid		
18	*Mycena*	Red	Mycearubin A	*Mycena rosea*	Peters and Spiteller [49]
19			Mycearubin B	*M. haematopus*	
20		Blue	Sanguinones A and B	*M. sanguinolenta*	Peters and Spiteller [50]
21		Red	Haematopodins	*M. haematopus*	Peters et al. [51]

(continued)

Table 1 (continued)

Serial #	Fungal genera	Color	Pigments responsible	Fungal species	References
22	*Tapinella*	Orange-yellow	Flavomentins A-D	*Tapinella* spp.	Béni et al. [8]
		Violet	Spiromentins A-D		
23	*Boletopsis*		Thelephoric acid	*Boletopsis leucomelaena*	Jägers et al. [32]
24	*Gomphidius*	Yellow	Gomphidic acid	*Gomphidius* spp., e.g., *G. glutinosus*	Knight and Pattenden [37]
25				*Suillus* spp.	
26	*Suillus*	Orange	Decarboxylated pulvinic acids	*S. grevillei*	Nelsen [47]
		Red			
27		Orange to red	Grevillins A, B, and C	*S. luteus*	Besl et al. [9]
28				*S. grevillei*	
29			Grevillins B, C, and D	*S. granulatus*	
30		Red	Tridentoquinone	*S. tridenticus*	Lang et al. [42]
31			Boviquinone-4	*S. bovinus*	
32	*Boletus*	Chocolate brown	Badione A	*Boletus badius*	Aumann et al. [4]
33	*Pisolithus*	Golden yellow	Norbadione A	*Pisolithus arrhizus*	Winner et al. [69]
34	*Scleroderma*	Bright yellow	Sclerocitrin	*Scleroderma citrinum*	
			Norbadione A		
35	*Chamonixia*	Deep brown and bright blue color	Chamonixin	*Chamonixia* spp.	Hanson [28]
36	*Gyrodon Paxillus*		Involutin	*Gyrodon* spp., *Paxillus* spp.	
37	*Gyroporus*		Gyrocyanin	*Gyroporus* spp.	
38	*Melanogaster*		Gyroporin	*Melanogaster* spp.	
39	*Paxillus*		Involutin	*P. involutus*	
40	*Chamonixia*		Chamonixin	*C. caespitosa*	
41	*Gyroporus*		Gyrocyanin	*G. cyanescens*	
42	*Suillus*	Oxidized into brown	Suillin	*S. granulatus*	Mühlbauer et al. [45]
43	*Cantharellus*	Pink Red-orange	β-carotene Lycopene	*Cantharellus cibarius* *C. cinnabarinus*	Hanson [28]
			α-carotene		
44			Xanthophyl canthaxanthin		
45	*Inonotus*	Red-brown	Bisnoryangonin	*Inonotus hispidus*	Ali et al. [3]
			Hypholomin B		
46		Yellow	Hispolon		
		Red-brown	Bisnoryangonin Hypholomin B	*Inonotus* spp.	Velíšek and Cejpekn [66]
47	*Phellinus*			*Phellinus* spp.	
48	*Hypholoma*	Yellow	Hispolon	*Hypholoma* spp.	

(continued)

Table 1 (continued)

Serial #	Fungal genera	Color	Pigments responsible	Fungal species	References
49	*Lenzites*	Red	Betulinans A and B	*Lenzites betulina*	Gill [21]
50	*Hapalopilus*	Dark red	Polyporic acid	*Hapalopilus nidulans*	
51	*Pycnoporus*	Red	Cinnabar	*Pycnoporus cinnabarinus*	Gripenberg [25]
52	*Terana*	Oxidized into indigo-blue	Corticins A, B, and C	*Terana caerulea*	Briggs et al. [12]
53	*Phanerochaete*	Dark red	Xylerythrin	*Phanerochaete sanguine*	Gripenberg and Martikkala [26]
54	*Laetiporus*	Yellowish or orange	Laetiporic acid A	*Laetiporus sulphurous*	Davoli et al. [16]
55	*Lactarius*	Carrot-colored into green	Blue lactarazulene	*Lactarius deliciosus*	Zhou and Liu [72]
56		Pale peach into greyish green	Red-violet lactaroviolin	*L. deterrimus*	
57		Orange	Dihydroazulen-1-ol	*L. deterrimus*	
58		Turns white to yellow	Lactarane and secolactarane	*L. scrobiculatus*	Caro et al. [14]
59		White to violet exposed to air	Uvidin A and B	*L. uvidus*	De Bernardi et al. [17]
60		Pink-red	Variety of benzofurans and chromenes	*L. fuliginosus*	
61	*Lactarius*	Pale green	Aminobenzoquinone blennione	*L. blennius*	Spiteller and Steglich [61]
62		Red	Lilacinone	*L. lilacinus*	Spiteller et al. [62]
63		Green-brown	Necatarone	*L. turpis*	Klamann et al. [35]
64	*Russula*	Yellow into red	Ochroleucin A changes into ochroleucin B	*Russula ochroleuca*	Sontag et al. [60]
65				*R. viscida*	
66	*Albatrellus*	Purple	Grifolinone B	*Albatrellus confluens*	Yang et al. [71]
		Red	Ketone albatrellin		
67	*Hydnellum*	Dark red	Terphenylquinone atromentin	*Hydnellum aurantiacum*	

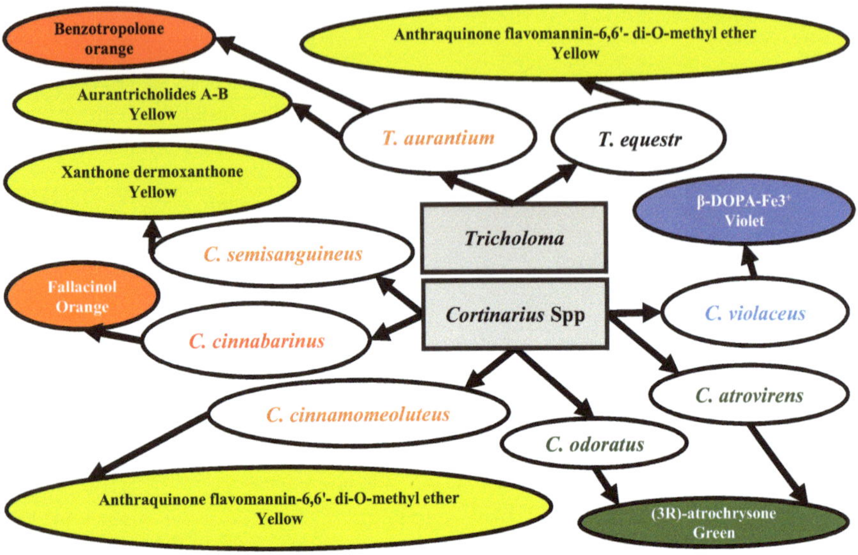

Fig. 2 Biological dyes produced by different species of *Tricholoma* and *Cortinarius* mushroom genera (Table 1 Serial # 6–13)

Tricholoma equestre can biosynthesize anthraquinone flavomannin-6,6′-di-O-methyl ether metabolite (Fig. 2), which exhibits a bright yellow color under fluorescence UV light [20, 22].

In contrast, various genera within the *Strophariaceae* family, including *Gymnopilus*, *Hypholoma*, and *Pholiota*, have the ability to biosynthesize yellow-brown pigments. The precursor of this pigment is styrylpyrone pigments, namely bisnoryangonin and hispid, which are widespread among these genera [20]. *Oudemansiella melanotricha* has been found to produce orange-yellow pigments in the form of polyenes amino acid derivatives such as dihydroxerulin, xerulin, and xerulinic acid [39]. Lang et al. [41] described the carrot truffle (*Stephanospora caroticolor*) as containing a tuber-like fruiting body that biosynthesizes stephanosporin compounds, resulting in a bright orange pigment. Similarly, Peters and Spiteller [49, 50] extracted pyrroloquinone alkaloids, specifically mycearubin A and mycearubin B, which serve as precursors for red pigments, from *Mycena* and *Roussel* mushroom species. Similar biological compounds were isolated from other species such as *M. rosea* and *M. haematopus*, contributing to their red coloration. Additionally, *M. sanguinolenta* and *M. haematopus* synthesize blue and red pigments, respectively (Fig. 3).

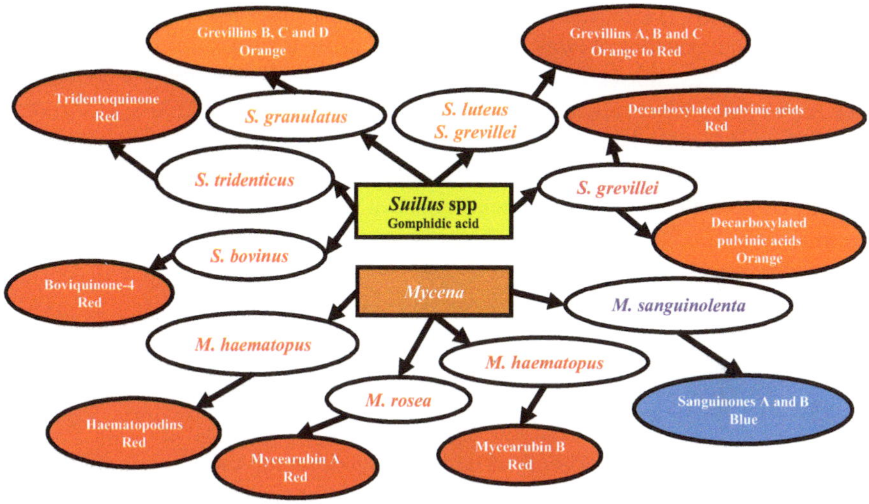

Fig. 3 Mushroom genera *Suillus* and *Mycena* biologically synthesized textile dyes (Table 1 Serial # 18–31)

4 Biopigment Synthesis in Boletales

All mushroom genera belonging to the order Boletales exhibit a diverse range of biopigments in various colors. These pigments are derived from the biosynthesis of terphenylquinones and pulvinic acids [65, 68]. The *Tapinella* genus, found within the Boletales order, produces trace quantities of orange-yellow flavomentins A-D and violet spiromentins A-D [10].

In the Boletales order, the primary pigment is yellow pulvinic acid, which is produced through lactone synthesis by opening the oxidized ring of terphenylquinones. Pulvinic acid is particularly prevalent in mushrooms belonging to the *Gomphidiaceae* and *Suillaceae* families, specifically in the *Gomphidius* and *Suillus* genera. The yellow gomphidic acid pigment was first extracted from the slimy spike-cap mushroom *G. glutinosus* [37]. Similarly, a well-known species in the *Suillus* genus, *Suillus grevillea*, synthesizes a variety of pigments in different colors, such as yellow, orange, and red. These pigments are extracted from the yellow-brown cap of *S. grevillea* and are produced through the decarboxylation of pulvinic acids into trihydroxypulvinone [47]. Other species within the *Suillus* genus produce grevillins A, B, C, and D, which range in color from orange to red (refer to Fig. 3). The grevillins A, B, C, and D are synthesized through the lactonization of atromentin and are extracted solely from the caps or fruiting bodies of *Suillus* mushrooms. *S. grevillea* and *S. luteus* have grevillins A, B, and C, while *S. granulatus* has grevillins B, C, and D, with grevillin D considered the major biopigment responsible for the color biosynthesis [9, 47].

Several species within the *Boletus* genus have been found to possess various radionuclides, which aid in their survival in radioactive-polluted environments.

Cesium-137 (Cs137), attributed to the Chernobyl nuclear reactor catastrophe, is bound to naphthalenoid pulvinic acids, existing as their potassium salts in the cap skin layers of these mushrooms. The primary substances responsible for the chocolate brown and golden yellow hues in the cap skin layers of these bolete mushrooms are badione A and norbadione A [4]. Norbadione A, in particular, is the main compound and is found in higher concentrations compared to badione A and other related compounds. Elevated concentrations of norbadione A have been isolated from the mushroom *Pisolithus arrhizus*, where it constitutes 25% of the mushroom's dry weight [69].

4.1 Colorful Biopigments in Earth Ball and Peppery Bolete

Previous studies have demonstrated that the common earth ball mushroom (*Scleroderma citrinum*) biosynthesizes the bright yellow pigment sclerocitrin along with norbadione A in its cap of the fruiting body. Sclerocitrin is produced by the combination of two molecules of xerocomic acid through oxidative coupling. Similarly, sclerocitrin is also naturally produced in the peppery Bolete mushroom (*Chalciporus piperatus*), giving it a lemon yellow color in varying shades [69].

4.2 Terphenylquinones and Pulvinic Acids Bio-colorants in Boletales

Terphenylquinones such as chamonixin, involutin, highly oxidized gyrocyanin, and gyroporin are synthesized through the decarboxylation of pulvinic acids. These pigments are produced by various genera of fungal families within the Boletales order, including *Chamonixia* and *Leccinum* (Boletaceae), *Gyrodon* and *Paxillus* (Paxillaceae), *Gyroporus* (Gyroporaceae), and *Melanogaster corda* (Melanogasteraceae). When the fruiting bodies of these mushrooms are crushed, a distinct deep brown and brilliant blue color can be extracted [18, 68]. Mühlbauer et al. [45] investigated the acetylated compound suillin, which is biosynthesized in the fruiting cap of *S. granulatus* and later oxidizes to give the brown color to the mushroom's fruiting body.

5 Carotenoid Pigments in Mushroom Species

Carotenoid pigments are commonly found in plants and higher groups of algae but are infrequently produced in fungi. However, certain mushroom species such as *Cantharellus cibarius* and *C. cinnabarinus* have the potential to synthesize carotenoid compounds including β-carotene, lycopene, α-carotene, and xanthophyll and canthaxanthin in the form of xanthophyll. These carotenoid compounds, upon oxidation, produce pink and orange-red colors [28].

6 Pigments in Cobalt Crust Fungus and Saprophytic Fungi

Lawrinowitz et al. [43] reported that the plant pathogenic fungus *Terana caerulea*, commonly known as cobalt crust fungus and belonging to the family Phanerochaetaceae, produces corticins A, B, and C through the polyporic acid pathway. These corticins give an ink-blue color when exposed to natural light and UV light. The compound xylerythrin is biologically synthesized in the saprophytic fungal strain *Phanerochaete sanguine* from various molecules of arylpyruvic acids, resulting in a red color. When *P. sanguine* infects a plant, its mycelia grow in a red color, turning the infected area red [26]. *Laetiporus sulphureus*, an edible wood-rotting fungus, contains non-isoprenoid polyene compounds known as laetiporic acid A and 2-dehydro-3-deoxylaetiporic acid and exhibits primary colors [16].

7 Terpenoid Pigments in Mushroom Species

After carotenoid-based biological pigments, several mushroom species produce terpenoid-based biopigments that can be used in the textile industry. The *Lactarius* genus of the Russulaceae family biosynthesizes terpenoid pigments of diverse colors (Fig. 4). Over time, these pigments convert into sesquiterpenoids, resulting in a change in color shade. Young fruiting bodies of *Lactarius deliciosus* have a carrot shade color, but as they age, the color changes from carrot to green. The responsible pigments for this color change are blue lactarazulene and lipophilic red-violet lactaroviolin (Fig. 4). Recent investigations have also revealed that *L. deliciosus* synthesizes three different naturally red azulene compounds [72]. Young fruiting bodies of *L. deterrimus* have a pale peach color and produce orange latex, but as they age, they turn apricot in color due to the presence of azulene pigment-derived sesquiterpenoids [72]. Velíšek and Cejpek [66] reviewed that *L. sanguifluus* produces 1,3,5,7(11),9-pentaenyl-14-guaianal and a lipophilic compound known as 1-hydroxymethyl-4-methyl-7-(1-methylethenyl) azulene stearate, which has been extracted from the caps of *L. indigo*'s fruiting bodies. Both isolated terpene compounds from *L. sanguifluus* and *L. indigo* give red and indigo color pigments, respectively. Bosetti et al. [11] isolated a variety of lactarane and secolactarane sesquiterpenoids, which are different from azulene terpenes. Lactarane and secolactarane compounds are synthesized by *L. scrobiculatus*, initially appearing white and later turning yellow upon oxidation when exposed to air. *L. uvidus* produces a variety of uvidin terpenoid compounds that are whitish in color but turn violet upon oxidation. The molecular structure of uvidin A and B is similar to a broad range of terpenoid compounds found only in *L. uvidus* [11, 66].

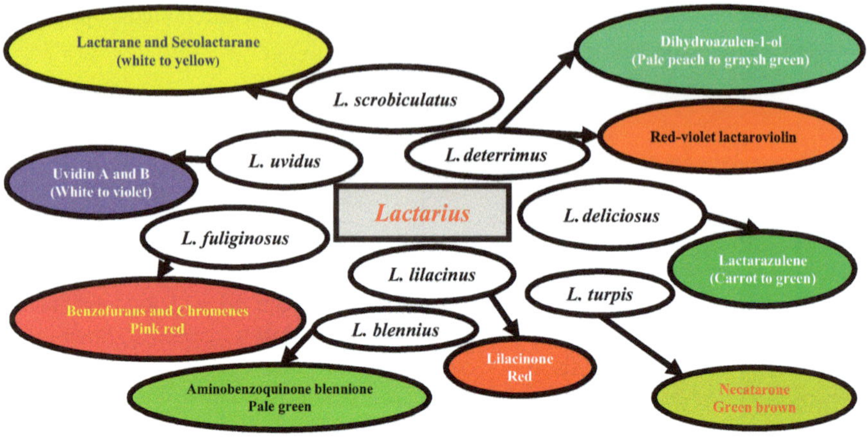

Fig. 4 Textile dyes produced by *Lactarius* genera (Table 1 Serial# 55–63)

7.1 *Isoprenoid Quinol Stearate in* Lactarius *Species*

Lactarius fuliginosus and *L. picinus* produce the isoprenoid quinol stearate of 4-methoxy-2 phenol derivatives, which results in a red-pink stain and an unpleasant odor (Fig. 4). This stain and acrid flavor develop in bruised toadstools due to hydrolysis followed by oxidation by red benzofurans and chromenes. In the fruit bodies of the slimy milkcap *Lactarius blennius*, the aminobenzoquinone blennione is found, giving a pale green hue. Its biosynthesis is likely influenced by 3,6-dihydroxyanthranilic acid. Another pigment, lilacinone, shares a close biological relationship with Bennione and is responsible for the violet color of *Lactarius lilacinus* [62].

7.2 *Alkaloidal Biopigment in* Lactarius turpis

Lactarius turpis has a green-brown flesh and cap skin characterized by an alkaloidal pigment known as necatarone (Fig. 4). The dehydrodimer form of necatarone is predominant in old fruiting bodies, while the 10-deoxydeoxydehydrodimer form is dominant in young fruiting bodies [19, 35]. Various biopigments obtained from different mushrooms (fungal fruiting bodies) are listed in Table 1.

8 Extraction Procedure for Color from Mushrooms

There are various methods used for the extraction of pigments from mushrooms, depending on the type of pigment and the nature of the mushroom's cell wall. Common extraction methods include solvent extraction, Soxhlet extraction, and hot water extraction [23]. Since mushrooms do not perform photosynthesis and lack photosynthetic pigments, extracting pigments from mushroom biomass is generally easier compared to plants. Researchers have developed preparative-scale methods for isolating pigments from wild mushrooms, making the extraction process more efficient and effective.

8.1 Preparative-Scale Isolation Method

This method, developed by Hynninen et al. [31], is specifically designed for extracting naturally occurring anthraquinone dyes. These dyes are commonly found as glycosides in various living organisms. Despite their water solubility, these pigments are unsuitable for dyeing fibers as they remain in the solution instead of binding to the fibers. The isolation method involves the utilization of fresh fruiting bodies of mushrooms and the application of an enzyme called β-glucosidase. This enzyme breaks the glucoside linkage present in the mushroom cell walls, enabling the extraction of free anthraquinone color pigments. Subsequently, these extracted pigments can be employed for dyeing fibers.

9 Modern Extraction Methods of Mushroom Pigments

9.1 Enzyme-Assisted Extraction

One modern method used to extract mushroom pigments is enzyme-assisted extraction. This method utilizes hydrolytic enzymes to break down the cell wall components of mushrooms, such as chitin and glucans. Enzymes like chitinase and glucanase are capable of effectively degrading these polysaccharides, thereby facilitating the release of pigment compounds. For instance, the application of carbohydrase enzymes derived from *Trichoderma harzianum* has proven effective in extracting the bioactive compound eritadenine from Shiitake mushrooms. The optimal conditions for enzyme activity usually involve a pH of 4.8 and a temperature of 50 °C. It has been demonstrated that pretreating mushroom biomass with hydrolytic enzymes enhances the extraction of mushroom pigment compounds [70].

9.2 Pulsed Electric Fields

Another modern extraction method is pulsed electric fields (PEF), which is used for extracting intracellular compounds from mushrooms, algae, and higher plants (Fig. 5). PEF involves the application of short-duration, high-voltage pulses to cells. This process temporarily permeabilizes cell membranes, facilitating the movement

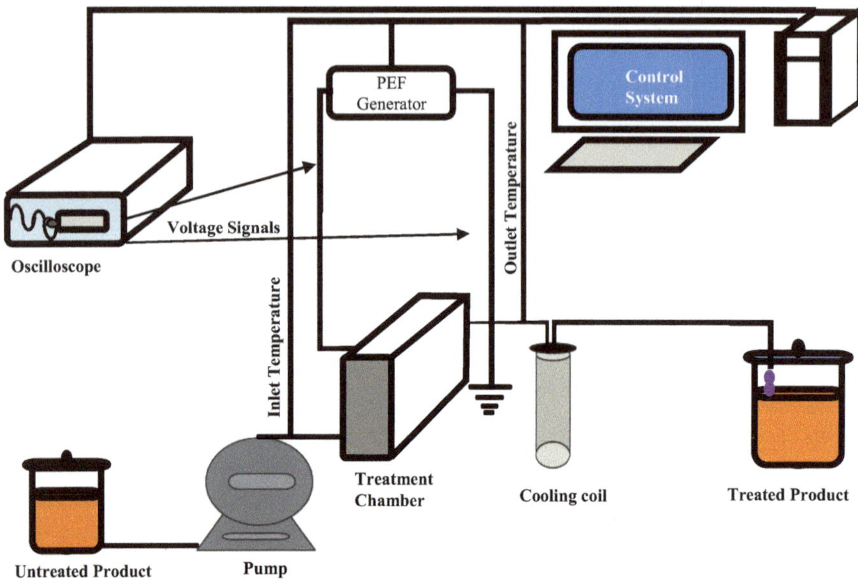

Fig. 5 Schematic representation of pulse electric fields (PEF) for the extraction of intracellular pigmented compounds. (Redrawn from Roselló-Soto et al. [56])

of charged species across cellular compartments. Consequently, cell damage or the induction of sub-lethal stress may occur. The PEF extraction system typically consists of a pulse generator, a treatment chamber, and a pumping system. The pulse generator delivers high-voltage pulses through electrodes in the treatment chamber, resulting in the extraction of targeted compounds. Short-duration direct current pulses, with voltages of up to 40 kV, can be selectively employed to extract high-value chemicals from complex biochemical compounds [52].

9.3 Ultrasound-Assisted Extraction

In recent years, ultrasound-assisted extraction (UAE) has garnered attention as a favorable method for extracting beneficial organic compounds from various sources, including mushrooms. This extraction technique offers several advantages over conventional methods. During UAE, the shock waves generated by cavitation bubbles in the ultrasound field disrupt mushroom cells, leading to cell rupture and enhanced mass transfer. Consequently, this improves both extraction yield and kinetics. UAE proves to be not only cost-effective and simple but also an efficient technique for isolating valuable compounds from mushrooms. When combined with solvents, UAE enables the reduction of extraction time and temperature, which is particularly advantageous for preserving thermolabile compounds. Research indicates that UAE facilitates the extraction of larger compounds such as polysaccharides from mushrooms, resulting in a significant increase from 52% to 129% in extraction yields.

9.4 Sub- and Supercritical Fluid Extraction

In recent years, subcritical and supercritical fluid extraction methods have gained significance as alternatives to traditional extraction techniques. These methods are considered more environmentally friendly and have the potential to yield nutritionally valuable compounds from various sources, including mushrooms. Subcritical fluid extraction involves using a solvent at temperatures and pressures below its critical point, while supercritical fluid extraction utilizes a solvent above its critical point. In both methods, a fluid such as carbon dioxide is employed as the extraction medium. The unique properties of subcritical and supercritical fluids enable the efficient extraction of target compounds from mushrooms. These extraction techniques offer advantages such as high selectivity, shorter extraction times, and the ability to extract compounds that are sensitive to heat or easily degraded by traditional extraction methods.

Furthermore, subcritical and supercritical fluid extraction methods are considered environmentally friendly due to the use of non-toxic solvents and the potential for solvent recovery and recycling. This makes them attractive alternatives for extracting nutritionally valuable chemicals from mushrooms and other natural sources.

9.5 Subcritical Water Extraction

Subcritical water extraction (SWE), also known as hot water extraction, is a method that utilizes water as the solvent at temperatures ranging from 100 to 374 °C (H_2O critical temperature) under high pressure. This unique condition enables water to remain in its liquid state without vaporization. It is important to consider the change in dielectric constant, which is influenced by temperature, during the extraction process. At low temperatures, compounds with high polarity can dissolve easily in water, while at higher temperatures (above 100 °C), water behaves more like an organic solvent and can dissolve organic compounds and compounds with low polarity. The use of SWE in mushroom extraction is an area that has not been extensively explored. Studying the kinetics between pressure, temperature, and the extraction of bioactive compounds in SWE is of particular interest as it indicates changes in water polarity (Fig. 6).

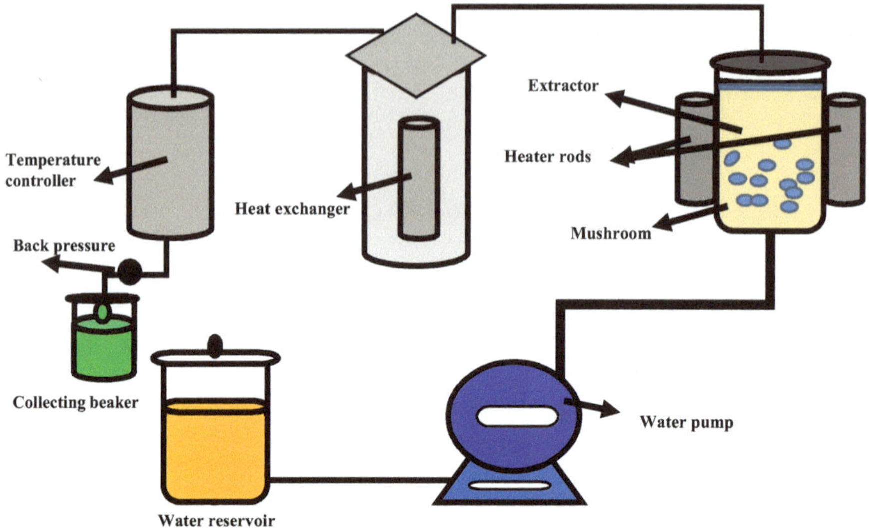

Fig. 6 Schematic representation of subcritical water extraction (SWE). (Redrawn from Roselló-Soto et al. [56])

9.6 Supercritical Fluid Extraction

Substances that reach a supercritical point of pressure and temperature are referred to as supercritical fluids. In this state, there is no distinction between the liquid and gas phases due to the temperature and pressure being above their critical point. Supercritical fluids exhibit a density similar to liquids and viscosities similar to gases. However, their diffusivity falls between the two states, allowing them to effectively extract and purify intracellular compounds despite their similar densities. Goto et al. [24] developed a mathematical model to calculate diffusivity (D), which was later modified by Mhemdi et al. [44] and the over all process is shown in Fig. 7. Supercritical fluid extraction (SFE) has a rich history and has been widely used in various industries, notably in food, agriculture, and fuel industries. It has also found applications in analysis, chromatography, pharmaceuticals, environmental contaminants, metal-ion extraction, and pesticide industries. In recent decades, there has been a particular emphasis on extracting highly nutritious substances from diverse sources, including mushrooms. Fig. 7, adapted from Roselló-Soto et al. [56], illustrates SFE utilizing CO_2 as a co-solvent in the extraction process. CO_2 acts as a supercritical fluid with limited solubility under critical conditions. By employing co-solvent extraction with organic solvents, low-polarity solvents can be extracted using low-polarity solvents, thereby providing a wider range of extraction options. Initially, CO_2 in SFE was primarily used to rapidly and conveniently obtain samples from *A. bisporus* for profiling carboxylic and fatty acids. At that time, the application of SFE-CO_2 for extracting chemical compounds from mushrooms was still in its early stages.

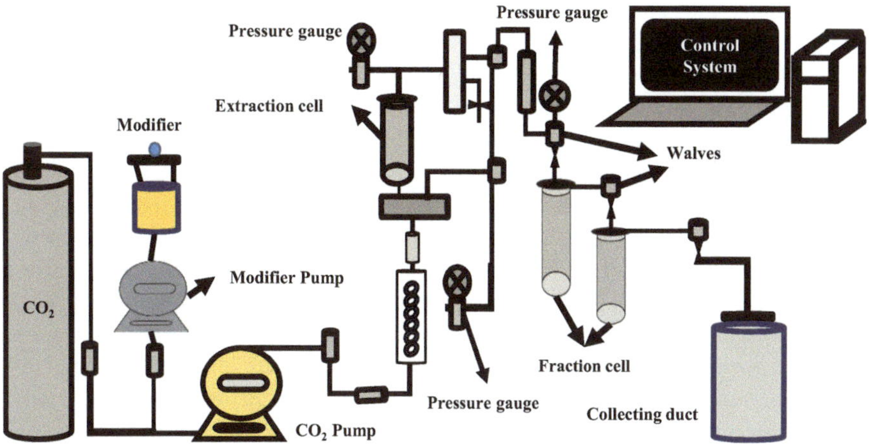

Fig. 7 Schematic representation of supercritical fluid extraction (SFE) for biological pigment extraction from mushrooms. (Redrawn from Roselló-Soto et al. [56])

10 Dyeing with Mushrooms: Exploring Natural Bio-colorants on Different Fibers

In 1804, Palmstruch's work Svensk Botanik [48] mentions the utilization of mushrooms primarily for food and medicinal purposes. However, there were also instances where various mushrooms were used as biological dyes to color cotton and linen. This unconventional use of mushrooms as dyeing agents stemmed from a lack of knowledge about fungi. Many species remained unrecognized, leading to doubt and confusion. During World War II, some studies in Europe examined the use of fungi as sources of textile colors [63]. However, it wasn't until the 1970s, when Miriam C. Rice, an American author, wrote a book on dyeing with mushrooms that these natural colorants gained popularity among artists, primarily due to their potent antibacterial properties [55].

With the growing concerns about fossil fuel depletion and ecological design, there has been an increased interest in natural dyes. Extensive research has been conducted to support the application of natural dyes in various industrial sectors, including food, cosmetics, and textiles (Table 1).

One of the primary bio-colorants used in mushroom-based dyeing methods to achieve a blue shade is terphenylquinone compounds. Sources of blue pigments include *Sarcodon*, *Phellodon*, *Hydnellum*, and *Telephora* spp. The fresh weight of *Hapalopilus nidulans* contains 20–40% polyporic acid. *Paxillus atrotomentosus* and *Telephora palmata* also possess pigments, albeit in lower concentrations. While *T. palmata* has trace amounts of telephoric acid (less than 1% of its dry weight), *P. atrotomentosus* contains a combination of atromentin and telephoric acid ranging from 2% to 4% [22, 58].

Differentiating between the two frequently mistaken *Sarcodon* species, *S. imbricatus* and S. *squamosus*, was previously challenging. It was believed that *Sarcodon imbricatus* imparted a blue color to wool fibers. However, achieving the desired blue hue was often perplexing for dyers. Current DNA studies have revealed that the *Sarcodon* genus is divided into two distinct species, with one growing in pine trees (*Pinus sylvestris*) and the other in spruce trees (*Picea abies*). Mushrooms harvested from pine woods yield more intense blue colors compared to those from spruce forests [33].

Producing vibrant blue hues necessitates a substantial quantity of fresh mushrooms, often exceeding the weight of the textile material used in the manufacturing process by 10–20 times. In some cases, this may result in the production of grayish or light blue colors. To achieve stronger colors, a quantity of mushrooms at least 30 times the weight of the fiber is required [29]. However, the low levels of colorants present in mushrooms make large-scale dyestuff manufacturing economically challenging, despite the mushrooms' ability to produce unique hues in the realm of natural dyes. In evaluating the future potential uses of terphenylquinones, their antibacterial properties should be taken into account [59].

Notable colorants found in important *Boletales* sp. Include grevillins, terphenylquinones, and benzoquinone derivatives. The grevillins from these mushrooms generate yellow to crimson colors [13, 59].

10.1 Anthraquinones: Vibrant Dyes from Cortinarius *Mushrooms*

Cortinarius mushrooms, belonging to the Cortinarius subgroup of *Dermocybe* species, have gained recognition as excellent sources for dyeing. The striking colors displayed by the fruit bodies are indicative of their high pigment content. *Cortinarius* mushrooms exhibit a wide range of colors, including brown, red, olive green, and even violet color.

Anthraquinones, the primary pigments found in *Cortinarius* mushrooms, hold significant importance among all categories of pigments used in industrial textile colorants. They are known for producing brilliant colors and demonstrating relative stability and light-fastness. In fact, ancient dyestuffs such as madder, cochineal, and kermes were renowned for their anthraquinone bases (*Handbook of Natural Colorants*, [6]). Anthraquinones derived from natural sources have also found applications as food colorings in recent times, including beverages, desserts, and other food items [15, 57]. Furthermore, natural colorants are increasingly being utilized in the cosmetic industry, including hair colorants [27, 57].

One notable example of a *Cortinarius* mushroom is the blood-red webcap, scientifically known as *Cortinarius sanguineus*. This species contains two prevalent anthraquinones: emodin and dermocybin-1-b-D-glycopyranosides. In young fungi, up to 90% of the pigments are in the form of glycosides. The biosynthesis of anthraquinones involves the proper folding and condensation of a polyketide chain derived from acetate units, which occurs through the acetate-malonate cycle. This pathway serves as a central route for anthraquinone biosynthesis. The acetate-malonate pathway reveals the biosynthetic relationships, where the structurally diverse red compounds such as dermocybin, dermorubin, and 5-chlorodermorubin are synthesized toward the end of the pathway, while the yellow compounds including emodin, physcion, and endocrine occur at the beginning of the pathway [53].

Color tests have shown that young *Cortinarius* mushrooms tend to produce more yellowish and orange hues, whereas older specimens yield dark red and reddish-brown tones. This observation has been confirmed through dyeing experiments.

11 Conclusion: Embracing the Potential of Mushroom-Based Dyes

The widespread reliance on synthetic dyes has posed significant environmental and health concerns. In response, the use of natural dyes obtained from various living resources, including plants, algae, microorganisms, fungi, and mushrooms, has gained substantial importance. Among these natural sources, mushrooms exhibit huge potential for the biosynthesis of a diverse range of color pigments that can be utilized in diverse industries including textiles. By harnessing the power of mushroom-based dyes, we can significantly reduce the environmental impact associated with synthetic dyes in a sustainable manner. It is crucial to raise awareness about the benefits and viability of mushroom-based dyes as a more sustainable and eco-friendly option for the textile industry.

Research studies have shown that numerous mushroom species, widely consumed as food and found abundantly in nature, can serve as valuable sources for creating a natural palette of dyes. In addition to their potential as colorants, mushrooms offer a wide array of metabolites that possess antimicrobial properties and have been explored for their applications in treating cardiovascular disorders. Additionally, these versatile compounds find utility in various industries, including pharmaceuticals, nutraceuticals, and cosmetics.

As we strive toward a more sustainable future in the modern world, it is essential to embrace the potential of mushroom-based dyes and explore their multifaceted benefits. By adopting these natural alternatives, we can foster environmentally conscious practices in the textile industry while promoting innovation and the efficient sustainable utilization of renewable resources.

12 Recommendations

Based on the significant potential and diverse applications of mushroom pigments, further exploration and investigation in this field are recommended. The study of biosynthetic pathways, genetic regulation, and chemical properties of mushroom pigments can provide valuable insights into their production and potential uses.

To advance the field, interdisciplinary collaborations among biologists, chemists, and industry professionals are encouraged. This collaboration can facilitate the development of sustainable and eco-friendly alternatives to synthetic dyes in industries like textiles, food, and cosmetics. Additionally, exploring the therapeutic potential of mushroom pigments can lead to the discovery of novel natural compounds with beneficial properties for medicinal applications. On the other hand, understanding the ecological functions and roles of mushroom pigments in attracting pollinators, deterring predators, and aiding in fungal communication is crucial. This knowledge can contribute to our understanding of fungal ecology and ecosystem dynamics. Academic institutions should promote interdisciplinary research programs and encourage collaborations across different fields. Overall, investing in and promoting research on mushroom biopigments will contribute to scientific advancements, sustainable practices, and the development of innovative applications in various industries, especially textiles.

References

1. Aishwarya, A. D. (2014). Extraction of natural dyes from fungus–An alternate for textile dyeing. *Journal of Natural Science Research, 4*(7), 1–7.
2. Akilandeswari, P., & Pradeep, B. V. (2016). Exploration of industrially important pigments from soil fungi. *Applied Microbiology and Biotechnology, 100*, 1631–1643.
3. Ali, N. A. A., Jansen, R., Pilgrim, H., Liberra, K., & Lindequist, U. (1996). Hispolon, a yellow pigment from Inonotus hispidus. *Phytochemistry, 41*, 927–992.
4. Aumann, D. C., Cloth, G., Steffan, B., & Steglich, W. (1989). Complexation of cesium 137 by the cap pigments of the bay boletus (Xevocomus badius). *Angewandte Chemie, International Edition English, 28*, 453–454.
5. Bayer, E., & Kneifel, H. (1972). Isolation of amavadin, a vanadium compound occurring in Amanita muscaria. *Zeitschrift für Naturforschung, 27B*, 207–210.
6. Bechtold, T., & Mussak, R. (2009). Natural colorants–quinoid, naphthoquinoid and anthraquinoid dyes. In *Handbook of natural colorants* (pp. 151–182). John Wiley & Sons Inc.
7. Bechtold, T., & Mussak, R. (Eds.). (2009). *Handbook of natural colorants* (Vol. 8). John Wiley & Sons.
8. Béni, Z., Dékány, M., Kovács, B., Csupor-Löffler, B., Zomborszki, Z. P., Kerekes, E., Urbán, E., Hohmann, J., & Ványolós, A. (2018). Bioactivity-guided isolation of antimicrobial and antioxidant metabolites from the mushroom. Tapinella atrotomentosa. *Molecules, 23*(5), 1082.
9. Besl, H., Michler, I., Preuss, R., & Steglich, W. (1974). Pigments of fungi, XXII Grevillin D, the main pigment of Suillus granulatus, S. luteus and S. placidus (Boletales). *Zeitschrift für Naturforschung, 29C*, 784–786.
10. Besl, H., Bresinsky, A., Geigenmiiller, G., Herrmann, R., & Steglich, W. (1989). Pilzfarbstoffe, 61 Flavomentine und Spiromentine, neue Terphenylchinon-Derivate aus Paxillus atrotomentosus und P. panuoides (Boletales). *Liebigs Annalen der Chemie, 1989*, 803–810.
11. Bosetti A., Fronza G., Vidari G., Vita-Finzi P. (1989). Norlactarane and lactarane sesquiterpenes from Lactarius scrobiculatus. *Phytochemistry, 28*, 14.
12. Briggs, L. H., Cambie, R. C., Dean, I. C., Hodges, R., Ingram, W. B., & Rutledge, P. S. (1976). Chemistry of fungi. XI. Corticins A, B, and C, benzobisbenzofurans from Corticium caeruleum. *Australian Journal of Chemistry, 29*, 179–190.
13. Cardon, D. (2007). Natural dyes. In *Sources, tradition, technology and science* (p. 268). Archetype.

14. Caro, Y., Venkatachalam, M., Lebeau, J., Fouillaud, M., & Dufossé, L. (2017). Pigments and colorants from filamentous fungi. In *Fungal metabolites* (pp. 499–568). Springer.
15. Carvalho, P. R. N., & Collins, C. H. (1997). HPLC determination of carminic acid in foodstuffs and beverages using diode array and fluorescence detection. *Chromatographia, 45*(1), 63–66.
16. Davoli, P., Mucci, A., Schenetti, L., & Weber, R. W. S. (2005). Laetiporic acids, a family of non-carotenoid polyene pigments from fruit-bodies and liquid cultures of Laetiporus sulphureus (Polyporales, Fungi). *Phytochemistry, 66*, 817–823.
17. De Bernardi, M., Vidari, G., Vita, F. P., & Fronza, G. (1992). The chemistry of Lactarius fuliginosus and Lactarius picinus. *Tetrahedron, 48*, 7331–7344.
18. Feling, R., Polborn, K., Steglich, W., Mühlbacher, J., & Bringmann, G. (2001). The absolute configuration of the mushroom metabolites involution and chamonixin. *Tetrahedron, 57*, 7857–7863.
19. Fugmann, B., Steffan, B., & Steglich, W. (1984). Necatarone, an alkaloidal pigment from the gilled toadstool Lactarius necator (Agaricales). *Tetrahedron Letters, 25*, 3575–3578.
20. Gill, M. (1994). Pigments of fungi (Macromycetes). *Natural Products Reports, 11*, 67–90.
21. Gill, M. (1999). Pigments of fungi (Macromycetes). *Natural Products Reports, 16*, 301–317.
22. Gill, M., & Steglich, W. (1987). Pigments of fungi (Macromycetes). In *Fortschritte der Chemie organischer Naturstoffe/Progress in the Chemistry of Organic Natural Products* (pp. 1–297).
23. Gong, P., Wang, S., Liu, M., Chen, F., Yang, W., Chang, X., et al. (2020). Extraction methods, chemical characterizations and biological activities of mushroom polysaccharides: A mini-review. *Carbohydrate Research, 494*, 108037.
24. Goto, M., Sato, M., & Hirose, T. (1993). Extraction of peppermint oil by supercritical carbon dioxide. *Journal of Chemical Engineering of Japan, 26*, 401–407.
25. Gripenberg, J. A. R. L. (1958). Fungus pigments VIII: The structure of cinnabarin and cinnabarinic acid. *Acta Chem. Scan, 12*, 603–610.
26. Gripenberg, J., & Martikkala, J. (1970). Fungus pigments. XX. On the structure of peniophorin. One of the pigments produced by Peniophora sanguinea Bres. *Acta Chemica Scandinavica, 24*, 3444–3448.
27. Grollier, J. F., Rosenbaum, G., & Cotteret, J. (1986). *U.S. Patent No. 4,602,913*. U.S. Patent and Trademark Office.
28. Hanson, J. R. (2008). Pigments and odours of fungi. In J. R. Hanson (Ed.), *The chemistry of fungi* (pp. 127–142). The Royal Society of Chemistry. Thomas Graham House.
29. Hiltunen, S. (2005). *Blue Colours from Mushrooms (in Finnish)*. Master's Thesis, Department of Home Economics and Craft Science, University of Helsinki.
30. Hinsch, E. M. (2015). *A comparative analysis of extracted fungal pigments and commercially available dyes for colorizing textiles*. Oregon State University.
31. Hynninen, P. H., Räisänen, R., Elovaara, P., & Nokelainen, E. (2000). Preparative isolation of anthraquinones from the Fungus Dermocybe sanguined using enzymatic hydrolysis by the endogenous β-glucosidase. *Zeitschrift für Naturforschung C, 55*(7–8), 600–610.
32. Jägers, E., Hillen-Maske, E., & Steglich, W. (1987). Pilzfarbstoffe, 54. Inhaltsstoffe von Boletopsis leucomelaena (Basidiomycetes): Klärung der chemischen Natur von "Leucomelon" and "Protoleucomelon". *Zeitschrift für Naturforschung, B: Chemical Sciences, 42*, 1349–1353.
33. Johannesson, H., Ryman, S., Lundmark, H., & Danell, E. (1999). Sarcodon imbricatus and S. squamosus–two confused species. *Mycological Research, 103*(11), 1447–1452.
34. Karuppan, P. E. R. U. M. A. L., Sekarenthiran, S. C., Velusamy, K., Sadasivam, M. O. N. A., Mahalingam, P. O. N. S. U. G. U. M. A. R. I., Ramayanam, B. M., & Devi, S. (2014, November). Prospective aspects of myco-chrome as promising future textiles. In *Proceedings of the 8th international conference on mushroom biology and mushroom products (ICMBMP8)* (pp. 412–416).
35. Klamann, J.-D., Fugmann, B., & Steglich, W. (1989). Alkaloidal pigments from Lactarius necator and L. atroviridis. *Phytochemistry, 28*, 3519–3522.
36. Klostermeyer, D., Knops, L., Sindlinger, T., Polborn, K., & Steglich, W. (2000). Novel nebzotropolone and 2H-furo[3,2b]benzopyran/2-one pigments from Tricholoma aurantium (Agaricales). *Europen Journal of Organic Chemistry, 4*, 603–609.

37. Knight, D. W., & Pattenden, G. (1976). Synthesis of the pulvinic acid pigments of lichen and fungi. *Journal of Chemical Society, Chemical Communications, 16*, 660–661.
38. Kremer, G., & Chemiker, D. (2004). *The history of pigments production- Use*. Possibilities. www.kremer-pigmente.com
39. Kuhnt, D., Anke, T., Besl, H., Bross, M., Herrmann, R., Mocek, U., Steffan, B., & Steglich, W. (1990). Antibiotics from Basidiomycetes. XXXVII. New inhibitors of cholesterol biosynthesis from cultures of Xerula melanotricha Dörfelt. *Journal of Antibiotics, 43*, 1413–1420.
40. Lagashetti, A. C., Dufossé, L., Singh, S. K., & Singh, P. N. (2019). Fungal pigments and their prospects in different industries. *Microorganisms, 7*(12), 604.
41. Lang, M., Spiteller, P., Hellwig, V., & Steglich, W. (2001). Stephanosporin, a "traceless" precursor of 2-chloro-4- nitrophenol in the abdominal fungus Stephanospora caroticolor. *Angewandte Chemie, 113*(9), 1749–1751.
42. Lang, M., Mühlbauer, A., Gräf, C., Beyer, J., Lang-Fugmann, S., Polborn, K., & Steglich, W. (2008). Studies on the structure and biosynthesis of tridentoquinone and related meroterpenoids from the mushroom Suillus tridenticus (Boletales). *European Journal of Organic Chemistry, 2008*, 816–825.
43. Lawrinowitz, S., Wurlitzer, J. M., Weiss, D., Arndt, H. D., Kothe, E., Gressler, M., & Hoffmeister, D. (2022). Blue light-dependent pre-mRNA splicing controls pigment biosynthesis in the mushroom Terana caerulea. *Microbiology Spectrum, 10*(5), e01065–e01022.
44. Mhemdi, H., Rodier, E., Kechaou, N., & Fages, J. (2011). A supercritical tuneable process for the selective extraction of fats and essential oil from coriander seeds. *Journal of Food Engineering, 105*, 609–616.
45. Muhlbauer, A., Beyer, J., & Steglich, W. (1998). The biosynthesis of the fungal meroterpenoids boviquinone-3 and-4 follows two different pathways. *Tetrahedron letters, 39*(29), 5167–5170.
46. Narsing Rao, M. P., Xiao, M., & Li, W. J. (2017). Fungal and bacterial pigments: Secondary metabolites with wide applications. *Frontiers in Microbiology, 8*, 1113.
47. Nelsen, S. F. (2010). Bluing components and other pigments of boletes. *Fungi, 3*(4), 11–14.
48. Palmstruch, J. W. (1829). *Svensk botanik* (Vol. 10).
49. Peters, S., & Spiteller, P. (2007). Mycenarubins A and B, red pyrroloquinoline alkaloids from the mushroom Mycena rosea. *European Journal of Organic Chemistry, 2007*, 1571–1576.
50. Peters, S., & Spiteller, P. (2007). Sanguinones A and B, blue pyrroloquinoline alkaloids from the fruiting bodies of the mushroom Mycena sanguinolenta. *Journal of Natural Products, 70*, 1274–1277.
51. Peters, S., Jaeger, R. J. R., & Spiteller, P. (2008). Red pyrroloquinoline alkaloids from the mushroom Mycena haematopus. *European Journal of Organic Chemistry, 2008*, 319–323.
52. Puertolas, E., Luengo, E., Álvarez, I., & Raso, J. (2012). Improving mass transfer to soften tissues by pulsed electric fields: Fundamentals and applications. *Annual Review of Food Science and Technology, 3*, 263–282.
53. Räisänen, R. (2019). Fungal colorants in applications–focus on Cortinarius species. *Coloration Technology, 135*(1), 22–31.
54. Ramesh, C., Vinithkumar, N. V., Kirubagaran, R., Venil, C. K., & Dufossé, L. (2019). Multifaceted applications of microbial pigments: Current knowledge, challenges and future directions for public health implications. *Microorganisms, 7*(7), 186.
55. Rice, M. (1980). *Mushrooms for color* (2nd ed.). Mad River Press.
56. Roselló-Soto, E., Parniakov, O., Deng, Q., Patras, A., Koubaa, M., Grimi, N., et al. (2016). Application of non-conventional extraction methods: Toward a sustainable and green production of valuable compounds from mushrooms. *Food Engineering Reviews, 8*, 214–234.
57. Saastamoinen, O. (1999). Forest policies, access rights and non-wood forest products in northern Europe. *UNASYLVA-FAO, 50*, 20–26.
58. Schweppe, H. (1993). *Handbook of natural dyes*. Nikol-Verl.-Ges.
59. Singh, R., Jain, A., Panwar, S., Gupta, D., & Khare, S. K. (2005). Antimicrobial activity of some natural dyes. *Dyes and Pigments, 66*(2), 99–102.
60. Sontag, B., Rüth, M., Spitteler, P., Arnold, N., Steglich, W., Reichert, M., & Bringmann, G. (2006). Chromogenic meroterpenoids from the mushroom Russula ochroleuca and R. viscida. *European Journal of Organic Chemistry, 2006*, 1023–1033.

61. Spiteller, P., & Steglich, W. (2002). Blennione, a green aminobenzoquinone derivative from Lactarius blennius. *Journal of Natural Products, 65*, 725–727.
62. Spiteller, P., Arnold, N., Spiteller, M., & Steglich, W. (2003). Lilacinone, a red aminobenzoquinone pigment from Lactarius lilacinus. *Journal of Natural Products, 66*, 1402–1403.
63. Sundström, C., & Sundström, E. (1983). *Sienivarjays*. Otava.
64. Téllez-Téllez, M., & Díaz-Godínez, G. (2022). Mushroom pigments and their applications. In *Biomolecules from natural sources: Advances and applications* (pp. 82–100). John Wiley & Sons.
65. Thoen, E., Harder, C. B., Kauserud, H., Botnen, S. S., Vik, U., Taylor, A. F., Menkis, A., & Skrede, I. (2020). In vitro evidence of root colonization suggests ecological versatility in the genus Mycena. *The New Phytologist, 227*(2), 601–612.
66. Velíšek, J., & Cejpek, K. (2011). Pigments of higher fungi-A review. *Czech Journal of Food Sciences, 29*(2), 87–102.
67. Walton, K., Coombs, M. M., Walker, R., & Ioannides, C. (2001). The metabolism and bioactivation of agaritine and of other mushroom hydrazines by whole mushroom homogenate and by mushroom tyrosinase. *Toxicology, 161*, 165–177.
68. Weijn, A., van den Berg-Somhorst, D. B., Slootweg, J. C., Vincken, J. P., Gruppen, H., Wichers, H. J., & Mes, J. J. (2013). Main phenolic compounds of the melanin biosynthesis pathway in bruising-tolerant and bruising-sensitive button mushroom (Agaricus bisporus) strains. *Journal of Agricultural and Food Chemistry, 61*(34), 8224–8231.
69. Winner, M., Gimenez, A., Schmidt, H., Sontag, B., Steffan, B., & Steglich, W. (2004). Unusual pulvinic acid dimers from the common fungi Scleroderma citrinum (common earthball) and Chalciporus piperatus (peppery bolete). *Angewandte Chemie, International Edition English, 43*, 1883–1886.
70. Xu, C., Tian, Z., Zhang, C., et al. (2014). Optimization of enzyme assisted extraction of polysaccharides from Ganoderma lucidum. *Engineering and Science, 12*, 17–20.
71. Yang, X.-L., Qin, C., Wang, F., Dong, Z.-J., & Liu, J.-K. (2008). A new meroterpenoid pigment from the Basidiomycete Albatrellus confluens. *Chemistry & Biodiversity, 5*, 484–489.
72. Zhou, Z. Y., & Liu, J. K. (2010). Pigments of fungi (macromycetes). *Natural Product Reports, 27*(11), 1531–1570.

Extraction and Application of Plant-Based Tannins as Sources of Natural Colourants

Shahid Adeel, Muhammad Yameen, Fariha Asghar, Nimra Amin, Meral Ozomay, and Somayeh Mirnezhad

Abstract With the progress of humankind, the world's favourable opinion of synthetic products has increased. Their hazardous effects are among the causes of environmental degradation, so these products are now subject to regulations. People are demanding ecofriendly methods, which has reinvigorated the use of plants. Among these plants, tannin-based functional molecules have excellent biological characteristics. Tannins are a class of polyphenol colorants which have various functional applications for colouring foodstuffs. Also, the plant-based tannins have antioxidant, antimicrobial and anticancer characteristics. This chapter will explore using plant-based tannins with modern and conventional methods and their isolation and applications in various fields, particularly textiles, fashion, food flavours, solar cells, etc. This brief review aims to help to provide a comprehensive knowledge base for the utilization of plant-based tannins in applied fields for the betterment of humanity.

Keywords Gallotannins · Majith · Microwave radiation · Oak · Ultrasonic radiation · Sustainability · Walnut

S. Adeel · N. Amin
Department of Applied Chemistry, Government College University Faisalabad, Faisalabad, Pakistan

M. Yameen (✉) · F. Asghar
Department of Biochemistry, Government College University Faisalabad, Faisalabad, Pakistan

M. Ozomay
Department of Textile Engineering, Marmara University, Istanbul, Turkey

S. Mirnezhad
Department of Carpet, University of Sistan and Baluchestan, Zahedan, Iran

S. S. Muthu (ed.), *Natural Dyes and Sustainability*, Sustainable Textiles: Production, Processing, Manufacturing & Chemistry,
https://doi.org/10.1007/978-3-031-47471-2_6

1 Introduction

The world is struggling to find the developmental processes that provide high efficiency and comfort. But this struggle also has side effects from the uncontrolled discharge of pollutants from all types of industries, especially those using dangerous and active compounds. Dyeing and tanning industries use excessive amounts of water and routinely generate hazardous wastewater [1]. A statistical analysis has shown that approximately 10,000 varieties are being utilized in different production units. The global demand for synthetic dyes is about 700,000 tons/year [2]. Among these, about 200,000 tons/year are discarded because of ineffective dyeing procedures in the textile industry. These waste dyes are treated as textile effluents, and discharging them into the environment causes degradation [3].

1.1 Dyes

Dyes are compounds which impart colour to materials such as textiles, fibres, papers, cosmetics, food, leather, pharmaceuticals, etc. Before the nineteenth century, these colours were derived from natural sources. Increasing public demand for dyed products encouraged the innovation of synthetic dyes. Dyes are applied in aqueous solutions and need mordants to boost their colour fastness on textile fibres. Finally, some colours and pigments absorb more light than others do.

There are some basic differences between dyes and pigments [4]. Dyes are organic, have small particle sizes and are soluble in water, whereas pigments are mostly inorganic, are water insoluble and form a suspension in water [5]. Dyes have short life spans and are combustible, whereas pigments have long life spans and are not combustible. Dyes are widespread throughout nature, and they impart colour via selective absorption, whereas pigments have less variety, and their colouring behaviour is due to the scattering of light [6].

1.2 Types of Dyes

Various synthetic and natural dyes are available. Synthetic dyes that are organic in nature, such as reactive dyes, processing dyes, etc., are commonly used [7]. To produce new shades, different chemicals and dyes are used in several combinations, and they render the market more competitive. Because of environmental issues, concerns have risen, and the changes in the production of dyestuffs has changed the textile industry [8]. Natural dyes include those from plants, animals and minerals. Natural dyes are nonsubstantiative and must be applied with chemicals or biomordants that have affinity for both dyes and fabrics [9].

1.2.1 Synthetic Dyes

Currently, the use of synthetic dyes is increasing day by day in many industries, like textiles, food, cosmetics and pharmaceuticals [10]. These widely used synthetic dyes impart solid hues to different materials, which appeals to people around the globe, but discharging these dyes is dangerous to the environment and human health [11]. These dyes and their intermediates, as effluents, are carcinogenic in nature and pose serious health problems for aquatic and terrestrial animal life as well. A popular but hazardous synthetic dye, azo dye, which is used in textile industry, produces 22 known carcinogens. These carcinogens are produced by bacteria on skin and are absorbed dermally. Azo dye effluent is one of the main causes of water pollution, so it represents a major concern [12].

1.2.2 Natural Dyes

In recent times, researchers have been struggling to find the best possible way of dyeing, by using effective and cheap resources [13]. Dyes extracted from plants are being promoted as part of the green revolution in the textile industry. More than 10,000 species have been found as sources of natural colorants and are being used in different fields. Concerns for preserving and cleaning the environment have spurred a shift towards using plant-based ecofriendly colorants to impart effective and durable properties to fabrics in the textile industry [14]. Natural dyes are the oldest materials used for dyeing. Madder made from mohanjo daro and the red Indian velvet made during the Mughal dynasty illustrate the efforts and skills of Indian dyers [15]. Dyeing with plant-based sources is the new, emerging trend in the textile industry. The use of natural dyes could be expanded because the process is natural and ecofriendly, produces high yields and is available at cheap rates. By using such dues, the textile industry can meet the demand for using plant-based dyes on fabrics, especially brown-dyed fabrics [16].

1.3 Tannin

Tannin was first found as a substance in different vegetable extracts, which plays a significant role in the conversion of animal skin into leather [17]. Tannins (tannic acids) are water-soluble phenolic compounds present in different fruits, vegetables, plant roots, bark, leaves and wood. These compounds' molecular weights range from 500 to 3000 D. These have large numbers of hydroxyl groups, which makes cross-linkages with different macromolecules, such as proteins and glucose [18]. Tannins are a diverse category of polyphenols that have large molecular weights and are water soluble in nature [19]. They are naturally found in many fruits and vegetables, leguminous seeds and grains. A large family of polyphenol compounds is present in the plant vacuoles of vegetal cells. This family of chemicals has several characteristics in common, such as the ability to create colourful complexes, the ability to oxidise in basic environments, and the ability to establish persistent cross-linked associations with other various molecules, including cellulose and protein, owing to its functional groups of tannins [20]. Thanks to these properties, tannins are differentiated from other polyphenolic compounds. Tannins can lead to various negative health consequences, including an antinutritional impacts, decreased digestibility and the potential for producing mutagenic and carcinogenic effects. Numerous health benefits have been confirmed, including antioxidant, antiallergic and many other anti-inflammatory activities. Tannins also provide defences against numerous cancers and allergic reactions. Because of their high level of astringency, the food applications of these substances have been quite restricted. However, they are extensively used in the pharmaceutical industry [21].

1.4 Classification of Tannins

The two primary categories of tannins are hydrolysable tannins and condensed tannins. Tannins are produced by plants and are stored in their vegetal components, such as bark, flowers, roots, leaves and fruits. Tannins are produced by vascular plants containing vacuoles, and metabolites are stored in vacuoles. These storage sites provide environments for plants that protect against their predators and do not interact with plants' metabolisms. These compounds act as scavengers for plants. Additionally, tannins are commonly found in the secondary xylem ad phloem, which is in the growth regions of trees between the cortex and the epidermis. Their presence suggests that tannins play roles in the development of these tissues. In general, the quantity of tannins is higher in bark than in the leaves, roots and branches of a tree. According to the chemical units involved, tannins can be divided into two main categories: hydrosylable and non hydrosylable tannins [22].

1.4.1 Hydrolysable Tannins

These compounds are mixtures of simple phenols and can be hydrolysed by weak acids and bases. Upon hydrolysis, carbohydrates and phenolic acids are produced. These compounds decompose at high temperatures. They produce pyrogallol, which is a hepatotoxic and highly irritant compound [23]. There are two different types of hydrolysable tannins, as follows:

1. Gallotannins
2. Ellagotannins

1.4.1.1 Gallotannins

Phenolic compounds form gallotannins when they are (partially or wholly) esterified with gallic acid [24].

Types of Gallotannins

1. Turkish tannins are obtained from *Quercus infectora*. Tannins in galls cause the coagulation of albumin, which is useful for skin conditions such as intertrigo, impetigo and eczema [25]. Diarrhoea, leucorrhoea and other vaginal discharges gleet and persistent gonorrhoea can all be treated with gall powder. This powder is applied directly to the skin to reduce excessive perspiration, eliminate odours and speed up the healing of wounds, thanks to its astringent nature and antioxidant qualities [26].
2. Chinese tannins are water soluble. These substances serve as agents in beer and wine that clarify the liquid, enhance flavours, modify colours and control the pH.
3. Tara tannins are extracted from *Caesalpinia spinosa*. The antioxidant properties of tara pod powder increase the shelf life and quality of pork products [27].
4. Acer tannins are extracted from *Acer ginnala*.

1.4.1.2 Ellagotannin

Hexahydroxydiphenic acid esterified with the central core of polyhydric alcohol forms an ellagotannin. Upon hydrolysis, it undergoes lactonization [28]. Corilagin, the simplest form of ellagotannin, has been found to possess great antibacterial properties. Ethanol extracts of *Terminalia chebula* fruits were examined and found to have a broad spectrum of antimicrobial and antibacterial activities against clinically important bacterial strains such as *Salmonella typhi* SSFP 4s, *Staphylococcus epidermis* MTCC3615 and *Staphylococcus aureus* ATCC 25923 [29].

1.4.2 Nonhydrolysable Tannins

These compounds are more complex than hydrolysable tannins. Owing to their structure complexity, the exact structure of condensed tannins has yet to be determined. They are mainly polymers of flavan-3-ol and galloyl units [30]. Polymerization occurs thanks to the condensation of two to ten catechin units. Further modification reactions, such as methylation, hydroxylation and glycosylation, increase the complexity of tannin molecules [31].

2 Extraction Methods

Tannins are abundant in nature and have garnered a lot of attention thanks to their wide range of possible uses. The quantity present in a plant varies depending on factors such as the geographic location, biological source, specific species, population size and location within the tree (whether it is in the inner or outer wood). Research on tannins has improved markedly in light of the perspective applications, but because tannins are heterogeneous, their extraction still represents the biggest obstacle to their commercialization. The extraction method for tannins exhibits significant variation and a still lacks a standardized process. The extraction procedure leaves several contaminants in the extracted tannin, such as sugars, minerals and stilbenes. For example, the inclusion of carbohydrates in tannins diminishes their ability to combat fungal growth, and it also poses challenges when attempting to impregnate them into wood for the preservation process. Moreover, processing variables, including particle size, temperature, pressure, duration, solvent type and solid-to-solvent ratio, affect the quantity of contaminants. Thus, the production variables must be carefully regulated when extracting tannins. Tannins are usually extracted from plant parts such as bark, leaves, stems and wood by using hot water or water mixed with different solvents.

Tannins were obtained from pine (*Pinus radiate*) bark at 90 °C, whereas the tannins in spruce (*Picea abies*) bark were extracted by using a hot-water extraction method at 85 °C. The temperature can reach up to 120 °C, depending on the type of solvent employed. The extraction of red angico (*Anadenanthera macrocarpa*), jabuticaba (*Myrciaria jabuticaba*) and umbu (*Spondias tuberosa*) was conducted. A higher temperature (105 °C) was employed even with water as the solvent. In contrast, the optimal temperature for extracting tannins from manjakani galls (*Quercus infectoria*) was found to be 75 °C. Depending on the availability of the solvent and the type of tannin, different extraction methods are described here. The duration of the extraction determines the effectiveness of the process and the quality of the resulting tannins.

2.1 Solid–Liquid Extraction Method

Historically, the solid–liquid extraction technique has been the most straightforward approach to extract tannins. In this method, a solvent is applied to a finely ground tissue, allowing it to pass through and dissolve the soluble matric compounds, all without the need of additional assisting mechanisms [32]. To investigate the influencing factors of the process, plant material and a mixture of solvent are used. The extraction of the desired compound is heavily affected by the particle size of the material and the solid–liquid ratio [33]. Catechins exhibit restricted solubility in polar organic solvents, whereas water and ethanol are frequently used solvents for extracting hydrolysable tannins [34]. In the past few years, different solvents have been tested. To remove chlorophyll and lipids, dichloromethane and hexane are used. The second step is to select a tannin extraction method. Depending on the solubility of the tannins, solvents with different polarities have been chosen. Condensed tannins exhibit a restricted ability to dissolve in a polar organic solvent. The hot-water extraction method has key benefits such as reduced costs, simplicity and productivity. The major drawbacks of using this method include the higher amount of solvent used and the longer extraction time [35].

2.2 Microwave-Assisted Method

Microwave-assisted extraction (MAE) includes the use of solvents and heating with a microwave [36]. Microwave radiation causes the breakdown of cell walls by creating pressure on the intracellular content. The intracellular content is released into the solvent [37]. Different solvents are used, depending on their availability. Water, ethanol and methanol are commonly used solvents for microwave-assisted extraction [38]. The microwave frequency rages from 300 MHz to 300 GHz, and the power plays an important role in the extraction of tannins. Usually, microwave power ranging from 100 to 900 W causes an incremental and efficient extraction of tannins. By using MAE, the extraction of tannins as colouring agents from mangrove biomass needed 30 min at 60 °C [39]. The results showed that the optimal extraction yield was up by 27.5% thanks to using an ethanol solvent at a material–solvent ratio of 0.04 g/mL.

Different yields were obtained under different conditions during extraction. A higher yield was obtained in hydroalcoholic solvent than in distilled water. Aqueous extracts showed a higher antioxidant property, but hydroalcoholic extracts showed overall better resistance against the tested bacterial strains [40]. The microwave technique is advantageous because it requires a low amount of solvent and provides the best recovery rates. On the other hand, the equipment required for this method is expensive for industrial applications, and the thermal degradation of tannins has also been reported [41].

2.3 Ultrasonic Extraction Method

During the ultrasonic extraction (UAE) method, microbubbles are created inside the liquid by using ultrasonic cavitation with frequencies of 20 kHz or above. Thanks to vibrations, microbubbles grow in size and cell walls break down. It causes the penetration of solvent in vegetal tissues for the extraction of tannins [20]. The cell assembly of the plant is damaged over time in the solvent, and the tannin extraction yield can be optimized; however, the quality might decrease with time. Another crucial element that controls the tannin extraction process is the solvent-to-solid ratio. Efficient extraction can be achieved through the continuous blending of solvents and solute particles. Optimal mixing occurs when there is a greater ratio of solvent to solid. This ratio transforms the particles, but not the solvent, leading to the tannin extraction. Solvents such as ethanol may serve as critical factors for increasing tannin content during ultrasonic extraction. The yield of tannins extracted through the ultrasonic method increases by 17.6% and is higher than that extracted through traditional methods. The efficiency of the method depends on the sound waves, which mechanically vibrate the particles. Following the optimum global processing conditions (18.7 min, 327.4 W and 69.3% ethanol) predicted by polynomial models, 178 mg of total tannin per g of extract was recovered by using UAE. Polar solvent – i.e. methanol or ethanol – yields higher recovery rates: 27.3 mg/g of dried extract. When these solvents are combined with ionic solvents, their extraction rate increases by up to 630.2 mg/g of dried extract [42]. Higher sonication power results in increased tannin quantity, up to a certain limit, beyond which tannin decomposition may occur. This extraction is characterized by its speed, taking no longer than 30 minutes, according to the protocols [43].

2.4 Supercritical Extraction Method

When substances are subjected to high temperature, they reach their critical point and coexist in a gas–liquid equilibrium. These are known as supercritical substances [44]. Carbon dioxide (CO_2) is used mostly because of its nontoxic and inert properties and existence as a supercritical fluid at relatively low temperatures and pressures [45]. Supercritical extraction is a contemporary method employed to extract tannins, utilizing supercritical solvents such as carbon dioxide, butane, pentane, fluorinated hydrocarbons, sulphur hexafluoride and nitrous oxide. A cosolvent or modifier is employed to enhance the selectivity of carbon dioxide. Supercritical carbon dioxide was used for the extraction of hydrolysable tannins from *Quereus infectoria*. Gall extracts have bioactivities for healing wounds. Using supercritical carbon dioxide under the optimal conditions for the pressure, temperature and particle size of the sample, the extraction yield was optimized by up to 6.02% in terms of tannin acid concentration – in this case, a 6149.71 mg/g sample and a gallic acid concentration of 96.85 mg/g of sample [46]. *Terminalia bellirica* fruit powder was investigated for the efficiency of extracting chebulagic acid through the supercritical extraction method, both with and without a modifier. Using 50% ethanol as a modifier was the most efficient method, more efficient than all the other conventional methods. Chebulagic acid was found to be present in the final product, which confirmed the DPPH free radical scavenging activity, and no more purification was required. An HPLC analysis was carried out to confirm the purity of the compound, and Nuclear magnetic resonance (NMR) spectra were used to confirm the chemical structure [47].

Assam tea seeds were employed in the supercritical CO_2 extraction method. Optimized extraction conditions provided highly purified oil, which had low acid values, few free fatty acids, reactive substances, high iodine values, high phenolic content and low values of DPPH [48]. *Peganum harmala* was considered for the extraction of phenolic compounds, employing the BOX-BEHNKEN statistical design at a fixed flow rate of carbon dioxide at 50 g/min and over an extraction time of 80 min. The maximum extraction yield was achieved at 300 bar and 55 °C. The particle size of the finely ground powder was 0.3 mm. The maximum total phenolic content was found to be 79.04 mg GAE/gE [49].

Several conditions are involved in the supercritical extraction method, such as temperature, pressure and cosolvents, which increase the solubility of the compound [50]. Higher values of pressure produce higher yields of tannins because the solubility of the compound increases [51]. In order to prevent the thermal degradation of tannins, lower temperature values should be selected. Cosolvents are used to enhance the solubility of tannins and thus produce high yields [52]. The supercritical extraction method has many benefits, including operating at low temperatures and pressures and having a short extraction time for yielding high amounts of tannin. But a heavy financial investment is required to install the equipment [53].

2.5 Pressurized Water Extraction

Pressurized water extraction is considered as a high-yield and ecofriendly extraction method. Bioactive compounds such as flavonoids, polyphenols and tannins from plant sources are extracted through this method. Different parameters are involved in the performance efficiency of this extraction method, such as the solvent, adsorbent, pressure and temperature [54]. The principle of extraction centres on the use of high temperatures and pressures under the critical point to maintain the liquid state of the water [55].

Phyllanthus tenellus was examined for the extraction of hydrolysable tannins. The pressurized hot-water extraction method was employed at 121 °C and 15 psi. The yield of this extraction method was almost double that of the methanol extraction method [56]. The extraction rates of tannins using this method vary significantly, spanning a range from 52.9 to 381.9 mg/g of dried extract [57]. The main advantages of this method are that it requires a short extraction time and that it uses water, eliminating its toxicity. The temperature and pressure can be adjusted, and the choice of cosolvent depends on the type of tannin being extracted [58].

3 Use of Plant-Based Tannins in the Tanning Industry

3.1 Cassia alata *(Candle Bush)*

Cassia alata is commonly known as candle bush and is a member of the Leguminosae family. The adaptation of animal skin into leather is known as tanning. Tannins obtained from plants play important roles in improving the ecofriendliness of the tanning process. *Cassia alata*, *Euphorbia hirta*, *Thespesia populnea* and *Wrightia tinctoria* leaves were examined to obtain tannin through water extraction. Tannin and tannin-induced silver nanoparticles (TISNPs) were found to be crystalline. The maximal number of tannins was obtained from *Cassia alata*, and TISNPs used in combination with chromium produced the optimal results. The highest tensile strength and the highest tear strength of leather were obtained [59]. Bertrand et al. investigated the antioxidant and anti-inflammatory properties of *Cassia alata* bark extract. The ethanolic extract of bark showed higher scavenging properties against hydrogen peroxide. Moniruzzaman et al. reported that 20% *Cassia alata* leaf paste is helpful in curing ringworm disease in calf hides. This is because of the bioactive compounds in the leaf powder. The alkaloids, polyphenols and tannins have a scavenging effect.

Another name for *Cassia alata* is ringworm cassia. Phytochemicals in the plant may have a variety of noteworthy pharmacological effects. *Cassia alata* has been applied as a treatment for atopic dermatitis. It has also conventionally been used to treat various contagious infections. Numerous studies have acknowledged its remarkable antimicrobial efficacy against *S. aureus*. *S. aureus* is subjected to moderate antimicrobial activity when exposed to alcoholic and aqueous extracts derived from the roots, barks, stems and leaves of *Cassia alata*. This is a potential source for the removal of bacteria from hides during leather processing.

Applying *Cassia alata* leaf paste and salt on goat skin improves the tanning of the skin. This is a greener approach and a revolutionary step in the tanning industry. Marudhamuthu et al. investigated the effects of salt and *Cassia alata* leaf powder applied onto goat skin. When *Cassia alata* leaf paste and salt are used at different concentrations, their antibacterial activities were found to be good on skin. The samples were treated with 50% salt + 5% paste; 10% + 10% paste; and 15% salt + 15% paste and were set to 21 days. Hair slip, smell and putrefaction were observed. The use of salt was found to be a greener approach to tanning, and reducing the amount of salt was preferably because the discarding of salt into water bodies after the tanning process increases their salinity.

3.2 Quercus robur *(Oak)*

The common name for *Quercus robur* is oak, which is a member of the Fagaceae family. Experimental studies have been conducted on extracting tannins from *Quercus robur* bark. According to researchers, bark contains various bioactive compounds, including derivatives of ellagic acids, gallotannins and some condensed tannins. When reactive oxygen atoms interact with glucose units and make ester bonds, these phenolic entities are formed. When these tannins are applied onto animal skin, they impart pink, red or dark-brown shades to leather. When these samples are treated with iron, dark-green spots are created. It makes leather more solid and less purified. Several extraction solvents, such as ethanol, ethyl acetate and diethyl ether, have been used and Gas chromatography–mass spectrometry (GC–MS) analyses was conducted. From them, 25 bioactive compounds were found, and the main compounds were phytol, 1-docosene, and 1,3-dioxolane. A Fourier transform infrared (FTIR) analysis showed the presence of carboxylic acids, alkanes, phenols, alcohol and ketones. Thanks to these findings, tannins from oak bark are considered as fine alternatives to chromium in leather tanning and are plant-based sources of tannins for the tanning industry [60]. Oak tanning is a labour-intensive process, and unlike many other vegetal materials, oak bark tannins do not permeate and interact with the hide as quickly. For this reason, the one-year procedure was made into law and distinguished the UK tanning business from others. The colour of oak bark–tanned leather is a highly distinctive medium brown. When oak bark extract is used for the tanning of leather, the final product is a light-coloured leather that ranges in hue from creamy to yellowish to light brown. When in contact with iron, pyrogallols produce bluish-black patches and do not react well to pH changes. Pyrogallols are found in oak galls and oak wood. When boiled with acid, this form of tannin generates tannin reds.

3.3 Tessmannia burttii

Condensed tannins obtained from *tessmannia burttii* (casalpiniaceae) are potential sources as ecofriendly tanning agents for the leather industry. Various extraction parameters were optimized, and the cross-linking activities between tannin powder and hide were analysed by using the differential scanning colorimetry method. Cross-linking occurs among basic amino groups and is vulnerable to acid hydrolysis. Stem bark produces higher extraction yields than root bark does. One study has shown that chromium can be replaced in the tanning industry by ecofriendly plant-based tanning agents [61].

3.4 Olea europaea *(Olive)*

Olea europaea is commonly known as olive. The underused residue of the extraction of olive oil, known as olive pomace (OP), contains polyphenols, which can be used as sources of antioxidants in the nutraceutical industry. The synchronized use of enzymes and microwave processing was assessed for its phenolic extractability and composition. By utilizing water as the solvent, a greater solid-to-liquid ratio and a faster reaction time, microwave-assisted extraction was able to achieve phenolic extractability equal to that of traditional solvent extraction. The use of enzymes increases the extractability of the phenols and tannins. During the processing of hides, for the removal of heavy metals, an adsorbent was prepared. In order to increase the removal effectiveness of Fe^{2+} and Mn^{2+}, batch adsorption experiments were carried out, and the impacts of the adsorbent chemical structure, dose, pH, contact duration and starting ion concentration were explored. Olive mills produce wet waste containing 60% water, and its management presents an emerging environmental issue. An HPLC analysis confirmed the presence of polyphenols, which exhibit high antioxidant activity. Solubilizing the olive pulp is more convenient when using it as raw material for leather tanning. It is a modern, bio-based, and ecofriendly alternative resource [62].

3.5 Coriaria neplansis *(Masuri Berry)*

Coriaria neplansis is also known as Masuri berry, and it belongs to the Coriariaceae family of shrubs. Tannins from *Coriaria neplansis* are renewable resources that have high biodegradability and produce few pollutants. *Coriaria nepalensis* bark (CNB) was investigated for tannin extraction and its application in converting skin into leather. CNB was utilized to extract a valuable tannin resource, and tannin extraction with sodium hydroxide was improved. Improvements in other physical characteristics, such as break elongation and tear strength, were found – up to 75.8% and 38.2 N mm^{-1}, respectively – while the tensile strength was measured at 6.1 MPa. It was similar to those attained by using commercial valonia tannin reagents. CNB tannins penetrate into the collagen fibres of leather to produce permanent cross-linkages, resulting in hard and finely grained surfaces, as demonstrated via scanning electron and atomic force microscopy. Other optimal conditions for extractions were found: 82 °C over 63 min with 0.22% NaOH. Using NMR spectroscopy confirmed the presence of tannins. It is used primarily to tan the sheep skin [63].

3.6 Castanea sativa *(Chestnut)*

Chestnut wood (*Castanea sativa*) tannins are of the pyrogallol type that belong to a class of glucosidic tannins that are simple to hydrolyse. Chestnut extracts are highly astringent and have the potential to repair large areas of hide. They include a suitable amount of acid groups and natural organic acids. Because of these characteristics, chestnut extract is particularly well suited for the tanning of thick skins to create leather soles. It is feasible to produce leather that is compact, firm, flexible and waterproof at high yields in weight. The leathers that have been chestnut tanned are elastic, lightfast and traction and abrasion resistant, and they exhibit good colours.

A combination of tanning treatments was employed to find effective procedures that use chestnut powder and aluminium. The skin was pretreated with 2% aluminium and then 20% chestnut powder. The resultant shrinkage temperature reached up to 100 °C. When aluminium was employed, followed by chestnut powder, the shrinkage temperature increased to 94 °C. The Al–chestnut combination treatment imparted good organoleptic properties to the leather. The combination is a natural, ecofriendly method for tanning [64].

3.7 Xylocarpus granatum *(Cannonball Mangrove)*

Mangroves are ancient plants and important thanks to their medicinal properties. These components provide protection against fungi, bacteria and other microbes. Owing to the poor solid waste management of chemicals in tanning, experts are encouraging the use of ecofriendly alternatives. A novel vegetable tanning agent obtained from *Xylocarpus granatum* bark is being used for this reason. An FTIR analysis showed the excellent tanning activity of extracts from *Xylocarpus granatum*. It is also a good dyeing agent for shoe leather [65].

3.8 Coffee

Used coffee grounds (UCGs) are being used to reduce chemical contamination and for other environmental purposes. The tanning effect of used coffee grounds was investigated in comparison to leather tanned with synthetic agents. UCGs have sufficient concentrations of tannins – i.e. 14.92%. The optimal concentration of UCG extract was found to be 40 g/100 mL, which provides a good shrinkage temperature and high tensile strength to leather after tanning. UCG extracts with 10% sulphonates provides good durability and fullness to leather. This vegetable-tanned leather can be used in making bags and shoes [66]. The hide powder approach was utilized to determine the tannin content in coffee arabica pulp, resulting in a measured concentration of 5.04% and a tanning strength of 2.26. In comparison, mimosa exhibited a tannin content of 63.56% and a tanning strength of 2.82. Using 40% UCG extract can achieve the requisite shrinking temperature. UCGs can effectively serve as tanning agents in vegetable tanning processes, yielding long-lasting leathers with excellent fullness.

3.9 **Hygenia abyssinica** *(African Redwood)*

A new plant-based tannin material was extracted from *Hagenia abyssinica* bark powder. FTIR and UV-visible spectroscopic analyses confirmed that higher amounts of condensed tannins were present in the bark powder [67]. It is a potential tanning agent against chromium. It is also used in sheep skin tanning. The performance of the leather produced by using *Hagenia abyssinica* and leather produced with commercial mimosa was evaluated after the sheepskins were turned into crust leathers. *H. abyssinica*–tanned leather exhibited the highest shrinkage temperature: 78.5 °C. There have also been assessments of the masses in tanning liquid effluents.

3.10 **Trema orientalis** *(Charcoal Tree)*

Trema orientalis belongs to the flowering family Cannabaceae (hemp) and is commonly known as the charcoal tree. Given the harmful effects of chromium on the environment, researchers substituted it with the stem bark powder of *Trema orientalis* because it is an ecofriendly plant-based tanning agent. Extracts were prepared in hot distilled water at different temperatures. Extracts were prepared under different parameters for temperature, time and chemical additive concentration to check the efficiency of the extracts and their chemical compositions. An increase in temperature had positive impact on tannin extraction. Different additives, such as sodium sulphite and sodium hydroxide, reduced extract efficiency. Gallic acid was found in the extract (32.12%). The extract exhibited excellent leather retanning ability, higher than that of commercially available tannins [68].

4 Applications for Tannin-Containing Plants in Textile Dyeing

4.1 Acacia nilotica *(Babul)*

Dye was extracted from *Acacia nilotica* bark, and shoe leather was dyed using this extract. Pretreated goatskin (leather) was employed as a dyeing sample: A colour analysis, its colour fastness and its dye exhaustion rate were examined. It imparted an amber colour to the surface of leather; upon mordanting it with potash alum and ferrous sulphate, it turned brown and grey. Premordanted samples obtained with extract dye through mordanting technique had better colour fastness, shades and colour strength [69]. *Acacia nilotica* was used for the extraction of natural dyes from henna leaves to dye wool fabrics and to increase those fabrics' fibre–dye affinity and colour intensity. Henna dye and henna–*Acacia* dye were applied to wool fibres under various dyeing conditions. Modified samples contained higher colour strength and yielded better fastness results.

Dhanania et al. explored ecofriendly dye extraction for cotton and silk. A Fourier transform infrared (FTIR) analysis and ultraviolet-visible (UV-vis) spectroscopy were used to identify the extracted colouring ingredient. The maximal UV-vis absorbance, which was determined to be 250 nm, indicated the presence of phenolic chemicals. On silk and cotton textiles, an alkaline seed extract was applied both with and without a mordant. For the purpose of determining the fastness qualities of these dyed fabrics, the impact of washing and sunshine on coloured silk and cotton fabrics has been researched. The dyed silk textiles had better fastness qualities than the dyed cotton materials had.

Bioactive wool textiles dyed with *Acacia* leaf extract were determined to have colorimetric features such as UV protection and antioxidant qualities. How these qualities would fare when subjected to iron premordanting was also evaluated. They showed good results for fastness when subjected to washing and rubbing, and they produced a wide variety of colours. The UV-protection capabilities of the produced bioactive wool fabrics were evaluated in terms of the ultraviolet protection factor (UPF) and the UV transmittance within the ultraviolet range (UV-A and UV-B) and then compared with those values in undyed wool textiles.

4.2 Triadica sebifera *(Chinese Tallow)*

The dyeing properties of Chinese tallow (CT) leaves were explored by applying bio-mordents and chemical mordents to wool fabrics. Dyeing variables were optimized, and the extract was analysed through an FTIR analysis, which confirmed the presence of tannins and other phenolic entities. Wool fabrics that were dyed with CT leaf extract showed good antioxidant and antibacterial properties against *E. coli*. The fabrics possessed a good UV-protection capacity. The results encourage the future use of CT extract for dyeing in the medical textile industry [70].

The best results were from dyeing silk fabrics with Chinese tallow at 90 °C and a pH of 4 for 60 min. A silk fabric coloured without a mordant had a reddish-brown hue, but a silk fabric dyed with a mordant imparted a variety of mild- to dark-reddish colour hues. The ferrous sulphate mordant, however, produced a deeper and duller colour. While the colour fastness to light was poor to fair and the colour fastness to crocking was good to exceptional, the colour fastness to washing was fair to good. The SEM images study of coloured silk at temperatures close to boiling revealed that this caused very little damage. Silk dyed with a plant extract was investigated for its UV-protection and antibacterial properties. This silk had tremendous antimicrobial and antioxidant properties after the dyeing.

4.3 Cocos nucifera *(Coconut)*

Tannin was extracted via the ultrasonic-assisted extraction method to dye silk fabric. A physiochemical analysis of the fabric before and after the ultrasonic treatment showed no changes in the chemical structure of the treated silk fabric. Ultrasonic waves have great potential for extracting tannin colorants for dyeing and ecofriendly mordanting under optimized conditions. Natural dyeing processes are more sustainable and cleaner than conventional methods because the natural processes use bio-based mordants [71].

Coconut coir (*Cocos nucifera*) husk extract can also be employed for the dyeing of wool fabric via the ultrasonic method. Fabric and extract can be used under irradiated and unirradiated conditions. Ultrasonic radiation provides best way to explore the colouring efficiency of plants for dyeing protein-based fabrics. Biomordanting wool fabrics improves the shade quality and fastness of wool fabrics [72].

Tannin extracted from coconut husk by using the liquid–liquid extraction method was used to impart colours, especially brown, to cotton fabrics. When alum was used on cotton, the fabric showed good colour fastness properties, in contrast to the naphthol binding agent [73].

4.4 Camelia sinensis *(Black Tea)*

Black tea (*Camelia sinensis*) leaf extract was utilized for the dyeing of wool fabrics via the modern method of extraction. The radiation from microwave ovens increases the amount of extracted tannin and improves the colour strength of wool fabrics. The dyeing is carried out through a simple adsorption method. Premordanting and postmordanting were also carried out to improve the quality of the shades [74].

Waste tea leaf extract was utilized for the dyeing of silk fabric via the microwave-assisted extraction method. The use of low amounts of chemicals and biomordents has valorized the dyeing of silk and produced eye-catching and sustainable colours with good fastness properties. Tannin extraction in aqueous media is a good way of producing sources of natural brown dyes for the coloration of surface-modified silk fabric [75].

The microwave radiation process was utilized to explore the colouring behaviour of tannin in tea leaves. Cotton fabric was dyed, and tannin was used as a mordant both alone and in combination with a metal mordant for dyeing cotton [76].

4.5 Azadirachta indica *(Neem)*

Azadirachta indica, commonly known as neem, grows in tropical areas. The microwave-assisted method was utilized to extract tannin from *Azadirachta indica* to dye silk fabric. Radiating the extract in an acidic medium at a pH of 2 for 2 min imparts good colour to silk fabric. In one study, administering a microwave treatment on the extract and the silk not only improved the extraction of the tannins from neem bark but also provided good colour strength to the fabric [77].

Neem has been used as natural dye for adding a superhydrophobic coating onto cotton fabric to prevent the dye colour from fading. Neem acts as a self-cleaning agent, which reduces the among of synthetic dye needed. The concentration of the dye and the superhydrophobic coating on the cotton fabric has an inverse effect on the porosity of fabric [78].

Neem leaf extract can be used to dye jute-cotton (jutton) fabric blends. The results of using the disk diffusion method to conduct the antimicrobial and antibacterial tests have shown high resistance to microbes when dyed with neem leaf extract. Its rubbing fastness was found to be of grade 4, and treated jutton maintains high-tensile and elongation properties. Using neem leaf extract is an ecofriendly technique for carrying out antibacterial treatments on jutton yarn in the textile industry.

4.6 Carica papaya *(Papeeta)*

Carica papaya could be an efficient source of dyeing agents for protein fabrics. An attempt has been made to dye protein fabrics with *Carica papaya* leaf powder extract as an ecofriendly dye source. All three mordanting techniques, namely pre-, meta- and postmordanting, were employed. Chemicals and biomordents produce comparable results on silk and cotton. Biomordants produce different shades on dyed wool fabrics and have great potential to replace the metal salts used as mordants. Silk fabrics dyed with *Carica papaya* leaf powder extract exhibit good antioxidant and antibacterial properties and provide good UV protection. *Carica papaya* leaf powder has a promising future as a dye in the medical textile industry [79].

4.7 Rubia cordifolia *(Majith)*

In Hindi, the common name for *Rubia cordifolia* is majith. It is a member of the Rubiaceae family. In modern techniques, madder has been used to colour various textiles, so madder has been one of the crucial components of textile-production processes. Thanks to its herbal nature, it has indirectly improved the economic growth of the regions where it is cultivated.

Various plants, including madder, pomegranate and raseda, are sources for extracting the tannin colourants used to dye silk. The microwave-assisted extraction method used for this process requires hot water as a solvent. The dyeing conditions were optimized, and an FTIR analysis of the dyed silk samples was carried out to check the changes in the fabric. The K/S (absorption and scattering) coefficients of the dyed samples directly corresponded to the dye concentrations [80].

Both synthetic and natural dyes can be used to colour wool garments. Madder has been used in the topping procedure (i.e. dyeing already-dyed wool) as a natural dye source to provide a different angle on the colouring of wool, and a spectrophotometer has been used to analyse the colour changes. In natural dyeing, this topping procedure can be used to create various hues and colours with adequate/good fastness values.

The dyeing of banana fibres using madder as a natural dye source has produced an adequate lightfastness property after mordanting. These fibres have shown an increase in colour strength (K/S) after chemical mordants, such as ferrous sulphate, alum and stannous chloride, had been applied.

4.8 Juglans regia *(Persian Walnut)*

A bio-dye was extracted from the bark residues of *Juglans regia*, commonly known as Persian walnut. Ticha et al. explored the colouring behaviour of Persian walnut shells in wool dyeing. Different parameters for extraction and dyeing were optimized, and the Persian walnut shells' concentrations of biocomponents such as phenols, flavonoids and tannins were estimated. Through the microwave extraction method, a high yield of tannins was obtained, and a good K/S value, up to 15, and a brown shade resulted. The waste of *Juglans regia* bark has a promising future in textile dyeing, especially for wool fabric [81]. The goal of this work [81] was to improve the microwave extraction procedure that produces *Juglans regia* bark residues that colour acrylic textiles. Hence, the microwave power, pH, extraction time and dry mass concentration were initially explored as extraction conditions. Each time, the numbers of flavonoids and tannins were measured. Moreover, acrylic fibres were dyed using the obtained extracts and a microwave, and the appropriate colour yield (K/S) was assessed. The results showed that the dyed acrylic fabric colour strength (K/S) obtained a value of 17, and the rubbing fastness results were satisfying.

Juglans regia has been used for the screen printing of handknit samples of wool yarn. The fluidity increased in acidic media, but the thickness and elasticity of the samples decreased. A 45-minute fixing process of steaming at 105 °C delivered the best results. Mordanting with iron sulphate improved the rubbing fastness, and pH had no effect on the thickness of the sample printing.

4.9 Arachis hypogaea *(Peanut)*

The skin of peanuts exhibits great antibacterial and antimicrobial properties. A peanut powder was exposed to microwave radiation, and this extraction method was found to be effective. It imparted variable shades to wool fabric. The quality of the shades varies when treated with biomordants. Premordanted fabrics provide great antibacterial protection, and postmordanted fabrics possess good UV protection. Peanut powder has great potential to be used as a bio-dye at the industrial level [82].

The biological properties of peanut shells allow their use as biomordants for dyeing natural fabrics. Peanut shells have great potential for imparting antibacterial and antimicrobial properties to dyed fabrics. They increase fabrics' resistance to microbes and act as natural wound-healing agents. They also provide good UV protection to fabrics.

4.10 Terminalia catappa *(Sea Almond)*

Terminalia catappa, also known as sea almond or ketapang, is effective in treating fish skin conditions. The antimicrobial tannin in ketapang leaves destroys viruses, fungi and bacteria. It is a superior substitute for fish antibiotics. Ketapang leaves have been used as tannin-based bio-dye extraction materials. In one study, the leaves were soaked in distilled water. The immersion method was used to extract the tannin and create a solvent. Ready-to-dye cotton fabric was immersed into a dye solution for up to 60 min to increase the fabric's dye uptake. Different chemical mordents were applied on the dyed fabric, and then its rubbing fastness was checked. The fabric mordanted with alum produced optimal rubbing fastness results [83].

5 Future Aspects

The textile industry is a large expanding industry which has direct impacts on the environment and human health. The world is turning its priorities towards green approaches. Plant-based tannins used as natural sources of colorants produce fabrics that are ecofriendly and safe for human skin. Tannins impart antimicrobial properties to fabrics, and fabrics dyed with tannin-containing extracts provide good shades and UV protection – all of which are promising for the medical textile industry.

Market demand for tannins will increase as awareness increases among consumers and other stakeholders, and governments are enacting policies to make the environment safe and clean. The textile industry can follow suit by using these plant-based products instead of chemically synthesized ones.

The tanning of leather produces large water effluent. Water from tanneries has high salinity and chromium concentrations. Using plant sources in tanning is an ecofriendly approach for converting animal skin into leather. It may require a high processing time, but the raw material being used for such leather tanning is easily available in nature. The use of these naturally available plant materials help protect the environment human health. Moreover, tannin-based coagulants are being used in the production of cleaner water and for water treatment, to improve water availability and reduce the risk of contamination. Market competition in the production of conventional coagulants should inspire new applications and technologies featuring tannins.

The current weaknesses of dyeing methods could be minimized by optimizing tannin extraction methods [84]. The addition of condensed tannin causes the nitrification and mineralization of temperate soil. Using condensed tannins could also mitigate greenhouse gas emissions [85]. Tannin-based adhesives can be produced, and ethylene carbonate can be used as a cross-linking agent. These could be used as alternatives to formaldehyde-based natural adhesives. Environmental assessments can be considered as starting points in the overall assessment of the environmental sustainability of the production system for tannin-based natural adhesives [86].

Tannins may result in a network structure, a decrease in free volume, and an increase in vapour passageways within films. Moreover, Soy protein isolate (SPI)-based films have retained the strong antioxidant capacity of the bayberry, showing that certain free sites on BT remained accessible to contribute to the antioxidation action. Chinese barberry tannins are therefore often used as cross-linkages and for other purposes, particularly packaging applications [87]. Interest in natural goods is growing across a variety of industries, encouraging collaboration across diverse production and marketing disciplines [88].

6 Conclusion

Tannins, found abundantly in various plant sources, are an important class of natural compounds. They come in multiple forms – condensed and hydrolysable tannins being the primary types. These tannins are predominantly found in the barks, leaves, fruits, fruit shells, seeds, shoots and stems of plants. Different solvents are used to extract tannins from various plant parts by using water alone or in combinations also using water. Common solvents include ethanol, sodium hydroxide and methanol. The purity and the yield of extracted tannins depend on the extraction method and solvent used.

Various conventional and modern technologies are used for tannin extraction. Moreover, achieving superior tannin quality can be facilitated by managing variables such as the temperature, pressure, solid-to-solvent ratio, particle size, source and extraction duration (time). Commonly, the presence of carbohydrate impurities in the extraction process limits the potential applications of tannins. Applications may vary depending on how the tannins are used – i.e. for tanning leather, dyeing fabric or preserving animal skins. Antibacterial and antifungal properties increase their value to textile industry and make their contact with human skin safe. Moreover, tannins have additional cutting-edge uses in the production of biomedical devices, bio-based foams, dietary supplements and pharmaceuticals. However, further research is required for each use, which might lead to new opportunities for tannin applications.

This project aimed to provide support for, or to at least offer recommendations on, a study aiming to reduce or eliminate the obstacles in the way of finding advanced applications for tannins. Finally, it addressed the extraction of tannins in a broad context.

References

1. Ismail, G. A., & Sakai, H. J. C. (2022). Review on effect of different type of dyes on advanced oxidation processes (AOPs) for textile color removal, *291*, 132906.
2. Wan, H., et al. (2021). Bioreduction (AuIII to Au0) and stabilization of gold nanocatalyst using Kappa carrageenan for degradation of azo dyes, *176*, 282–290.
3. Verma, R. K., et al. (2021). Eradication of fatal textile industrial dyes by wastewater treatment. *Biointerface Res. Appl. Chem, 12*, 567–587.
4. Do, K. L., et al. (2022). From historical dye to bio-colourant: Processing, identification in historical textiles and potential applications of anthraquinone-based morindone, *205*, 110482.
5. Li, L., et al. (2022). Influence of indigo-hydroxyl interactions on the properties of sepiolite-based Maya blue pigment, *200*, 110138.
6. Gürses, A., Güneş, K., & Şahin, E. (2021). Removal of dyes and pigments from industrial effluents. In *Green chemistry and water remediation: Research and applications* (pp. 135–187). Elsevier.
7. Hanafi, M. F., & Sapawe, N. (2020). A review on the water problem associate with organic pollutants derived from phenol, methyl orange, and remazol brilliant blue dyes. *Materials Today: Proceedings, 31*, A141–A150.
8. Jamee, R., & Siddique, R. (2019). Biodegradation of synthetic dyes of textile effluent by microorganisms: an environmentally and economically sustainable approach. *European Journal of Microbiology and Immunology, 9*(4), 114–118.
9. Zhang, Y., et al. (2022). Sonochemical mordanting as a green and effective approach in enhancing cotton bio natural dye affinity through soy surface modification. *Journal of Cleaner Production, 336*, 130465.

10. Khan, S., et al. (2021). Synergistic role of bacterial consortium to biodegrade toxic dyes containing wastewater and its simultaneous reuse as an added value, *284*, 131273.
11. Habib, N., et al. (2021). Environmental friendly sustainable application of plant-based mordants for cotton dyeing using Arjun bark-based natural colorant, *28*(38), 54041–54047.
12. Kapoor, R. T., et al. (2021). Exploiting microbial biomass in treating azo dyes contaminated wastewater: Mechanism of degradation and factors affecting microbial efficiency. *Journal of Water Process Engineering, 43*, 102255.
13. Mariotti, N., et al. (2020). Recent advances in eco-friendly and cost-effective materials towards sustainable dye-sensitized solar cells. *Green Chemistry, 22*(21), 7168–7218.
14. Thakker, A. M. (2021). Sustainable processing of cotton fabrics with plant-based biomaterials Sapindus mukorossi and Acacia concinna for health-care applications. *The Journal of The Textile Institute, 112*(5), 718–726.
15. Kumari, P. (2021). *Innovation in design of traditional Mashru textile for product diversification*. Maharaja Sayajirao University of Baroda (India).
16. Kumbhar, S., et al. (2019). Eco-friendly dyeing of cotton with brown natural dye extracted from Ficus amplissima Smith leaves, *17*(2), 1161–1166.
17. Muna, F., Tanzania's leather value chain: A review of the literature. Trade and investment in East Africa: Prospects, challenges and pathways to sustainability, 2022: p. 241-266.
18. Ghosh, S., et al. (2020). Molecular and macromolecular crowding-induced stabilization of proteins: Effect of dextran and its building block alone and their mixtures on stability and structure of lysozyme. *International Journal of Biological Macromolecules, 150*, 1238–1248.
19. Smeriglio, A., et al. (2017). Proanthocyanidins and hydrolysable tannins: occurrence, dietary intake and pharmacological effects, *174*(11), 1244–1262.
20. Fraga-Corral, M., et al. (2020). Technological application of tannin-based extracts, *25*(3), 614.
21. Sieniawska, E., & Baj, T. (2017). Tannins. In *Pharmacognosy* (pp. 199–232). Elsevier.
22. Lamy, E., et al. (2016). *Determinants of tannin-rich food and beverage consumption: oral perception vs. psychosocial aspects*.
23. Jiménez, N., et al. (2014). Tannin degradation by a novel tannase enzyme present in some Lactobacillus plantarum strains, *80*(10), 2991–2997.
24. He, H.-F. (2022). Gallotannins. In *Handbook of food bioactive ingredients: Properties and applications* (pp. 1–16). Springer.
25. Pal, S., Sengupta, S., & Osta, M. (2021). Role of C-reactive protein, serum ferritin, albumin, and lactate dehydrogenase in hospitalized coronavirus disease-19 infected patients-A study in tertiary care coronavirus disease-19 center in Eastern India. *National Journal of Physiology, Pharmacy and Pharmacology, 11*(4), 441–443.
26. Akbar, S. (2020). Quercus infectoria G. Olivier (Fagaceae). In *Handbook of 200 medicinal plants* (pp. 1505–1511). Springer.
27. Skowyra, M., et al. (2015). Effect of tara (Caesalpinia spinosa) pod powder on the oxidation and colour stability of pork meat batter during chilled storage, *53*(4), 419.
28. Lalchandani, D. S., et al. (2022). Significance of stressor media on the stability of statins: A critical assessment. *Chemical Papers, 76*(10), 6047–6071.
29. Kannan, P., Ramadevi, S., & Waheeta, H. J. A. J. O. M. R. (2009). Antibacterial activity of Terminalia chebula fruit extract, *3*(4), 180–184.
30. Suvanto, J., et al. (2017). Variability in the production of tannins and other polyphenols in cell cultures of 12 Nordic plant species, *246*(2), 227–241.
31. Garcia, D. E., et al. (2016). Modification of condensed tannins: From polyphenol chemistry to materials engineering, *40*(1), 36–49.
32. Zhang, Q.-W., Lin, L.-G., & Ye, W.-C. J. C. M. (2018). Techniques for extraction and isolation of natural products: A comprehensive review, *13*(1), 1–26.
33. Simeonov, E., Koleva, V. J. C., & Quarterly, B. E. (2012). Solid-liquid extraction of tannins from Geranium Sanguineum L.–experimental kinetics and modelling, *26*(3), 249–255.
34. Ramos, S., et al. (2020). Cleaner production strategies for the food industry. In *The interaction of food industry and environment* (pp. 1–34). Elsevier.

35. Dhawale, P. V., et al. (2022). *Tannin as a renewable raw material for adhesive applications: A review*.
36. Bagade, S. B., & Patil, M. J. C. R. I. A. C. (2021). Recent advances in microwave-assisted extraction of bioactive compounds from complex herbal samples: A review, *51*(2), 138–149.
37. Patra, A., et al. (2021). Microwave-assisted extraction of bioactive compounds from cashew apple (Anacardium occidenatale L.) bagasse: Modeling and optimization of the process using response surface methodology, *15*(5), 4781–4793.
38. Pinto, D., et al. (2021). Microwave-assisted extraction as a green technology approach to recover polyphenols from Castanea sativa shells, *1*(2), 229–241.
39. Handayani, P. A., et al. (2019). Microwave-assisted extraction (MAE) process of tannin from mangrove propagules waste as natural dye for coloring Batik tulis. In *Key engineering materials*. Trans Tech Publ.
40. Nisca, A., et al. (2022). Optimization of microwave-assisted extraction conditions to improve phenolic content and in vitro antioxidant and anti-microbial activity in quercus cerris bark extracts, *11*(3), 240.
41. Arshanitsa, A., et al. (2022). Advantages of MW-assisted water extraction, combined with steam explosion, of black alder bark in terms of isolating valuable compounds and energy efficiency, *181*, 114832.
42. Poveda, J., et al. (2018). Revalorization of winery by-products as source of natural preservatives obtained by means of green extraction techniques, *112*, 617–625.
43. Ribeiro, M. M., et al. (2022). Recent advances in the recovery of tannins from natural matter, *75*, 289–328.
44. Talmaciu, A. I., et al. (2016). Isolation of bioactive compounds from spruce bark waste using sub-and supercritical fluids, *117*, 243–251.
45. Carlqvist, K., et al. (2022). Life cycle assessment for identification of critical aspects in emerging technologies for the extraction of phenolic compounds from spruce bark, *333*, 130093.
46. Mohd-Nasir, H., et al. (2021). Optimization of the supercritical carbon dioxide extraction of Quercus infectoria galls extracts and its bioactivities, *45*(2), e15156.
47. Guggilla, M., et al. (2022). Supercritical fluid extraction of chebulagic acid from Terminalia bellirica fruits (Baheda), *46*(4), e16455.
48. Muangrat, R., & Jirarattanarangsri, W. J. J. O. F. P. (2020). Physicochemical properties and antioxidant activity of oil extracted from Assam tea seeds (Camellia sinensis var. assamica) by supercritical CO_2 extraction, *44*(3), e14364.
49. Melloul, S., et al. (2022). Supercritical CO_2 extraction of bioactive compounds from local Peganum Harmala plant seeds and optimization of the extraction yield and the antioxidant activities, *28*, 100729.
50. Sadegh, N., et al. (2022). *Recent advances and perspectives of tannin-based adsorbents for wastewater pollutants elimination: A review* (p. 100763).
51. Mansouri, F. E., et al. (2022). Evaluation of different extraction methods on the phenolic profile and the antioxidant potential of Ceratonia siliqua L. pods extracts, *27*(19), 6163.
52. Zhang, X., et al. (2021). Extraction and application of tannin in the medical, food and chemical industries, *3*(1), 62–75.
53. Hassim, N., et al. (2020). Effect of static extraction time on supercritical fluid extraction of bioactive compounds from Phyllanthus niruri, *17*(2-3), 918–924.
54. Alvarez-Rivera, G., et al. (2020). Pressurized liquid extraction. In *Liquid-phase extraction* (pp. 375–398). Elsevier.
55. Pereira, D. T. V., et al. (2019). Pressurized liquid extraction of bioactive compounds from grape marc, *240*, 105–113.
56. Mohd Jusoh, N. H., et al. (2019). Pressurized hot water extraction of hydrosable tannins from Phyllanthus tenellus Roxb, *13*(1), 1–10.
57. Chao, B., et al. (2017). Tannin extraction pretreatment and very high gravity fermentation of acorn starch for bioethanol production, *241*, 900–907.
58. Cuong, D. X., et al. (2019). *Tannins: Extraction from plants*.

59. Raji, P., et al. (2019). Greener approach for leather tanning using less chrome with plant tannins and tannins mediated nanoparticles, *30*(6), 1533–1543.
60. Ayalew, A. A., & Wodag, A. F. J. C. A. (2023). *Extraction and chromatographic analysis of ethiopian oak bark plant for leather tanning applications* (pp. 1–10).
61. China, C. R., et al. (2021). Tanning capacity of Tessmannia burttii extracts: The potential eco-friendly tanning agents for the leather industry, *3*(1), 1–9.
62. Solé, M. M., et al. (2021). Characterization of wet olive pomace waste as bio based resource for leather tanning, *14*(19), 5790.
63. Guo, L., et al. (2020). Optimisation of tannin extraction from Coriaria nepalensis bark as a renewable resource for use in tanning, *149*, 112360.
64. Yahia, M., et al. (2019). Chestnut-Alu minium combination tanning system for high stability leather, *6*(5), 1–6.
65. Das, R. K., et al. (2022). Application of indigenous plant-based vegetable tanning agent extracted from Xylocarpus granatum in semi-chrome and chrome retanned leather production, *32*(3), 258–264.
66. Nasr, A. I., El Shaer, M. A., & Abd-Elraheem, M. A. J. J. O. E. E. (2023). Potential application of used coffee grounds in leather tanning, *24*(3), 10–19.
67. Saratale, R. G., et al. (2021). Grape pomace extracted tannin for green synthesis of silver nanoparticles: Assessment of their antidiabetic, antioxidant potential and antimicrobial activity. *Polymers, 13*(24), 4355.
68. Chowdury, M. J., et al. (2022). Extract of Trema Orientalis (L.) stem bark: A potential source of environmentally friendly tanning agent for leather industry, *117*(1), 28–34.
69. Mahdi, M., et al. (2020). Dyeing of shoe upper leather with extracted dye from acacia nilotica plant bark-An eco-friendly initiative, *14*(4), 241–258.
70. Zhou, Q., et al. (2020). Environmental friendly bioactive finishing of wool textiles using the tannin-rich extracts of Chinese tallow (Sapium sebiferum L.) waste/fallen leaves, *176*, 108230.
71. Adeel, S., et al. (2022). Ecofriendly application of coconut coir (Cocos nucifera) extract for silk dyeing, *29*(1), 564–572.
72. Adeel, S., et al. (2020). Sustainable ultrasonic dyeing of wool using coconut coir extract, *90*(7-8), 744–756.
73. Kholil, A. S. A., et al. (2021). Utilization of old coconut (Cocos nucifera) husk waste as potential source of natural dye and its dyeing properties on cotton cloth. In *Key engineering materials*. Trans Tech Publ.
74. Duraisamy, R., et al. (2020). Extraction, screening and spectral characterization of tannins from Acacia xanthophloea (Fever Tree) bark. *Research Journal of Textile and Leather, 1*(1), 1–10.
75. Adeel, S., et al. (2023). Sustainable application of microwave-assisted extracted tea based tannin natural dye for chemical and bio-mordanted wool fabric, *20*(1), 2136322.
76. Win, K. H., et al. (2020). *Extraction of tannin as a natural mordant from tea leaf for dyeing of cotton cloth*. MERAL Portal.
77. Zuber, M., et al. (2020). Influence of microwave radiation on dyeing of bio-mordanted silk fabric using neem bark (Azadirachta indica)-based tannin natural dye, *17*(10), 1410–1422.
78. Singam, R. T., et al. (2019). The preparation and characterization on natural dyes based on neem, henna and turmeric for dyeing on cotton with superhydrophobic coating, *11*(7), 137–143.
79. Rani, N., Jajpura, L., & Butola, B. J. J. O. T. I. O. E. S. E. (2020). Ecological dyeing of protein fabrics with Carica papaya L. leaf natural extract in the presence of bio-mordants as an alternative copartner to metal mordants, *101*(1), 19–31.
80. Vadood, M., & Haji, A. J. C. (2022). Application of ANN weighted by optimization algorithms to predict the color coordinates of cellulosic fabric in dyeing with binary mix of natural dyes, *12*(10), 1519.
81. Ticha, M. B., et al. (2022). Bark residues recovery of Juglans regia. l for the dyeing of wool fabrics: Development of microwave-assisted extraction and dyeing processes, *19*(13), 5282–5296.

82. Rather, L. J., et al. (2021). Valorization of agro-industrial waste from peanuts for sustainable natural dye production: Focus on adsorption mechanisms, ultraviolet protection, and antimicrobial properties of dyed wool fabric, *1*(3), 427–442.
83. Faisal, R., & Chafidz, A. (2019). Extraction of natural dye from ketapang leaf (Terminalia catappa) for coloring textile materials. In *IOP conference series: Materials science and engineering*. IOP Publishing.
84. Tomasi, I. T., et al. (2022). *Tannin-based coagulants: Current development and prospects on synthesis and uses* (p. 153454).
85. Siniscalchi, D., et al. (2022). Effects of condensed tannins on greenhouse gas emissions and nitrogen dynamics from urine-treated grassland soil, *29*(56), 85026–85035.
86. Arias, A., et al. (2022). Tannin-based bio-adhesives for the wood panel industry as sustainable alternatives to petrochemical resins, *26*(2), 627–642.
87. Liao, J., et al. (2022). Investigation of potential use of soybean protein isolate–Chinese bayberry tannin extract cross-linked films in packaging applications, *15*(15), 5260.
88. Jose, S., et al. (2022). *Functional aspects of natural dyes* (p. 267–294).

Social and Environmental Impact of Natural Dyeing

Shanthi Radhakrishnan, Radhika Rajangam, and Praveena Peruran

Abstract Textile production is the prime reason for effluent creation, and excessive research is being done for efficient effluent management systems all over the world. Though regulations and standards have been devised by countries across the world, problems still exist as the sources of them have not been addressed. Synthetic materials and auxiliaries are being used extensively for textile production which has a bearing on the health of the industry personnel as well as the environment. Synthetic dyes are manufactured with a high degree of colour fastness for long-lasting effects. Even small quantities of leftover dyes contribute to the colour of the wastewater. Colour and toxicity removal require special processes after textile processing is completed. The consciousness of sustainable textile manufacturing has created the need for the selection of more safe alternatives. One such process is natural dyeing which provides innumerable opportunities for new colour combinations, safe throw-aways for agricultural use, using ingredients of natural origin and affording a non-hazardous subtle feel to human skin. This chapter studies the social and ecological impacts of natural dyeing giving rise to sustainable results. An analysis of a few case studies will help to prove that natural dyes taken from renewable resources will be the trend in the fashion industry for many years in the future.

Keywords Textile value chain · Cleaner production · Natural dyeing · Sustainability

S. Radhakrishnan (✉)
Department of Costume & Apparel Design, PSGR Krishnammal College for Women, Coimbatore, India
e-mail: deancad@psgrkcw.ac.in

R. Rajangam · P. Peruran
Department of Costume Design & Fashion, PSGR Krishnammal College for Women, Coimbatore, India
e-mail: radhikar@psgrkcw.ac.in; praveena@psgrkcw.ac.in

S. S. Muthu (ed.), *Natural Dyes and Sustainability*, Sustainable Textiles: Production, Processing, Manufacturing & Chemistry,
https://doi.org/10.1007/978-3-031-47471-2_7

1 Introduction

The digitalization of the textile market as e-commerce platforms creating connectivity and reach and covering longer distances and unique products has created a drive for extensive demand and growth. In 2021, the value of the global textile market is around USD 993.6 billion and the estimated compounded annual growth rate (CAGR) from 2022 to 2030 is 4.0% [1]. Apart from designing, production and distribution of textiles and apparel, the textile market includes technical and functional products for specific needs and requirements. The highest market share of 73% is attributed to fashion and clothing followed by the technical segment, household segment and others [1]. Natural fibres like cotton, wool and silk have taken the lead while the polyester segment includes polyethylene, polypropylene, aramid and polyamide due to the growth of the technical sector.

The fashion and clothing segment shows the demand for functional and quality-oriented formal and casual wear for different age groups, for example, crease-free shirting and suiting fabrics and high-quality dyed and printed fabrics, catering to the needs of the growing population and urbanization in emerging economies of the developing countries. High performance characteristics and the targeted end-user applications have been the basis for the demand in technical segments like transportation, construction, medical and protective clothing. Bedding, upholstery, carpets, kitchen, bath linen and others are products that stipulate the home furnishing segment, propelling the growth of the household application sector. Such driving forces call for an increase in production and consumption leading to the predicted growth of the textile market and volume of business turnover.

Despite the growing demand and being a growth factor to many economies, this industrial business model is highly polluting and devours many natural resources. Wet processing coupled with a plethora of chemicals in the production chain operations create havoc on the social and environmental equilibrium, resulting in various disastrous impacts that are compounded by the passage of time. Release of dust and particulate matter, nitrogen and sulphur oxides, and volatile organic compounds constitute air pollution; discarded packaging and textile and yarn scraps comprise the solid waste at the primary level; excessive volumes of unwanted heterogeneous compositions of textile sludge reveal loads of heavy metals, pathogenic microorganisms, micronutrients and organic matter while unwanted clothes end up in landfills. Since water is the main component which facilitates the application of any pretreatment, colouring and finishing substances to fabric, it is inevitable that effluents of different compositions are left behind after the process. One of the major damages to the environment and community is the discharge of untreated effluents into the soil or water bodies. It has been estimated that untreated hazardous effluents constitute 80% of total discharge produced by the textile industry's processing sector [2] which has high levels of biochemical oxygen demand (BOD) and chemical oxygen demand (COD) [3, 4]. It has been found that this condition is greatly due to the presence of organic compounds like textile dyes [5] being non-biodegradable in nature.

This chapter highlights the problems related to textile dyeing, proposes natural dyeing as a sustainable alternative to synthetic dyes and studies the impact of using natural dyes in the value chain by analysing a few case studies.

2 Key Issues in Textile Dyeing

Dyes are substances that impart colour to the substrate on which they are combined, for example, textiles, leather, food, cosmetics, pharmaceuticals and so on. Usually the dyeing process is carried out in a solution containing dyes and the materials to be dyed with the aim of developing colour with a certain degree of colour fastness. From the end of the nineteenth century, natural dyes were replaced by synthetic dyes. Generally dyes should be safe and free of toxins, allergens, mutagens and carcinogens, but there are many reports that speak about the adversity of dyes on textiles [6, 7]. Unlike cosmetics or food, it is difficult to detect the composition of the textile dyes as it is not mandatory to declare the dye components [8].

2.1 *Synthetic Dyes for Textiles*

Synthetic dyes are available commercially in a wide range of over one lakh compounds. The annual production of synthetic dyes is more than seven lakh tons of which around 15% are discharged into the water bodies and environment [9]. The colouration is due to the presence of chromophoric groups and the autotrophic groups are responsible for the fixation of colour to the polar groups of the textile yarns or fabric [10, 11]. There are three main categories of synthetic dyes: anionic, cationic and non-ionic dyes. Cationic (basic dyes) and anionic dyes (direct, acid and reactive dyes), which are classified under the head ionic dyes, are water-soluble organic compounds while non-ionic dyes (vat and disperse dyes) are water-insoluble but soluble in organic solvents. Water-soluble dyes are difficult to remove from wastewater and cannot be addressed by regular water treatment systems [12]. Reactive dyes are very stable, complex in structure and remain unaffected in the environment for long periods of time (maybe around 46 years), and these tend to accumulate in aquatic organisms causing toxicity [13, 14]. In the case of complex metal dyes, they exhibit a resistance of half-life of upto 13 years with a hazardous composition of nickel, copper, cobalt and chromium [15, 16].

2.2 *Environmental Problems*

Dyes in the dyebath are not completely utilized and the unutilized dyes enter the wastewater to be released into the environment, for example, about 15–50% of azo dyes, which are considered to be highly toxic, do not bind to the fabric during the dyeing process [17]. This proves to be negative to the structure and functioning of the ecosystem and microbial organisms of the soil which in turn affect the germination and growth of plants [18, 19]. The dye molecules absorb sunlight and prevent the formation of photosynthesis stunting the underwater plant growth. The presence of chromium in textile dyes creates oxidative stress, leading to considerable damage to plants in areas of photosynthesis and carbon dioxide assimilation [20]. Another example is the release of metal dyes into the environment, causing the assimilation of heavy metal cations by the fish gills due to the negative charges and accumulation into certain tissues. Once consumed by humans they cause a sequence of pathologies [7, 21].

When effluent is released after the dyeing process, the coloured effluent is usually visible, causing unpleasant and aesthetic damage to the rivers and lakes. They prevent the penetration of light through the water, leading to a reduction in photosynthesis and oxygen levels which affect the entire system and aquatic biota [3, 4, 12]. The dyes exist permanently in the environment and enter the food chains, moving from the basic level to the more complicated higher levels. As it gets transferred from one level to another, biomagnification of the pollutant occurs, leading to higher levels of contamination in the predator than the prey ([22–24]). The persistent nature of the dye's existence in aerobic environments causes bioaccumulation in sediments and soil which finally gets them into the public water supply systems, causing problems for humans [25]. In some cases there is a partial degradation or transformation of the azo-type compounds forming dangerous aromatic amines; dyes may also combine with other intermediate synthetic compounds or other degraded products to form new carcinogenic substances [26].

2.3 Hazards to Human Health

The manufacturing of synthetic dyes and intermediaries, like benzidine and 2-naphthylamine, has caused a high occurrence of bladder cancer; enzymatic co-factors may be substituted to cause inactivation of enzyme activities resulting in many disorders [20]; other problems due to textile dyes may be dermatitis and disorders of the central nervous system [27, 28]. Oral ingestion and inhalation of the dyehouse dust lead to irritation in the skin and eyes; contact dermatitis, allergic conjunctivitis, rhinitis, occupational asthma and allergic reactions are common among workers who work with or handle reactive dyes. Human serum albumin and reactive dyes combine to act as antigen (disease-producing starter) producing immunoglobulin E antibodies which combine with histamine, resulting in chronic diseases [29].

Textile dyes contain genotoxic substances that may directly or indirectly cause damage to enzymes responsible for DNA replication, causing mutations leading to cancer or inheritable damage such as birth defects. This long-term hazard was demonstrated in studies made in *Allium cepa* root cells that showed chromosomal irregularities [30, 31]. DNA changes caused by mutagens harm cells leading to cancer. Dyes like Azure-B reveal mutagenic potentiality [29] whereby it interrupts the helical structure of DNA and the duplex RNA [32, 33] and also indicates cytotoxic effects by inhibiting the enzymes in the central nervous system that dictate human behaviour and those that regulate biological responses and events [34].

Synthetic counterparts in textile dyeing create problems for human health and the environment. All living systems from the tiniest to the macro-level organisms constitute the environment and anything or any influences which tamper with their equilibrium will cause immense harm to the interlinking ecosystem. Hence, alternative technologies are essential for a better life and prosperity.

3 Sustainable Alternatives to Textile Processing

Sustainability is the most sought phenomenon that is in the current trend and the motive behind all enterprises and businesses. An achievement in this concept calls for appreciation and certification to be a role model for others and move the bandwagon towards the sustainable development goals (SDGs). Many apparel manufacturers are conducting research to fabricate their production processes to reduce their environmental footprint and cost and enhance the safety of the products to enable sustainable growth. The textile and apparel sectors are water, energy and chemical intensive, and green chemistry is promoted for resource efficiency and effluent reduction which in turn will have a positive impact on the environment. Green chemistry is also known as 'sustainable chemistry' with a greener framework across the entire life cycle of a product. It aims at a 'no harm' policy, and to achieve this motto its methodology works with eco-friendly alternatives, non-hazardous substances, low-risk production processes, prevention of waste, recycling, circular economy and research-based databank development initiatives.

A sustainable product/process has the least or minimal impact on the environment in terms of raw material usage, emissions and waste generation [35, 36]. Raw material or resource usage must be within certain limits so that it does not deplete or permanently damage the resource; every natural resource requires time to regenerate and this has to be taken into consideration while using the resource for production [37]. The cradle-to-grave concept has been replaced by the cradle-to-cradle concept where no material is wasted but may be refabricated to enter the next cycle of production along with the virgin material. Moreover, a sustainable product can be economically, environmentally and socially viable. Environmental product declarations (EDPs) are made by the manufacturers by using the concept of cradle-to-gate where a partial life cycle assessment (LCA) is made from resource extraction (cradle) to the point before it is transported to the consumer (gate) called business-to-business EDPs [38].

The aesthetic, functional and comfort properties of textile fabrics are enhanced during wet processing operations. This step requires a huge amount of water as the medium for applying the chemical and for washing and rinsing operations to remove the excess chemicals and end products after the process is completed. The drain water will be loaded with the washing and finishing agents and the unfixed dyes which would contaminate the environment if discharged as untreated water. Green chemistry envisages alternative biodegradable chemicals for wetting, washing and finishing operations; more reactive and ecologically degradable dyes; and ionic liquids as the medium for wet processing operations, thereby ensuring sustainability [39].

Sustainable practices are to be incorporated generally in the manufacturing industries rather than on the basis of compulsion. Today's customer is very conscious about being eco-friendly and is willing to pay more for the same. The export market is regulation driven and without the eco-friendly certification, regular business cannot be undertaken. Resource scarcity has caused spiralling of prices, increasing the product cost and in turn affecting the prosperity of the organization. Demand for better products with reasonable costs is the market trend, and only continuous research and innovation will help to create maximum benefits at a minimum cost. Market competitiveness and innovation go hand in hand.

3.1 Substitution of Chemicals

To make a product or process sustainable, the process ingredients and components are to be made greener than the conventional pathway to minimize pollution. In many processes, solvents play an important role in enabling processes in terms of dissolving reactants, extraction, separation, dispersion, washing apparatus and so on. Many of the conventional solvents are toxic, volatile and highly inflammable. The greatest challenge is leakages in workplaces during storage and transportation causing photochemical smog. Safety in handling is essential to prevent fire and explosions. Alkyl phenol ethoxylates used in detergents, wetting agents and levelling agents are highly toxic to aquatic life and disturb the reproduction of aquatic species; polybrominated diphenyl ethers and chlorinated paraffins, halogenated phenols and benezenes used in the production of flame-retardant finishes have been indicated as 'priority hazardous substances' by Oeko-Tex; sequestering agents like ethylenediamine tetraacetic acid (EDTA), diethylenetriaminepentaacetic acid (DTPA) and nitrilotriacetic acid (NTA) form stable complexes with metals which is hazardous in nature.

Biosolvents, like ethanol, are produced from renewable sources. The biosolvent production process involves the fermentation of feedstock/starchy substances/lignocellulosic materials. This creates a greener pathway as fossil resources used in petrochemical solvents are avoided and the subsequent carbon dioxide emissions are prevented. Further CFCs, which cause the depletion of ozone, is replaced by the use of supercritical solvent $scCO_2$. Ionic solvents are derived from molten salts that have special solvating properties for polar materials and have low melting points. Ionic liquids from molten salts serve as alternatives to dimethyl formamide (C_3H_7NO), dimethyl sulphoxide (C_2H_6OS), etc. [40].

Non-volatile eco-friendly solvents of liquid portions of polyethers, for example, polyethylene glycol and siloxane polymers with a wide range of polarities, are biodegradable and inexpensive alternatives that are non-toxic to human and marine life. Apart from these many volatile and non-volatile ester solvents that are used in carrier oils, plasticizers and biodiesel like isopropyl laurate, rapeseed methyl ester, glycerol triacetate and dibasic ester (DBE) are being used in eco-friendly inks and are accepted as phthalate-free plasticizers and biodegradable carrier oils. All these alternatives are prepared from renewable sources through green processes, have a low tendency to cause violent reactions, are non-flammable, are low in toxicity and heavy metals, are biodegradable and are low in bioaccumulation in food chains.

In the case of dyeing, the dye bath is a conglomeration of chemicals used to enhance/counteract the effects of other chemicals. The auxiliaries used in the dyeing process are alkalis, salts, reducing and oxidizing agents, and fibre residual contaminants. Most of the dyeing processes cause pollution from the synthetic dyes used and the auxiliaries that constitute the dye bath or dye formulation like dispersing agents and anti-foaming agents.

3.2 Green Preparatory Processing

Preparatory processing is an essential step which sets the fabric for future processing like dyeing or finishing. Enzymes are being extensively used in the preparation of fabric which includes desizing, scouring and bleaching. Enzymes are biological catalysts consisting of proteins that are made up of polypeptide chains. Enzymes are from individual proteins with RMM (relative molecular mass) of around 13,000 which catalyses a single reaction up to multi-enzyme complexes [41]. Denaturing of the enzyme occurs when favourable conditions do not exist as these fragile living organisms need a specific temperature, pH and concentration to work on the substrate.

According to Presa and Tavcer [42], enzymes may be applied in various stages of textile processing, namely desizing – amylase, lipase; scouring – pectinase, cellulase, cutinase; bleaching – oxidoreductase, xylanase; finishing – cellulase, oxidoreductase and lipase; and textile waste composting – laccases, cellulase, protease, xylanase and polyesterases.

Enzyme usage as an alternative to chemicals converts the process to an eco-friendly one and is termed as 'bioprocess'. The first process in textile preparation is desizing where the warp sizes composed of starch, modified starch or organic polymers and polyacrylates are removed by hydrolysis with water or mineral acids, oxidation with per-oxy compounds or sodium bromite. Bio-desizing uses amylases, enzymes obtained from malt, pancreatic or bacterial sources, along with lipases (for fat removal). Regular amylase works in medium temperatures of 25–55 °C and pH 5.5–7.0 while industrial high-temperature amylases work at 95 °C. The end products after the enzymatic desizing process are sugars and dextrins which are not toxic and do not negatively impact the BOD of wastewater. The next process is scouring where the natural impurities in the raw fibres are removed before the colouration process. Pectate lyase, an alkaline pectinase, hydrolyses the salts of polygalacturonic acids (pectins) that are non-cellulosic impurities present in the primary wall matrix of the fibre. This reaction is carried out under slightly alkaline conditions of pH 8–9.5 and temperature around 55–60 °C in the presence of chelating agents [43]. The benefits of bioscouring are that the enzyme does not destroy the cellulose of the fibre, strength retention, lower weight loss leaves optimal wax in the fibre to maintain the fabric's softness, and works with low temperatures and less water. The BOD and COD of the bioscouring process are reduced by 55–80% when compared to conventional scouring with sodium hydroxide. The health risks of the operators are minimized, and it is also safe for the environment [44]. Alkaline pectinase is commercially useful when compared to conventional pectinases [45]. All round savings on all fronts have been achieved by simultaneous desizing and scouring with combined enzyme systems due to the specific target approach of the enzymes present in the combined systems [46].

An important step in the preparatory sequence is the bleaching of fabric to impart whiteness. It had been studied at Auburn University, USA, that there was an improvement in whiteness index with low strength loss while using glucose oxidase enzyme [47]. The reaction takes place in the presence of molecular oxygen where the enzyme converts glucose into gluconic acid and hydrogen peroxide; D-Gluconic acid acts as a sequestering agent during bleaching. Biobleaching of wool has also been carried out where H_2O_2 bleaching in the presence of protease enzyme (Bactosol SI from Clariant) helped to improve the whiteness index and hydrophilic properties. Compatibility of enzymes has also led to combined preparations of desizing, scouring and bleaching to save water and energy with the required properties essential for further processing. A very safe method of removing hydrogen peroxide after bleaching is the use of catalase enzymes which work in neutral pH and temperature 20–50 °C. The enzyme only attacks hydrogen peroxide and is changed to oxygen and water, removing all the hydrogen peroxide. The presence of H_2O_2 may cause problems and hence has to be removed before the dyeing process.

3.3 Eco-Friendly Processing

Other finishing processes where enzymes are used are in the decolourization of dyes by laccases in the dyeing and printing industries [48] and in the delignification of woody fibres in the pulp and paper industries [49]. Laccases are glycoproteins under the oxidoreductases called the blue-copper family of oxidases [50, 51] that catalyse the oxidation of substrates at the expense of molecular oxygen.

3.4 Natural Dyeing

From ancient times, dyes have been acquired from natural sources like plants, animals and minerals for colouration of fabric. Natural fibres like cotton, wool and silk were dyed with natural dyes on a wide scale but with the advent of synthetic dyes, its popularity lost ground.

Currently, the consciousness towards a safe environment has turned the interest of manufacturers and consumers from petrochemical-derived dyes towards the use of natural dyes. The major challenges in the use of naturally sourced dyes are their colour fastness, colour evenness, colour reproducibility, heavy metal salts as mordants and availability of natural resources to meet the demands of the ever-growing population. Two major classifications exist depending on the application method, namely substantive and non-substantive dyes. In the substantive class, the dyes are applied without mordant pretreatment while the non-substantive class requires dyeing after mordanting or in the presence of mordants in the dye bath. A mordant is a dye fixative used to set dyes on the fabric by forming a coordination complex with the dye thereby attaching the dye to the fabric [52]. The classification of mordants is given in Fig. 1. The commonly used fixing agents are Turkey red oil, ferrous sulphate, tartaric acid, stannous chloride and alum. Tannin-metal salt combinations also serve as mordants as they form metal chelates after pre-/posttreatment of the substrates with metal salts. Important sources of tannins available in nature are oak bark and wood (*Quercus infectoria*), pomegranate rind, cutch (*Acacia catechu*) and so on [53].

Sustainability in production and consumption can be realized when ecology is taken as a priority in decision-making by industrialists, retailers and consumers. Products can be manufactured with due interest and priority to the environment and safety. Some measures include using lesser resources; attaining better yields for economic viability; less or no pollution of water, soil and air; no unwanted by-products and less waste generation. An international set of standards under the ISO 14000 series helps in the assessment and improvement of environmental management in the industry [54].

A better understanding of the life cycle of chemicals, low process inefficiency, application of green chemistry and clean technologies can create more manufacturing systems that are environment-compatible, leading to more scope towards the 'Green Concept'.

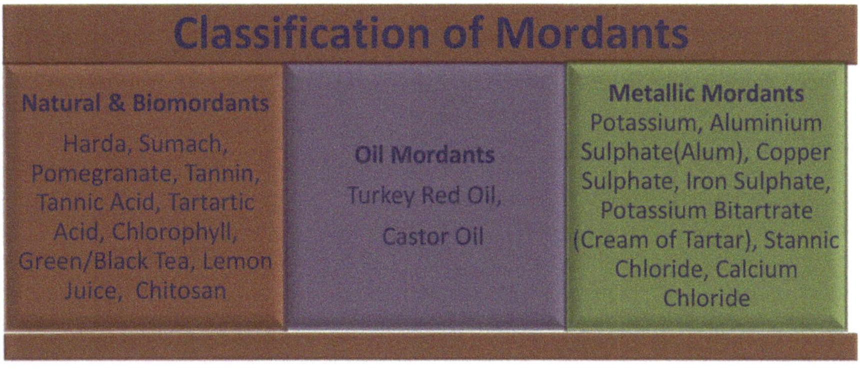

Fig. 1 Classification of mordants

4 Concerns and Remedies

4.1 Areas of Apprehension in Natural Dyeing

Issues related to natural dyeing are very many. Natural dyes have a low affinity to textile fibres and need huge doses of dye, leaving a lot of dyes in the bath due to low exhaustion from the bath to the fibre. Dyes are available in small quantities in plants and hence a lot of natural resources need to be dismantled to obtain small dye quantities, leaving huge biomass as waste. The cost of transportation of plant material from harvest to process centres is high, and the remains can be used as fuel or fertilizer. Indiscriminate cutting of trees for bark can lead to damage or deforestation; heavy usage of flowers can lead to disturbing pollination and reproductive cycle. This can lead to disruptions of the ecosystem and plants can become endangered [55].

The areas of concern regarding natural dyes are reproducibility, low fixation, colour fastness properties and cost efficiency, which can be overcome by using mordants to certain permissible limits [56, 57]. For example, improved colour fastness and a wide range of colour shades can be obtained with the help of alum and iron mordants.

The problems of using metallic mordants include the generation of wastewater with toxic metal ions which cause negative impacts on the environment and allergic and health-related problems [58, 59].

When taking sustainability into consideration, all waste cannot be used randomly. Wastes from agriculture and forests have inbuilt risks. Only certified organic farm waste can be used; there may be copper zinc and toxic substances in grape pomace as there are many insecticides sprayed on them during growth. Bioaccumulation of heavy metals, radioactive elements and pollutants are common in plants. In a study, the presence of hexavalent chromium was analysed in the bark of oak, based on the natural environment and urban deforestation for the setting up of metal-processing companies [60]. Further, wastes may contain good dyestuffs as well as instable colourants and segregation of the two will be a Herculean task.

However, natural dyes are expensive with issues on availability and have difficult operational methodologies. The conventional methods of extraction of natural dyes are soaking, stirring + heating, boiling, refluxing, heating at boiling, maceration and Soxhlet extraction. Water and energy consumption is high and only a few dyes can compete with synthetic dyes. Immense research has to be undertaken to find better working methods to make the process viable for commercial applications. Though natural dye-producing plants and trees can be grown on wastelands, as they need minimal resources, the yield will not be sufficient to cater to the large industrial needs and consumer demand. It can only cater to the small-scale sector and the cottage industries and cannot become a viable alternative to the synthetic dye industry.

4.2 Alternative Methodologies for Natural Dyeing

Extensive research and sustainability studies on plant-based extracts have been undertaken to replace conventional metallic mordants. Bio-mordants are plants with high tannin content and serve as alternatives to metallic mordants, for example, peel of pomegranate, acacia, henna, turmeric, thuja and rosemary leaves are considered alternatives to aluminium, iron sulphate, copper sulphate, stannous chloride and potassium dichromate [61–63]. Another source of metals is from plants called hyperaccumulators. They are unique and unusual plants that collect metals or metalloids in living tissues. The levels of metal content are usually hundreds to thousands of times more than that occurring in many normal plants [64–66].

Green and clean substitutes from biomass and eco-friendly alternatives with high tannin content and metal-accumulating compositions are being utilized to advance towards sustainability [67–70].

The traditional isolation method is used to isolate natural dyes from dye-yielding plants rich in colouring matter. This will be a solution to overcome the problems of low yield and poor fastness of natural dyes.

However, few modern methods are being adopted to align the fabric towards the dye, enhancing the colour yield as well as surface modification of the fabric. *Ultrasonic radiations* are rays used for the isolation of natural products with added advantages like time, solvent, cost and labour. *Microwave radiation* is a sustainable heating source used in domestic and industrial applications. This extremely fast process uses lesser solvent, lower temperature and lesser labour. Gamma radiation is a powerful heating source used for the isolation of colourant from natural dyes which is helpful in dye degradation (conversion of the dye into innocuous products). *Ultraviolet radiations* are radiations that help in dye isolation and surface modification of fabric. These electromagnetic rays within the range of 200–380 nm are found to be efficient in energy, time and cost [71, 72]. It has been reported that the colourant molecules from Arjun bark in various media were isolated after microwave treatment for 6 minutes to give enhanced colour depth. The irradiated dye powder was encapsulated by using the solvent-acidified methanolic extract, giving excellent colour yield when silk fabric was dyed in lower conditions for 65 min. at 65 °C using dye bath (1 pH) with 7% of Na_2SO_4 as exhausting agent [73]. Similarly, ultrasonic treatment was used to isolate the colourant in methanolic medium from neem bark for a time span of 30–60 minutes. A 30-minute stimulation was helpful to encapsulate the colourant from 4 grams of neem bark powder to be used for silk dyeing at 70 °C for 60 minutes [74].

Natural colour components like chlorophyll, carotenoids, flavonoids and betalains are being replicated in the laboratory and comparisons are being made with the natural one [75]. The extraction of natural dyes with the help of enzymes is also being researched. Xylanase enzyme is used to destruct the matrix of the epidermal layer of the root of *Lithospermum erythrorhizon*, a natural red dye, thereby improving the dye extraction [76]. Similarly the pigment extracted from Gardenia, a blue natural pigment, by chemoenzymatic reaction, proved to have good fastness properties in the absence of a mordant [77]. Research is being done on bacteria-derived pigments where microorganisms are used to produce pigments; the advantages being high production yield and feasibility for bioprocess development, for example, *Janthinobacterium lividum* XT1 is being reported to produce a violet pigment [78, 79].

The by-products of agricultural production create opportunities for fibre and natural dye production. Sustainability is enhanced when plants grown for a specific use can provide opportunities for other end uses eg. banana plant (pseudo stem) can provide fibers as well as an extract which helps in the colouration of natural dyes. Another source of natural dyes is from the wastes and by-products of the food, timber and beverage industries. These industries produce a large amount of biological waste, namely moulded fruits, distillation residues, pomace and by-products that possess a large amount of natural dyes. Pomace from fruits like cherries, red and black currants, and purple grapes are rich in anthocyanins; coloured vegetable wastes contain anthocyanins and betalains while timber harvest results in bark material waste rich in tannins [60, 80]. Segregation and use of these wastes are a pioneering task.

To enable the use of bio-resourced materials for textile dyeing and finishing [81], extensive research should be undertaken to attain technical, economic and ecological standards of the twenty-first century. Unused remnants can be sent back to agriculture for composting and biogas production. Further studies are to be done to understand and judiciously use bio-colourants for textile colouration. Hence bio-colourants may be the next-generation green alternative that can bring in colour to textiles.

Zero emission is the aim of the textile industry to attain sustainability. Waste from agriculture, forest and by products can be investigated to become the raw material for another production cycle. To lower the cost of natural dye production, the abundantly available ash tree bark and leaves of teak and eucalyptus can be used [82–84]. In the natural dyeing process, the wastewater contains unused dyes which has to be removed as per the standards of textile effluents [85]. Extraction by sonification followed by natural dyeing seems to be a better option than thermal exhaustion. This method helps to enhance dye exhaustion, resulting in reduction in effluent load and lower costs [68, 70, 86].

5 Analysis of the Impact of Natural Dyeing

5.1 Case Study I

This study analyses the energy consumption and greenhouse gas emissions of natural dyeing of 1000 kg of cotton fabric, using pomegranate rind and mordanting with eucalyptus bark, by the life cycle approach. The dyes and mordant were prepared from locally available pomegranate rind and eucalyptus bark. The preparation of aqueous solution in both dye and mordant was carried out by taking 5 g of powder +100 ml of distilled water boiled for one hour at 90 °C with stirring using a hot plate and magnetic stirrer. The solution is filtered and used for dyeing or mordanting. The pretreated sample was immersed in the mordant solution for 15 minutes with continuous stirring followed by shade drying. The mordanted fabric was soaked in the dye solution for 60 minutes at 50 °C with continuous stirring. The Material Liquor Ratio (MLR) was 1:20. This was followed by shade drying. Testing was carried out for the colour fastness of the dye.

Life cycle assessment studies on natural resources have been extensively studied [87, 88]. The LCA was based on the calculation process, system boundary factors, LCA factors and inventories for processes, as given in Table 1.

The impact analysis was done for different stages, namely raw materials, production, storage, use, transport, total impact, co-products and finally total + co-products. The figures in the case study are expressed in Table 2.

In all the parameters studied storage and use were nil so they have not been represented in the table. The principal results of the study are the total carbon footprint is 1.11×10^5 kgCO2eq/fu, total water usage 1.54×10^5 m^3 water/fu and total water footprint stress weighted is 1.47×10^5 m^3 water/fu.

5.2 *Case Study II*

The textile industry is known to be the most polluting industry and current practices are based on the sustainability aspects and the methods of reaching these goals [90, 91]. Synthetic dyes have a high environmental impact, but it has been reported that direct and reactive dyes have a lower negative impact when compared to vat or sulphur dyes. This trend may also be seen in natural dyeing where some dyes are more eco-friendly than the others, for example, the agricultural production of indigo dyes has a high environmental impact [92]. The aqueous solution extracted from the bark of *Acacia dealbata* tree, an agro-industrial waste, is being used for dyeing. In this study, the LCA was carried out to study the environmental impact of the natural dyeing process (Table 3).

Methodology

Direct dyeing was performed in a neutral or slightly alkaline bath with the addition of electrolyte. The dye uptake was assessed in terms of with/without electrolyte addition. The electrolyte used was NaCl at 5 g/L. The dyeing process was by exhaust method at 95 °C for one hour followed by two subsequent rinses. The colour yield was medium; the residual bath had a high colour saturation and was used for new dyeing.

Results

1. The colour coordinates of the dyed cotton with and without electrolytes are compared with synthetic reactive dyeing for the shade RGB 0÷255. A slightly darker shade was achieved with electrolyte addition and this residual bath was used for redyeing.

Table 1 Parameters for life cycle assessment (LCA) [89]

I. Calculation process for dyeing 1000 kg of cotton fabric	
1. Dye powder – pomegranate rind 2. Weight of one fruit rind = 40 gms 3. Rinds required = 25,000 nos. 4. Annual yield per five year tree = 100 fruits 5. Land requirement = 2.5 acres (100 trees per acre) 6. Irrigation water requirement per day = 32 litres 7. Annual irrigation water requirement = 11,680 litres/year/tree	8. Rinse water for mordanting and dyeing – 20,000 litres 9. Fertilizer requirement for five year tree = 600–700 g N, 200–250 g P_2O_5, 200–250 g K_20, 75 g ammonium sulphate and 10 kg of farmyard manure 10. CO_2 released by transport for 1 tonne km (standard is 100 km app. 12 kg CO_2) = 0.12 kg CO_2 released
II. Factors considered for system boundary	
Cultivation Harvesting Pomegranate rind cutting	Grinding Colour extraction Pretreatment as mordanting and posttreatments – colourant fixing, dyeing
III. Factors taken for LCA	
Impact categories chosen for the study: Global warming potential, eutrophication and human toxicity potential Cradle to gate (as usage of fabric excluded) Assumption – zero pesticide in the cultivation process Methods used for LCA – IMPACT 2002, CED, Recipe 2016 and Eco indicator 99	Tool used: Carbon footprinting tool (CCaLC2) to perform LCA made up of ecoinvent and CCaLC2 databases Midpoint impact categories – carbon footprint, acidification potential, water usage and footprint, eutrophication, photochemical smog potential, human toxicity potential, ozone layer depletion potential
IV. Inventories used for the various processes for LCA with name and functional units	
Sodium chloride = 40 kg Soap = 20 kg Tannic acid = 40 kg Urea = 339.062 kg Phosphoric acid = 120 kg MOP (muriate of potash) = 105.8 kg Alum/stannic chloride/stannous chloride = 40 g Water requirement for cultivation = 15,62,500 litres	Land use = 2.5 acres Energy requirement for dye extraction = 315 kWhr Energy requirement for dyeing = 420 kWhr Energy requirement for desizing and scouring = 10 GJ / ton Transport = 12 kg CO_2 eq Water requirement for mordanting, washing, desizing and cleansing with soap solution (scouring) = 100,000 litres
V. Midpoint impact categories and their functional units	
Carbon footprint – 1.11 × 105 kg CO_2 eq/fu Human toxicity potential – 4104 kg dichlorobenzene eq/fu Acidification potential – 68.2 kg SO_2 eq/fu Eutrophication – 22.5 kg phosphate eq/fu	Water usage – 1.54×10^5 m^3 water/fu Water footprint – 1.47×10^5 m^3 water eq/fu Photochemical smog potential – 0.621 kg ethene eq/fu Ozone layer depletion – 7.32×10^{-4} kg R11 eq/fy

Table 2 Particulars of the impact analysis and their values [89]

Sl. No.	Impact particulars	Raw materials	Production	Transport	Total	Co-products	Total + co-products
1.	Carbon footprint [kgCO2eq/fu]	1.00E+5	1.36E+4	1798	1.11E+5	2665	1.14E+5
2.	Photochemical smog potential [kg ethane eq/fu]	0.621	0	0	0.621	NA	NA
3.	Eutrophication [kg phosphate eq/fu]	6.32	14.0	2.26	18.2	4.37	22.5
4.	Human toxicity potential [kg dichlorobenzene eq/fu]	1057	2177	871	3239	865	4104
5.	Acidification potential [kg SO_2 eq/fu]	15.5	44.5	6.78	56.4	10.5	66.8
6.	Ozone layer depletion potential [kg R11 eq/fu]	2.75E-4	2.05E-4	2.52E-4	4.91E-4	2.41E-4	7.32E-4
7.	Cumulative energy demand (3 stage) [MJ/fu]	2.68E+4 2.45E+4 5.13E+4	5.83E+4 2117 6.04E+4	3.00E+4 787 3.08E+4	8.87E+4 1.58E+4 1.04E+5	2.64E+4 1.17E+4 3.81E+4	1.15E+5 2.74E+4 1.43E+5
8.	Water usage (2 stage) [m^3 water / fu]	2.89E+5 2.89E+5	0	0	1.53E+5 1.54E+5	1.35E+5 1.35E+5	2.89E+5 2.89E+5
9.	Water footprint stress weighted [m^3 water /fu]	2.78E+5	0	0	1.47E+5	1.31E+5	2.78E+5

Colour fastness to washing and wet rubbing was carried out and colour change was given degree 4 and for wet rubbing was degree 5. Degree 4 and 5 show that the colour fastness of the fabric was good.

2. The *Acacia dealbata* tree bark (ADTB) natural dyeing LCIA results are as follows:

- Water depletion (m^3) was identified within the range 1,OE+00 to 1,OE+02.
- Climate change (kg CO_2 eq) and human toxicity (1,4-DB eq) was found to be within the range of 1,OE+00 to 1,OE-02.
- Terrestrial acidification (kg SO_2 eq) and photochemical oxidant formation (kg NMVOC) were within the range 1,OE-02 to 1,OE-04.
- Freshwater eco-toxicity (kg 1,4-DB eq), marine eco-toxicity (kg 1,4-DB eq) and terrestrial eco-toxicity (kg 1,4-DB eq) were found to be in the range 1,OE-04 to 1,OE-06.
- Freshwater eutrophication (kg P eq) was found to be within the range 1,OE-06 to 1,OE-08.

Table 3 Parameters for LCA, dyeing and testing [93]

I. Factors taken for LCA	
Impact categories chosen for the study – goal and scope definition, life cycle inventory analysis (LCI), life cycle impact assessment (LCIA) and life cycle interpretation phase (LCI) Gate to gate (pretreatment and bath procurement excluded, only dyeing stage taken into account) Assumption – time frame is 100 years Methods used for # LCA – ISO 14040; #LCIA – GaBi software – database platform: education database 2014; # characterization methodology – ReCiPe 1:08 (H) Functional unit: 1 kg cotton fabric	LCIA inputs – water volume, energy consumption, mass of fabric and dye stuffs; LCIA outputs – mass of dyed fabric, numerical values of the negative effects of wastewater. Energy consumption calculation based on literature; gas emissions not taken into account in this study Midpoint impact categories – water depletion (WD), climate change (CC), terrestrial acidification (TA), freshwater eutrophication (FE), ozone depletion (OD), freshwater eco-toxicity (FET), marine eco-toxicity (MET), terrestrial eco-toxicity (TET), human toxicity (HT) and photochemical oxidant formation (POF)
II. Dyeing and testing particulars	
Cotton dyeing pH 4.5 Mordant dye Machine used for dyeing and rinsing: Datacolour Ahiba IRTM machine – dyeing pot capacity 150 ml Material liquor ratio – 1:10	Woven cotton fabric (desized, edges stitched) – 125 g/m^2 Testing: residual bath testing – standard methods of wastewater assessment; dyed sample testing – commercial spectrophotometer Datacolour SF 600 Plus CT

- Ozone depletion (kg CFC-11 eq) was seen within the range 1,OE-10 to 1,OE-12.

The results also highlighted that natural dyeing of 1 kg of cotton fabric showed 0,7 kg of CO_2 eq in climate change while 12,4 kg of CO_2 eq was released in the case of synthetic reactive dyeing. A medium colour was obtained with reduced environmental impact when compared to the reactive dyed sample. Further, agro-industrial wastes can be utilized for value-added natural dyeing process, providing a niche opportunity to textile products [93] (Table 4).

6 Sustainability and Natural Dyeing

6.1 Sustainability Considerations in Natural Dyeing

Sustainability of a process is based on economics, environment and society. The sustainability parameters and outcomes are shown in Fig. 2. The demand for natural dyes is moving towards greater heights, creating new markets for the textile industry; these efforts can be addressed by offering unique value-added products and standardizing the procedures can increase productivity [94, 95]. Natural dyes obtained from indigo, madder and weld have been considered economically viable due to modern cultivation techniques whereby the raw materials, production and

Table 4 Colour coordinates and shade of dyed cotton [93]

Colour coordinates	With NaCl 5 g/L	Without electrolytes	Synthetic reactive dyeing	Desized cotton fabric	First dyeing	Second redyeing
L	62. 05	63. 88	62.30	89, 47	64, 35	63, 32
a	9.95	9.39	5.10	0, 99	8, 33	8, 37
b	13.67	12.78	5.02	8, 37	10, 08	13, 28
RGB 0÷255						

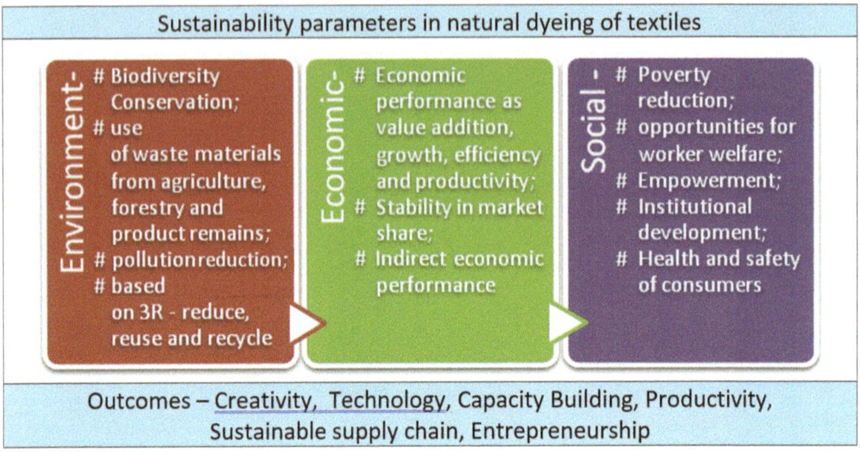

Fig. 2 Sustainability parameters in natural dyeing of textiles. (Created by author)

waste-handling costs are lower. This in turn reduces the consumer price and creates new SMEs to flourish. Further the fresh feel and odour-free properties help natural dyed fabrics to catch the eye of the customer [60, 96]. Another method of reducing production costs is to optimize natural dye production by combining extraction from cultivated plants and using natural dye producing waste. Dyes are said to be economically and technically viable when they provide the required colour with good strength, excellent affinity and compatibility with the fibre and superior fastness properties after dyeing.

Sustainability indicators for the environment can be analysed in terms of usage of resources, pollution and human behaviour activities. Though natural dyes are obtained from cultivated plants for large scale use, the plant/animal species can be preserved by biodiversity conservation. To make the raw material renewable, industries have to tie-up with farmers to grow the required plants eg. *Indigofera tinctoria*. The waste from natural dyes after extraction can be reused into derivative materials like animal feed, biogas and compost, fulfilling the concept of the 3R – reduce, reuse and recycle. Use of bio-mordants can ensure sustainability throughout the production process of natural dyeing. Two case studies report that the use of natural dyes would reduce additives by 50% and energy by 15% [97]; further, the adoption of plasma technology would eliminate the drying process and bring about reduction in energy usage [98].

Natural dye production will provide opportunities for entrepreneurship and poverty reduction, maintain cultural heritage and competitiveness, and guarantee industrial workers and consumers health and safety [99]. Natural dyes on natural fabrics would be a worthy combination for the eco-friendly trend. A joint effort by fashion designers, brand owners, natural dye producers and traders can help in promoting natural dyed textiles and apparels to build a green and sustainable industry.

6.2 Roadmap for Natural Dyeing (Table 5)

To propagate natural dyeing in the global market, research is required in all facets right from the sourcing of raw materials to the end product. Data is available about the existing sources of natural dyes, but we may not know if it will be feasible when undertaken at our end. Research on new natural resources for the identification of new dyes and detection of new dyes from waste materials may be the initial steps. While discussing waste materials, the resources can be by-products of agricultural production, for example, the remains or wastes of the food and beverage industry and timber trade. It is also advisable to explore the efficacy of bacteria-derived pigments, biomass-based pigments and bio-colourants. All dyes have components that produce colour; efforts are directed towards lab replication of natural colour components like chlorophyll, carotenoids, flavonoids and betalains. It has been reported that bacteria-derived pigments from Monascus species (Monascus sanguineus and Monascus purpureus) are capable of producing natural red pigment [100]; carotenoid has been obtained from *Phaffia rhodozyma*, *Micrococcus roseus*, *Brevibaacterium linens* and *Bradyrhizobium*, which is a yellow to orange colour [101, 102]. A study [103] reveals that enzyme *Aspergillus niger* was employed to convert tea polyphenols into precursor pigment for natural dyeing, leading to better dye uptake and fixation. Use of biomass pigments by bioconversion leads to ecological textiles.

Table 5 Roadmap for natural dyeing

Research focus				
Raw materials	Identification of dyes from new natural resources Use of waste materials for dye extraction	Bacteria-derived pigments Biomass-based pigments Studies on bio-colourants	By-products of agricultural production – food, timber and the beverage industries as raw materials for dye production	Lab-replicated natural colour components like chlorophyll, carotenoids, flavonoids and betalains
Colour	Colour engineering and combinations	New mix and match combos	Natural and special looks	Standardizing the recipe
Extraction and optimization	Extraction methodology of natural dyes Experimentation of optimum conditions for natural dyeing	Enzyme-extracted natural colours Enhanced extraction for more output Extraction by sonification	Extraction and reuse of unused dyes from natural dyeing wastewater New technologies for colour yield + surface modification of fabric	Ultrasonic, microwave, gamma and ultraviolet radiations
Assistive materials and technologies	Selection of the right mordant	Right method for the required colour	Biomordants and hyperaccumulators as replacement for metal mordants	Agriculture and industrial biotechnology and enzyme technology

(continued)

Table 5 (continued)

Assessment and evaluation	Colour identification and new colour coding	Optimization methods for extraction of the dye and application methods	Pre- and posttreatment for natural dyeing; colour fixative methods; testing for colour fastness	Techno-economic studies; life cycle analysis; sustainability studies
Role of stakeholders	Increasing consumer awareness for natural and sustainable products – preference for organic products – willingness to pay and procure – advent of digital marketing	Research propagation for natural dyes – sources, combination of colourants, new process sequence for creating new colours, trying mixing of colourants or process sequence from different sources, reliability and validity of processes – research lab trials to commercialization	Joint pacts between natural dye producers and traders; fashion designers, manufacturers and brand owners; forecasters, trend analysts, design houses, buying houses, exporters and importers, global market development	Government subsidy for natural dye research through the Indian Knowledge System (IKS); create data bank and documentation of research findings for natural dyeing; reporting and marketing of the findings for global use; development of domestic, export, international, multinational and global markets
Futuristic	Biotechnology-based agriculture with tools like cell and tissue cultures, genetic engineering – as replacement for conventional growing techniques	Natural dye waste reused into derivative materials like animal feed, biogas and compost Plasma technology can reduce energy consumption	Increase the number of natural dye sources by biotechnological tools like gene manipulation Modern methods for extraction, optimization, reduction in cost, energy, water and chemicals with better output	Enzyme technology – extraction of dye, pretreatment of fabric for better dye uptake and fastness properties; ultrasonic dyeing with natural dyes
Laboratory to market/consumer				

New colours and combinations can create an ocean of looks for different applications. This involves research for optimizing the combinations with greater advantages. New technologies like ultrasonic, microwave, gamma and ultraviolet radiations are being used for extraction of colour from the resource along with enzyme technology and plasma treatment. These provide immense opportunities for enhanced extraction and output. Bio-mordants and hyperaccumulators can serve as replacements for conventional metal mordants. All these efforts can be evaluated with techno-economic studies, life cycle assessments and sustainability estimations. Stakeholders play an important role as awareness and judicious consumption form the basis for sustainable efforts on part of the design, manufacture and retail teams at regional, national and global scales.

7 Conclusion

Many reports and estimates have stated that the global market for natural dyes and eco-textiles is growing by leaps and bounds. Awareness of the adverse effects of synthetic dyes and enhanced health consciousness have increased the demand for natural textiles and the adoption of natural methods of manufacture. The massive use of natural reserves can be avoided by the adoption of biotechnology tools for enhanced production of natural dyes with minimal resources. This can be taken forward with assistive technologies like enzyme treatments, plasma technologies and radiation methods. The revival of traditional cultures together with harmonious living with natural colours can bring about communal and social transformations. Openings are in plenty for small and medium enterprises, entrepreneurs, agriculturalists, textile traders and manufacturers. This creates a huge unexplored space to be researched as nature has unlimited possibilities which can be tapped to create a greener and healthier world.

To conclude we have a few quotes [104–106]:

> Mere colour, unspoiled by meaning and unallied with definite form can speak to the soul in a thousand different ways.
>
> —Oscar Wilde

> I am of the African race and in the colour which is natural to them of the deepest dye; and it is under a sense of the most profound gratitude to the Supreme Ruler of the Universe.
>
> —Benjamin Banneker

> We don't go 'NATURAL', We return. 'Natural, is where it began'.
>
> —Unknown

References

1. GVR Inc. (2023). *Textile market size, share & trends analysis report by raw material (cotton, wool, silk, chemical), by product (natural fibers, nylon), by application (technical, fashion), by region, and segment forecasts, 2022 – 2030*. https://www.grandviewresearch.com/industry-analysis/textile-market#:~:text=The%20global%20textile%20market%20size,4.0%25%20from%202022%20to%202030. Accessed 4 Dec 2022.
2. Bai, R., Yu, Y., Wang, Q., Yuan, J., & Fan, X. (2016). Effect of laccase on dyeing properties of polyphenol-based natural dye for wool fabric. *Fibers and Polymers, 17*, 1613–1620.
3. Setiadi, T., Andriani, Y., & Erlania, M. (2006). Treatment of textile wastewater by a combination of anaerobic and aerobic pro-cesses: A denim processing plant case. In S. Ohgaki, K. Fukushi, H. Katayama, S. Takizawa, & C. Polprasert (Eds.), *Southeast AsianWater environment 1* (pp. 159–166). IWA Publishing, Bangkok, Thailand.

4. Setiadi, T., Andriani, Y., & Erlania, M. (2006). Treatment of textile wastewater by a combination of anaerobic and aerobic processes: A denim processing plant case. In S. Ohgaki, K. Fukushi, H. Katayama, S. Takizawa, & C. Polprasert (Eds.), *Southeast Asian Water environment 1: Selected papers from the first International symposium on Southeast Asian Water environment (biodiversity and water environment), Bangkok, Thailand, October 2003* (pp. 159–166). IWA Publishing.
5. Berradi, M., Hsissou, R., Khudhair, M., Assouag, M., Cherkaoui, O., El Bachiri, A., & El Harfi, A. (2019). Textile finishing dyes and their impact on aquatic environs. *Heliyon, 5*(11), e02711.
6. Sharma, B., Dangi, A. K., & Shukla, P. (2018). Contemporary enzyme based technologies for bioremediation: A review. *Journal of Environmental Management, 210*, 10–22.
7. Khan, S., & Malik, A. (2018). Toxicity evaluation of textile effluents and role of native soil bacterium in biodegradation of a textile dye. *Environmental Science and Pollution Research - International, 25*(5), 4446–4458.
8. Nikfar, S., & Jaberidoost, M. (2014). Dyes and colorants. Reference module in biomedical sciences. In *Encyclopedia of toxicology* (Third ed., pp. 252–261). Elsevier.
9. Ajmal, A., Majeed, I., Malik, R. N., Idriss, H., & Nadeem, M. A. (2014). Principles and mechanisms of photocatalytic dye degradation on TiO_2 based photocatalysts : A comparative overview. *RSC Advances, 4*, 37003–37026. https://doi.org/10.1039/c4ra06658h. https://www.academia.edu/10156547/Principles_and_mechanisms_of_photocatalytic_dye_degradation_on_TiO2_based_photocatalysts_a_comparative_overview
10. Shamey, R., & Zhao, X. (2014). *Modelling, simulation and control of the dyeing process*. Elsevier.
11. Wardman, R. H. (2017). *An introduction to textile coloration: Principles and practice*. Wiley.
12. Hassan, M. M., & Carr, C. M. (2018). A critical review on recent advancements of the removal of reactive dyes from dyehouse effluent by ion-exchange adsorbents. *Chemosphere, 209*(1), 201–219. https://doi.org/10.1016/j.chemosphere.2018.06.043
13. Leena, R., & Selva, R. D. (2008). Bio-decolourization of textile effluent containing reactive black-B by effluent-adapted and non-adapted bacteria. *African Journal of Biotechnology, 7*(18), 3309–3313.
14. Siddiqui, S. I., Fatima, B., Tara, N., Rathi, G., & Chaudhry, S. A. (2019). *Recent advances in remediation of synthetic dyes from wastewaters using sustainable and low-cost adsorbents* (pp. 471–507). Woodhead Publishing. https://doi.org/10.1016/B978-0-08-102491-1.00015-0
15. Brock, T., Groteklaes, M., & Mischke, P. (2000). *European coatings handbook*. Vincentz Network GmbH & Co KG.
16. Christie, R. M. (2001). *Colour chemistry*. Royal Society of Chemistry.
17. Rehman, K., Shahzad, T., Sahar, A., Hussain, S., Mahmood, F., & Siddique, M. H. (2018). Effect of Reactive Black 5 azo dye on soil processes related to C and N cycling. *PeerJ, 6*, e4802.
18. Imran, M., Crowley, D. E., Khalid, A., Hussain, S., Mumtaz, M. W., & Arshad, M. (2015). Microbial biotechnology for decolorization of textile wastewaters. *Reviews in Environmental Science and Biotechnology, 14*(1), 73–92.
19. Rawat, D., Mishra, V., & Sharma, R. S. (2016). Detoxification of azo dyes in the context of environmental processes. *Chemosphere, 155*, 591–605.
20. Copaciu, F., Opriş, O., Coman, V., Ristoiu, D., Niinemets, Ü., & Copolovici, L. (2013). Diffuse water pollution by anthraquinone and azo dyes in environment importantly alters foliage volatiles, carotenoids and physiology in wheat (Triticum aestivum). *Water, Air, and Soil Pollution, 224*(3), 1478.
21. Vargas, A. M. M., Paulino, A. T., & Nozaki, J. (2009). Effects of daily nickel intake on the bio-accumulation, body weight and length in tilapia (Oreochromis niloticus). *Toxicological and Environmental Chemistry, 91*(4), 751–759.
22. Aquino, J. M., Rocha-Filho, R. C., Ruotolo, L. A., Bocchi, N., & Biaggio, S. R. (2014). Electrochemical degradation of a real textile wastewater using β-PbO2 and DSA® anodes. *Chemical Engineering Journal, 251*(2014), 138–145.

23. Sandhya, S. (2010). Biodegradation of azo dyes under anaerobic condition: Role of azoreductase. In H. A. Erkurt (Ed.), *Biodegradation of azo dyes. The handbook of environmental chemistry* (Vol. 9, pp. 39–57). Springer.
24. Newman, M. C. (2015). *Fundamentals of ecotoxicology: The science of pollution*. CRC Press.
25. Vikrant, K., Giri, B. S., Raza, N., Roy, K., Kim, K. H., Rai, B. N., et al. (2018). Recent advancements in bioremediation of dye: Current status and challenges. *Bioresource Technology, 253*(2018), 355–367.
26. Ito, T., Adachi, Y., Yamanashi, Y., & Shimada, Y. (2016). Long–term natural remediation process in textile dye–polluted river sediment driven by bacterial community changes. *Water Research, 100*, 458–465.
27. Christie, R. M. (2007). *Environmental aspects of textile dyeing*. Elsevier.
28. Clark, M. (Ed.). (2011). *Handbook of textile and industrial dyeing: Principles, processes and types of dyes*. Elsevier.
29. Hunger, K. (2003). *Industrial dyes: Chemistry, properties and applications*. Willey-VCH.
30. Thakur, I. S. (2006). *Environmental biotechnology: Basic concepts and applications*. I. K. International Pvt.
31. Tiwari, S., Tripathi, A., & Gaur, R. (2016). Bioremediation of plant refuges and xenobiotics. In R. L. Singh (Ed.), *Principles and applications of environmental biotechnology for a sustainable future* (pp. 85–142). Springer Science.
32. Li, H., Zhang, R., Tang, L., Zhang, J., & Mao, Z. (2014). Evaluation of Bacillus sp. MZS10 for decolorizing Azure B dye and its decolorization mechanism. Journal of the. *Environmental Sciences, 26*(5), 1125–1134.
33. Petzer, B. H., Harvey, G., & Wegener, J. P. P. (2012). Azure B, a metabolite of methylene blue, is a high-potency, reversible inhibitor of monoamine oxidase. *Toxicology and Applied Pharmacology, 258*(3), 403–409.
34. Couto, N., Wood, J., & Barber, J. (2016). The role of glutathione reductase and related enzymes on cellular redox homoeostasis network. *Free Radical Biology & Medicine, 95*, 27–42.
35. Cyprus, J. (2020). *What makes a product sustainable?* https://earth911.com/business-policy/what-makes-a-product-sustainable/. Accessed 15 Dec 2022.
36. Team H. (2022). *What is a sustainable product?* The Complete Guide. https://hivebrands.com/blogs/news/what-is-a-sustainable-product. Accessed 15 Dec 2022.
37. Gauntlett, E. (2021). *What makes a product sustainable?* https://bowercollective.com/blogs/news/what-makes-a-product-sustainable. Accessed 15 Dec 2022.
38. Molinare, J. C., Korre, A., & Durucan, S. (2014). 24th European symposium on computer aided process engineering. *Computer Aided Chemical Engineering, 33*(2014), 1861–1866.
39. Gulzar, T., Farooq, T., Kiran, S., Ahmad, I., & Hameed. (2019). *Green chemistry in the wet processing of textiles* (pp. 1–20). Woodhead Publishing. The Impact and Prospects of Green Chemistry for Textile Technology.
40. Iclisoy, J. (2012). *Green chemistry in textile*. https://textilevaluechain.in/in-depth-analysis/articles/textile-articles/green-chemistry-in-textile/. Accessed 18 Dec 2022.
41. Kiron, M. I. (2022). *Sustainability in textile processing*. https://textilelearner.net/sustainability-in-textile-processing/. Accessed 29 Jan 2023.
42. Preša, P., & Tavčer, P. F. (2009). Low water and energy saving process for cotton pretreatment. *Textile Research Journal, 79*(1), 76, 13p–88.
43. Lange, N. K. (2000). Biopreparation in action. *International Dyer, 185*(2), 18–22.
44. Pawar, S. B., Shah, H. D., & Andhorika, G. R. (2002). Man-made text India, *45*(4), 133.
45. Sundar, P. S., Bhatoye, S. K., Karthikeyan, N., & Prabhu, K. H. (2007). Enzyme applications in textiles. *Indian Text Journal, 117*(8), 25–31.
46. Karmakar, S. R. (1998). Application of biotechnology in the pre-treatment process of textiles. *Colourage Annual, 45*, 75–86.
47. Zia, M. A., Sheikh, M. A., & Khan, I. A. (2010). Chemically treated strain improvement of Aspergillus niger for enhanced production of glucose oxidase. *International Journal of Agriculture and Biology, 12*(6), 964–966.

48. Du, L. N., Wang, S., Li, G., Wang, B., Jia, X. M., Zhao, Y. H., & Chen, Y. L. (2011). Biodegradation of malachite green by Pseudomonas sp. strain DY1 under aerobic condition: Characteristics, degradation products, enzyme analysis and phytotoxicity. *Ecotoxicology, 20*, 438–446.
49. Galante, Y. M., & Formantici, C. (2003). Enzyme applications in detergency and in manufacturing industries. *Current Organic Chemistry, 7*(13), 1399–1422.
50. Barreca, A. M., Fabbrini, M., Galli, C., Gentili, P., & Ljunggren, S. (2003). Laccase/mediated oxidation of a lignin model for improved delignifi cation procedures. *Journal of Molecular Catalysis B: Enzymatic, 26*, 105–110.
51. Riva, S. (2006). Laccases: Blue enzymes for green chemistry. *Trends in Biotechnology, 24*(5), 219–226.
52. Dictionary.com. (2022). *Mordant*. https://www.dictionary.com/browse/mordant. Accessed 2 Feb 2023.
53. Roy Choudhury, A. K. (2014). Sustainable textile wet processing: Applications of enzymes, roadmap to sustainable textiles and clothing. In S. S. Muthu (Ed.), *Textile science and clothing technology series* (pp. 203–238). Springer Science+ Business Media Singapore. https://doi.org/10.1007/978-981-287-065-0_7
54. Boiral, O., & Roy, M. J. (2007). ISO 9000: Integration rationales and organizational impacts. *International Journal of Operations & Production Management, 27*(2), 226–247.
55. Saxena, S., & Raja, A. S. M. (2014). Natural dyes: Sources, chemistry, application and sustainability issues. In *Roadmap to sustainable textiles and clothing: Eco-friendly raw materials, technologies, and processing methods* (pp. 37–80).
56. Burkinshaw, S. M., & Kumar, N. (2009). The mordant dyeing of wool using tannic acid and $FeSO_4$, Part 1: Initial findings. *Dyes and Pigments, 80*(1), 53–6.
57. Yusuf, M., Ahmad, A., Shahid, M., Khan, M. I., Khan, S. A., Manzoor, N., & Mohammad, F. (2012). Assessment of colorimetric, antibacterial and antifungal properties of woollen yarn dyed with the extract of the leaves of henna (Lawsonia inermis). *Journal of cleaner production, 27*, 42–50.
58. Zheng, G. H., Fu, H. B., & Liu, G. P. (2011). Application of rare earth as mordant for the dyeing of ramie fabrics with natural dyes. *Korean Journal of Chemical Engineering, 28*(11), 2148–2155.
59. Li, Y. V., Malensek, N., Sarkar, A. K., & Xiang, C. (2016). *Clothing and Textiles Research Journal*. https://doi.org/10.1177/0887302X16647124
60. Shahid-ul-Islam, Shahid, M., & Mohammad, F. (2013). Perspectives for natural product based agents derived from industrial plants in textile applications – A review. *Journal of Cleaner Production, 57*, 2–18.
61. Prabhu, K. H., & Teli, M. D. (2011). Eco-dyeing using *Tamarindus indica* L. seed coat tannin as a natural mordant for textiles with antibacterial activity. *Journal of Saudi Chemical Society, 1*, 1. https://doi.org/10.1016/j.jscs.2011.10.014
62. Rane, S., Hate, M., Hande, P., Ajitkumar, B. S., & Datar, A. (2017). Dyeing of cotton with Tectona grandis leaves and Terminalia arjuna bark extracts. *International Journal of Textile Science, 6*(2), 72–77.
63. Haji, A. (2019). Natural dyeing of wool with henna and yarrow enhanced by plasma treatment and optimized with response surface methodology. *Journal of the Textile Institute*, 1–9.
64. Reeves, R. D. (2003). Tropical hyperaccumulators of metals and their potential for phytoextraction. *Plant and Soil, 249*, 57–65.
65. van der Ent, A., Erskine, P. D., & Sumail, S. (2015). Ecology of nickel hyperaccumulator plants from ultramafic soils in Sabah (Malaysia). *Chemoecology, 25*, 243–259.
66. Cunningham, A. B., et al. (2011). Hanging by a thread: Natural metallic mordant processes in traditional Indonesian textiles. *Economic Botany, 65*, 241–259. https://doi.org/10.1007/s12231-011-9161-4
67. Oda, H. (2012). Improvement of light fastness of natural dye: effect of ultraviolet absorbers containing benzotriazolyl moiety on the photofading of red carthamin. *Coloration Technology, 128*(2), 108–113.

68. Mansour, H. F., & Heffernan, S. (2011). Environmental aspects on dyeing silk fabric with sticta coronata lichen using ultrasonic energy and mild mordants. *Clean Technologies and Environmental Policy, 13*, 207–213.
69. Foo, K. Y., & Hameed, B. H. (2010). Insights into the modeling of adsorption isotherm systems. *Chemical engineering journal, 156*(1), 2–10.
70. Mansour, H. F., & Heffernan, S. (2011). Environmental aspects on dyeing silk fabric with Sticta coronate lichen using ultrasonic energy and mild mordants. *Clean Technologies and Environmental Policy, 13*, 207–213.
71. Adeel, S., Amin, N., Fazal-ur-Rehman, T. A., Batool, F., & Hassan, A. (2020). *Sustainable isolation of natural dyes from plant wastes for textiles*. Chapter June 2020. https://doi.org/10.1002/9781119620532.ch17
72. Adeel, S., Naseer, K., Javed, S., Mahmmod, S., Tang, R. C., Amin, N., & Naz, S. (2020). Microwave-assisted improvement in dyeing behavior of chemical and bio-mordanted silk fabric using safflower (Carthamus tinctorius L) extract. *Journal of Natural Fibers, 17*(1), 55–65.
73. Adeel, S., Rehman, F. U., Hameed, A., Habib, N., Kiran, S., Zia, K. M., & Zuber, M. (2018). Sustainable extraction and dyeing of microwave-treated silk fabric using arjun bark colorant. *Journal of Natural Fibers, 17*, 1–14.
74. Adeel, S., Zia, K. M., Abdullah, M., Rehman, F. U., Salman, M., & Zuber, M. (2018). Ultrasonic assisted improved extraction and dyeing of mordanted silk fabric using neem bark as source of natural colourant. *Natural Product Research, 33*(14), 2060–2072.
75. Dawson, T. L. (2009). Biosynthesis and synthesis of natural colours. *Coloration Technology, 125*, 61–73.
76. Kim, D. H., Kim, J. H., Bae, S. E., Seo, J. H., Oh, T. K., & Lee, C. H. (2005). Enhancement of natural pigment extraction using Bacillus species xylanase. *Journal of Agricultural and Food Chemistry, 53*, 2541–2545.
77. Cho, Y. J., Kim, S. Y., Kim, J., Choe, E. K., Kim, S. I., & Shin, H. J. (2006). One-step enzymatic synthesis of blue pigments from geniposide for fabric dyeing. *Biotechnology and Bioprocess Engineering, 11*, 230.
78. Lu, Y., Wang, L., Xue, Y., Zhang, C., Xing, X.-H., Lou, K., Zhang, Z., Li, Y., Zhang, G., & Bi, J. (2009). Production of violet pigment by a newly isolated psychrotrophic bacterium from a glacier in Xinjiang, China. *Biochemical Engineering Journal, 43*, 135–141.
79. Cho, Y. J., Park, J. P., Hwang, H. J., Kim, S. W., Choi, J. W., & Yun, J. W. (2002). Production of red pigment by submerged culture of Paecilomyces sinclairii. *Letters in Applied Microbiology, 35*, 195–202.
80. Punrattanasin, N., et al. (2015). Silk fabric dyeing with natural dye from mangrove bark (Rhizophora apiculata Blume) extract. *Industrial Crops and Products, 49*, 122–129.
81. Shabbir, M., Islam, S. U., Bukhari, M. N., Rather, L. J., Khan, M. A., & Mohammad, F. (2017). Application of Terminalia chebula natural dye on wool fiber—evaluation of color and fastness properties. *Textiles and Clothing Sustainability, 2*, 1–9.
82. Bechtold, T., Mahmud-Ali, A., & Mussak, R. A. (2007). Reuse of ash-tree (Fraxinus excelsior L.) bark as natural dyes for textile dyeing: Process conditions and process stability. *Coloration Technology, 123*, 271–279.
83. Prusty, A. K., Das, T., Nayak, A., & Das, N. B. (2010). Colourimetric analysis and antimicrobial study of natural dyes and dyed silk. *Journal of Cleaner Production, 18*, 1750–1756.
84. Mongkholrattanasit, R., Krystufek, J., & Wiener, J. (2010). Dyeing and fastness properties of natural dyes extracted from eucalyptus leaves using padding techniques. *Polymer Fiber, 11*, 346–350.
85. Mussak, R. A. M., & Bechtold, T. (2009). Natural colorants in textile dyeing. In T. Bechtold & R. A. M. Mussak (Eds.), *Handbook of natural colorants* (pp. 315–337). Wiley, Chichester, UK.
86. Sivakumar, V., Anna, J. L., Vijayeeswarri, J., & Swaminathan, G. (2009). Ultrasound assisted enhancement in natural dye extraction from beetroot for industrial applications and natural dyeing of leather. *Ultrasonics Sonochemistry, 16*, 782–789.
87. Kumar, S., Singh, J., Nanoti, S. M., & Garg, M. O. (2012). A comprehensive life cycle assessment (LCA) of Jatropha biodiesel production in India. *Bioresource Technology, 110*, 723–729.

88. Ajayebi, A., Gnansounou, E., & Raman, J. K. (2013). Comparative life cycle assessment of biodiesel from algae and jatropha: A case study of India. *Bioresource Technology, 150*, 429–437.
89. Anandhan, M., & Prabaharan, T. (2018). Environmental impacts of natural dyeing process using pomegranate peel extract as a dye. *International Journal of Applied Engineering Research, 13*(10), 7765–7771.
90. Roy Choudhury, A. K. (2014). Environmental impacts of the textile industry and its assessment through life cycle assessment. In S. Muthu (Ed.), *Roadmap to sustainable textiles and clothing - environmental and social aspects of textiles and clothing supply chain* (pp. 1–39). Springer.
91. Nieminen, E., Linke, M., Tobler, M., & Vander Beke, B. (2007). EU COST Action 628: Life cycle assessment (LCA) of textile products, ecoefficiency and definition of best available technology (BAT) of textile processing. *Journal of Cleaner Production, 15*(13–14), 1259–1270.
92. Vanessa, P., Sandrine, P., Anne, P., & Nemeshwaree, B. (2012). Comparison of methodologies for Lca processes: Application to the dyeing of cotton. In *Proc. 2nd LCA Conf.*, no. November, pp. 6–7.
93. Linharesa, T., & Pessoa de Amorima, M. T. (2017). LCA of textile dyeing with Acacia Dealbata tree bark: A case study research. *Procedia Engineering, 200*(2017), 365–369.
94. Arena, M., Ciceri, N. D., Terzi, S., Bengo, I., Azzone, G., & Garetti, M. (2009). A state-of-the-art of industrial sustainability: Definitions, tools and metrics. *International Journal of Product Lifecycle Management, 4*(1–3), 207–251.
95. Raju, A. P. (2013). *A value chain in Natural Dyes*. Final Report (India: National Agricultural Innovation Project) Indian Council of Agricultural Research.
96. Senthilkumar, R. P., Bhuvaneshwari, V., Sathiyavimal, S., Amsaveni, R., Kalaiselvi, M., & Malayaman, V. (2015). Natural colours from dyeing plants for textiles. *International Journal of Biosciences and Nanosciences, 2*(7), 160–174.
97. Stegmaier, T., Linke, M., Dinkelmann, A., Von Arnim, V., & Planck, H. (2009). Environmentally friendly plasma technologies for textiles. In *Sustainable textiles* (pp. 155–178). Woodhead Publishing.
98. Angelis-Dimakis, A., Alexandratou, A., & Balzarini, A. (2016). Value chain upgrading in a textile dyeing industry. *Journal of Cleaner Production, 138*, 237–247.
99. Elsahida, K., et al. (2019). Sustainability of the use of natural dyes in the textile industry. *IOP Conference Series: Earth and Environmental Science, 399*(2019), 012065. https://doi.org/10.1088/1755-1315/399/1/012065. IOP Publishing.
100. Dikshit, R., & Tallapragada, P. (2013). Comparative study of Monascus sanguineus and Monascus purpureus for red pigment production under stress condition. *International Food Research Journal, 20*, 1235–1238.
101. Vazquez, M., Santos, V., & Parajo, J. C. (1997). Effect of the carbon source on the carotenoid profiles of Phaffia rhodozyma strains. *Journal of Industrial Microbiology & Biotechnology, 19*, 263–268.
102. Chattopadhyay, M. K., Jagannadham, M. V., Vairamani, M., & Shivaji, S. (1997). Carotenoid pigments of an antarctic psychrotrophic bacteriumMicrococcus roseus: Temperature dependent biosynthesis, structure, and interaction with synthetic membranes. *Biochemical and Biophysical Research Communications, 239*, 85–90.
103. Gong, J., Wang, F., Ren, Y., Li, Z., Zhang, J., & Li, Q. (2018). Preparation of biomass pigments and dyeing based on bioconversion. *Journal of Cleaner Production, 182*. https://doi.org/10.1016/j.jclepro.2018.01.219
104. http://www.picturequotes.com/i-am-of-the-african-race-and-in-the-colour-which-is-natural-to-them-of-the-deepest-dye-and-it-is-quote-654634 (quotes for conclusion).
105. https://www.pinterest.com/KinksandChaos/ (quotes for conclusion).
106. https://www.pinterest.ie/vshirran/inspirational-quotes/ mere colour unspoiled (quotes for conclusion).
107. Shahid, M., & Mohammad, F. (2013). Recent advancements in natural dye applications: A review. *Journal of Cleaner Production, 53*, 310–331.

Classification of Natural Dyes for Sustainable Exploitation

Jamiu Mosebolatan Jabar

Abstract Synthetic dyes have often been used as coloring agents by many industries since the mid-nineteenth century. Nevertheless, the recent campaign for ecological sustainability and conservation is now reviving the interest of the global community in utilizing natural dyes in all sectors. Natural dyes can be isolated from mineral ore, animals, and plants. Each of the colorants obtained from these sources has a unique chemical structure, color, and application method. Plant-based natural dyes are the most abundant and sustainable dyes belonging to all classes of colorants, ranging from indigoids to terpenoids, quinonoids, and phenolics. Animal-based natural dyes are commonly found as quinonoids, while the mineral ore-based dyes are inorganic (metallic) colorants. Natural dyes are less toxic, biodegradable, sustainable, renewable, and cleaner water-soluble coloring substances for pharmaceuticals, foods, cosmetics, electronics, leathers, textiles, etc. Besides the adorable bright colors ranging from yellow to orange, red, blue, green, and black, natural dyes have functional properties that account for their general acceptability. The challenges facing natural dyes, which are low shade, uniformity, and color-anchor, are now being circumvented through the involvement of the mordanting agents during the application process. Frequently used metal salts as color-anchors are hazardous and have detrimental effects on the environment and health of living beings when they are discharged as wastewater. Therefore, for achieving green application of natural dyes, electrolyte mordants must be replaced with herbal-anchors that have potential color fastening, intensifying and new color-generating properties. The categorization of natural dyes in accordance with their sources, structures, applications, and colors has been discussed in this chapter.

Keywords Color-anchors · hazardous · natural dyes · sustainable · wastewater

---

J. M. Jabar (✉)
Textile and Polymer Research Laboratory, Chemistry Department, the Federal University of Technology, Akure, Nigeria
e-mail: jmjabar@futa.edu.ng

S. S. Muthu (ed.), *Natural Dyes and Sustainability*, Sustainable Textiles: Production, Processing, Manufacturing & Chemistry,
https://doi.org/10.1007/978-3-031-47471-2_8

1 Introduction

Natural dyes can simply be described as water-soluble colorants obtained from nature, which are sources of colors for human beneficiation. Human beings have started loving colors since creation and will continue to love them until living ceases [1]. As a result of the intimacy between man and colors, it can be said that mankind is the regular friend of colors and colors turn out to become human comforters [2]. The light which human beings see through is a combination of several colors; white light, in particular, is a combination of seven different colors [3].

The natural inorganic red ochre (Fe_2O_3) was the coloring matter first known and used during pre-historical times for coloring substrates. Later (about 15,000 BC), a number of natural inorganic colorants such as clay, carbon black, red, and yellow ochre were ground together at varying proportions (depending on the desired color), dissolved in water, and applied with brushes for coloration of substrates. The first organic colorant named pitch was used by Noah as an ark surface–coating agent [4].

The Egyptian civilization (about 5000 BC) came with massive utilization of natural organic colorants for the beatification of clothes and other substrates. The clothes colored with indigo leaf dye extract (the foremost known plant-derived dye) were discovered in the crypts of ancient Egypt. Another natural dye discovered in the Middle East identified as Tyrian purple (a derivative of indigo dye) was obtained from mollusk species and used for dyeing royal clothes. These marked the beginning of natural dye exploration and exploitation in all sectors of life. Prior to the nineteenth century, natural dyes were used for coloring foods, drinks, leather, cosmetics, clothes, etc., and many used these naturally extracted substances as drugs. After that era, the natural dyes' general acceptance faded away due to the discovery of man-made dyes that were considered to be superior [5].

Ecological attentiveness has rekindled the lost glory of natural dyes to synthetic dyes due to the peculiar eco-demerits associated with synthetic dyes. Several research studies have been done and are still being carried out to circumvent the degradation of the environment from hazardous synthetic dyes by their partial or complete replacement with natural dyes. The reinforcement of the natural dyes enhances their attributes to meet the qualities desired in domestic and industrial coloring applications. Such reinforcements were done through the involvement of modern heating tools in the extraction process and the exploration of mordanting agents in the application process [6].

Overall, natural colorants can be broadly classified into two, namely schemochromes and biochromes. The schemochromes are optical phenomena of the colorants that are based on the monochromatization of incident light by the structural cells or tissues (chromatophores) of the specific species of animals. Unfortunately, this class of colorants can't easily be extracted. A number of specific groups of animals developed variable schemochrome purposely for some of the reasons highlighted below:

- Luring the opposite sex for mating
- Warning one another against perceived danger

- Hiding from predators
- Adapting to environmental change and others

The process of altering the color appearance is stimulated by motor neurons and relies on the animals' developed cells and tissues called chromatophores. Some of the animals with such traits and their appearances are presented in Table 1. Others that are not presented in the table are certain species of snakes, grasshoppers, zebra, and human hair to mention a few [7].

The biochromes are the biochemical compounds (mostly secondary metabolites) in the cells of the sources of the colorants that interact with the electromagnetic spectrum of the visible light and transmit or/and reflect the unabsorbed (complementary) component(s) in the form of color sensation. Although some species of

Table 1 Adorable color appearance in the animal realm

Animal	Visual color	Associated biochemical compounds
 Cock	Muticolor	Iridophore, xanthophone
 Butterfly	Wings' adorable color	Scale structure
 Chameleon	Camouflage	Melanophore, xanthophore
 Crustacean	Humidity and temperature-dependent color	Astaxanthin (carotenoid)
 Bullfrog	Humidity and temperature-dependent color	Melanophore, xanthophore
 Agama lizard	Orange, white, and black	Lipochrome (carotenoid), pheomelanin and eumelanin

animals are also found in this group of colorants, virtually all plants belong to this group and the colors on these living things are easily extractable in aqueous medium and other suitable solvents [8].

This chapter will discuss biochromes (extractable natural dyes) in detail, under the following headings:

(i) Sources
(ii) Chemical structures
(iii) Methods of application
(iv) Colors

2 Natural Dyes Classification

Although there exist many ways of classifying natural dyes, in this chapter, they are grouped based on: (i) sources, (ii) chemical structures, (iii) methods of application, and (iv) colors [9].

3 Natural Dyes Source-Based Classification

Up to the present moment, only inorganic/mineral ores, certain species of animals, and nearly all plants have been identified as origins of natural dyes. Each of the origins is discussed comprehensively in this chapter.

3.1 Natural Dyes from Inorganic/Mineral Ores

The dyes obtained from this source are called mineral-based natural dyes; a few members of this group of dyes are water-soluble inorganic colorants. They were the first set of colorants discovered and used for aesthetic purposes by man [10]. Some of the dyes from this source, their colors, and chemical formulae are listed in Table 2.

3.2 Natural Dyes from Specific Species of Animals

Animal-based dyes were obtained from certain species of insects and mollusks. For instance, archeological evidence showed that red-hued kermes was discovered from dried bodies of female insects called *Kermes ilicis*. Natural dyes obtained from other insects are cochineal that was extracted from female insects called *Coccus ilicis* and *Dactylopius coccus*, and laccaic acid from lac insect (*Kerria lacca*) [11].

Table 2 Mineral-based natural dyes

Dye	Visual color	Chemical formula/structure
Ochre	Red/yellow	$Fe_2O_3 \cdot nH_2O$
Clay	Red/orange	$Al_2O_3.2SiO_2 \cdot 2H_2O$
Red lead	Red	Pb_3O_4
Ruby sulfur	Yellow	As_2S_3
Talc	White	Ti_2O
Chalk	White	$CaCO_3$
Ultramarine-blue	White	$Na_5Al_3Si_3SO_{12}$
Zinc-white	White	ZnO
Lead-white	White	$PbCO_3$
Verdigris	Green	$Cu(CH_3COO)_2$
Malachite	Green	$Cu_2CO_3(OH)_2$
Ultramarine-blue	Green	$Na_5Al_3Si_2S_2O_{12}$
Ultramarine-blue	Blue	$Na_5Al_3Si_2S_3O_{12}$
Bone-black	Black	C and other elements in trace quantity
Graphite	Black	C and other elements in trace quantity
Charcoal-black	Black	C and other elements in trace quantity
Lamp-black	Black	C and other elements in trace quantity

The color of the cochineal is crimson or scarlet, depending on the source insect species. Its chemical structure is closer to kermes with the exception of the presence of glycosides in its structure, while laccaic acid inserts a dark red hue onto the substrate [10].

The chemical structure of the purple-colored natural dye called Tyrian purple obtained from different species of mollusks (snails) is akin to that of the first discovered natural organic dye called indigo with the exception of the presence of bromine at positions 6 and 6′ in its structure. Some of the mollusk species that yield Tyrian purple are *Murex brandaris, Murex trunculus, Purpura haemastoma, Purpura aperta,* and *Nucella lapillus*. Table 3 shows the source, class, color, and chemical structures of animal-derived natural dyes. Natural dyes obtained from the same group of animals (irrespective of their species) belong to the same class as displayed in the table. For instance, all insect-based natural dyes belong to the anthraquinone class of dyes, while those obtained from mollusks do belong to the same class called indigoid [9].

3.3 Natural Dyes from Plants

Most of the vegetables are dye-yield materials and, fantastically, dyes can be obtained from all their parts. As a result of this unique trait, the bulk of the available natural dyes come from the plant. The parts of the plant from which dyes have been extracted by the researchers are roots, tubers, barks, stems, flowers, fruits, seeds,

Table 3 Animal-based natural dyes

Source	Class	Color	Chemical structure and name [9–11]
Kermes ilicis	Anthraquinonoids	Red	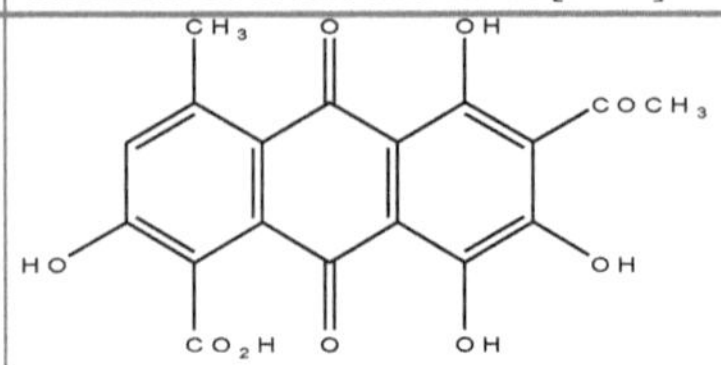 Kermesic acid
Dactylopius coccus	Anthraquinonoids	Crimson/scarlet	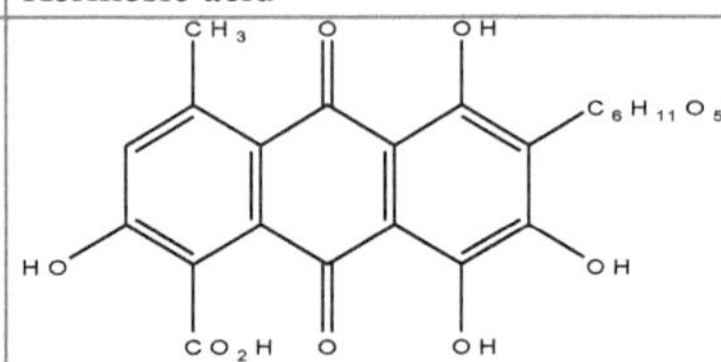 Carminic acid
Kerria lacca	Anthraquinonoids	Dark red	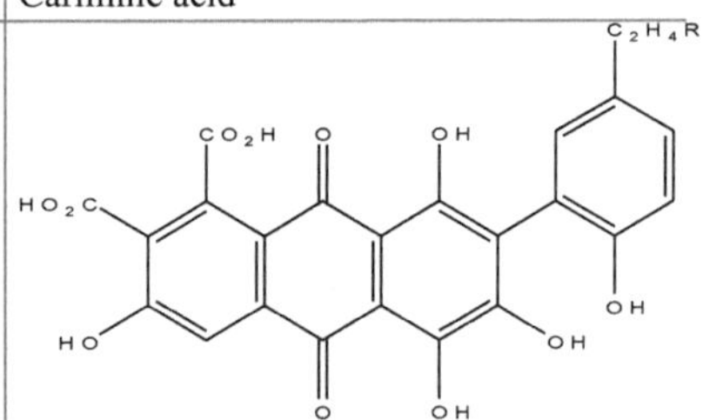 Laccaic acid
Mollusks	Indigoids	Purple	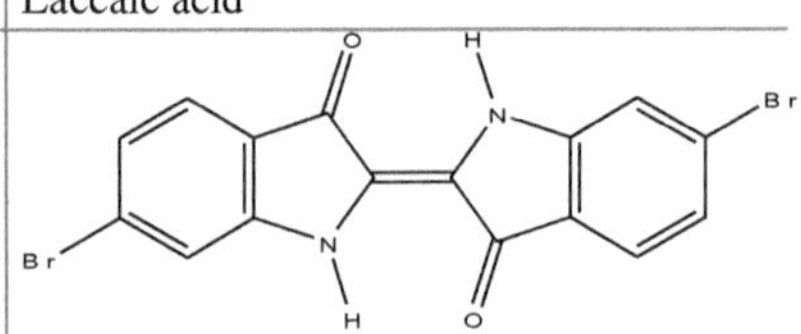 Tyrian purple

and leaves. The latest is the most abundant and sustainable source that can be obtained from the plant without causing any harm to the plant [5]. The dye-yielding plant's parts and a detailed discussion of their attributes are presented in this section.

3.3.1 Plant Roots and Tubers

The roots and tubers are the parts of the plant beneath the ground that absorb minerals and water from the soil for the survival and proper growth of the plant. They are also responsible for firm anchoring of the plant to the ground.

Various plant roots and tubers (Fig. 1) have been used as foods, antibiotics (for treatment of various biotic ailments), anti-malaria (for herbal malaria treatment) and for management of medical illness like diabetics and hypertension and in particular as natural dye precursors since creation [12]. Some of the plant roots and tubers

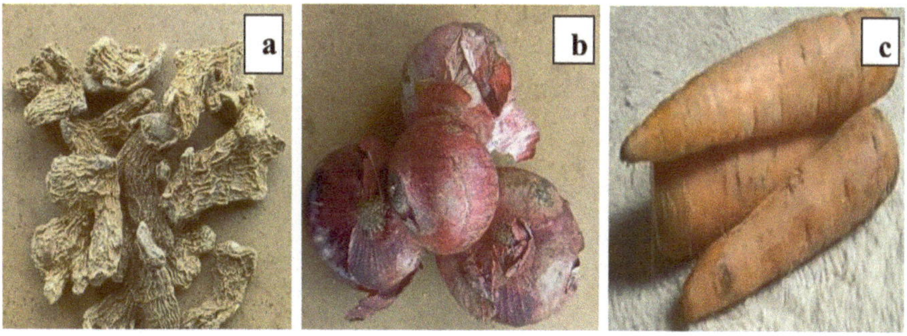

Fig. 1 Natural dye-bearing (**a**) ginger roots, (**b**) onion bulbs' scale, and (**c**) carrot tuber

(that have been used previously by some researchers for obtaining of natural dyes), their colors, class and chemical structures are listed in Table 4.

3.3.2 Plant Bark

The bark is the exterior part of the plant covering the stem, which largely contains dead tissues, where secondary metabolites are mostly deposited. It is directly connected to the leaves and, in many plants, it links the fruits and stems together. Apart from the aforementioned role of plant bark, in Nigeria, it is also used for many biomedical functions (Fig. 2). The presence of secondary metabolites is responsible for the natural dye-yielding ability of the plant barks [13]. Some of the plant barks with their color extracts, class, and chemical structures of major coloring constituents are presented in Table 5.

3.3.3 Plant Stem and Wood

The stem is the wooden part of the plant that serves as the structural framework (skeletal) for the plant. It is otherwise known as the plant's trunk, which connects the roots to the leaves. The stem (Fig. 3) assists in transporting the absorbed nutrients and water from the soil to different parts of the plant. Therefore, it is essential for the survival of the plant. Many plants acquire functional stems used as coloring and health-care agents. Jiang et al. [14] were some of the researchers to extract biodyes from plant stems for coloring textiles.

Table 6 displays some potential dye-yielding plant stems. Although a range of hues is obtainable from the stems of different plants, an interesting aspect of stem-based natural dyes is their common phenolic class, as can be seen from the table.

Table 4 Plant root-based natural dyes

Source	Class	Color	Chemical structure and name
Rubia tinctorum	Anthraquinonoids	Red	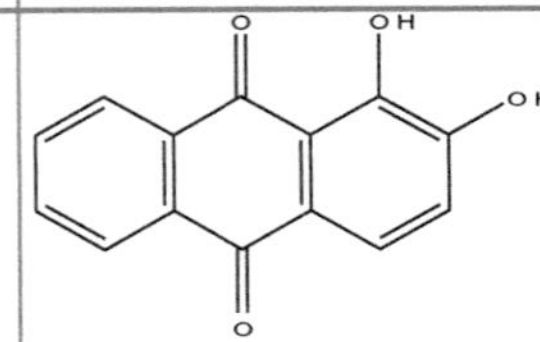 Alizarin
Moringa oleifera	Terpenoids	Yellow	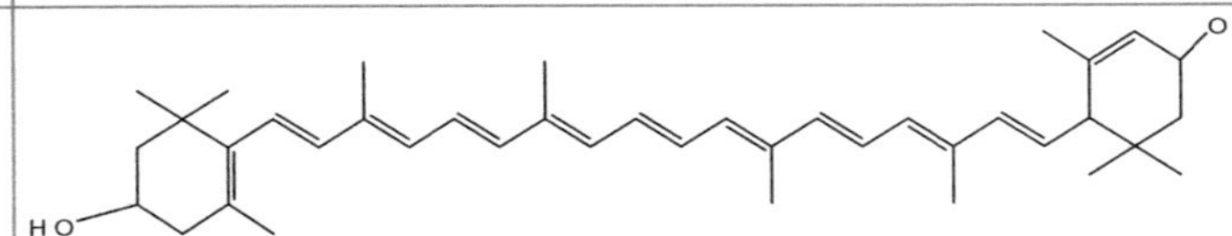 Lutein
Curcuma tinctoria	Terpenoids	Bright yellow	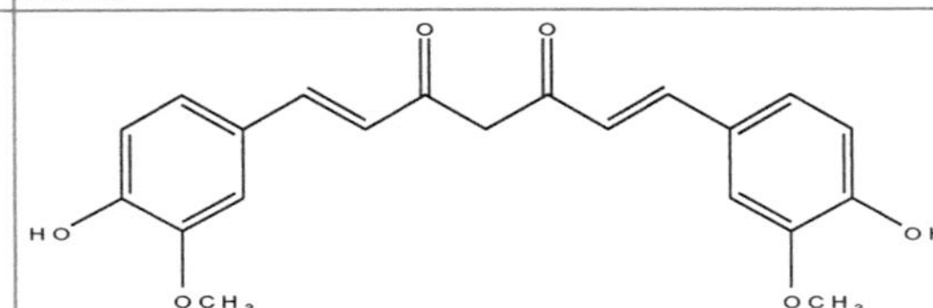 Curcumin
Galium verum	Anthraquinonoids	Red	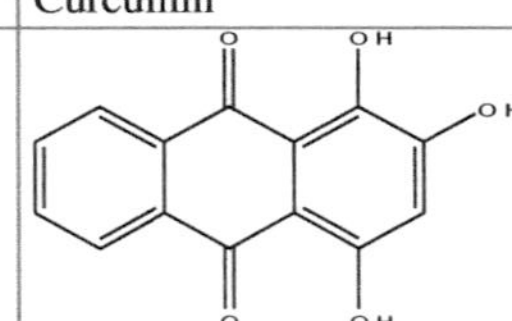 Purpurin

Alkannin tinctoria	Naphthoquinonoids	Red	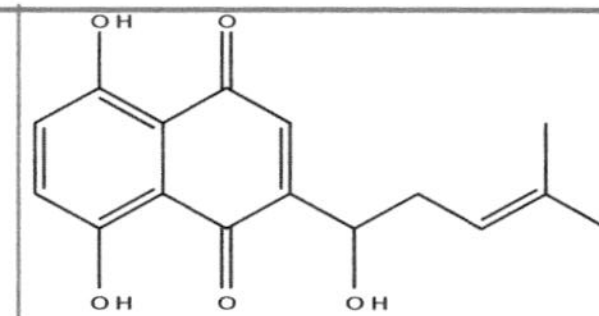 Alkannin

Fig. 2 Natural dye-bearing (**a**) mango, (**b**) cashew, and (**c**) neem bark

3.3.4 Plant Flowers and Petals

The flowers and petals (Fig. 4) are brightly colored plant parts that survive for only a short period of time. Within their short period of existence, they often engage in sexual reproduction to form seed-bearing fruits. They are admired by humans and animals (particularly butterflies and honeybees) for their sweet odor, juice, and attractive colors. Sinha et al. [15] extracted different kinds of colored dyes from different varieties of plant flowers and petals for beautifying substrates. Table 7 lists some plant flowers with extractable natural dyes and the class and chemical structures of the colorants.

3.3.5 Plant Fruits, Fruit Peels, and Pods

The fruits (Fig. 5) are soft, fleshy, and pulpy parts of the plant that enclose the seeds. Fruit develops from the ovaries of the plant's angiosperms above the ground with an adorable color(s), sweet odor, and taste. The primary objective of planting fruit-bearing plants is for the provision of food since fruit is one of the nutritive diets for proper human growth and wellness. The developmental stages of fruits are associated with color change. Table 8 shows some classes of fruit-based dyes extracted and used by some researchers for staining and functionalizing substrates [15].

3.3.6 Plant Seeds and Seed Husks

The matured ovules of the plant found in fruits are popularly called seeds (Fig. 6). It is made up of three parts: seed coat, endosperm, and embryo. Plant seeds consist of minute undeveloped plants and in most plants they act as food reservoirs. Haque et al. [16] are some of the researchers that used seed dye extract for coloring textiles. Some of the plant seeds containing extractable natural dyes are presented in Table 9.

Table 5 Plant bark-based natural dyes

Source	Class	Color	Chemical structure and name [5]
Bridelia ferruginea	Flavonoids	Yellow to orange	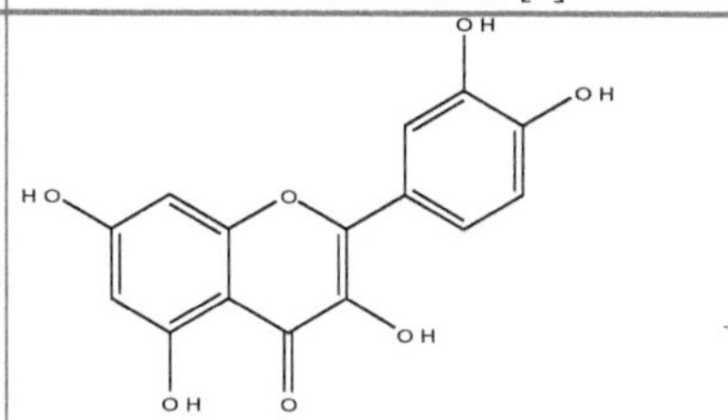 Quercetin
Azadirachta indica	Tannins	Orange	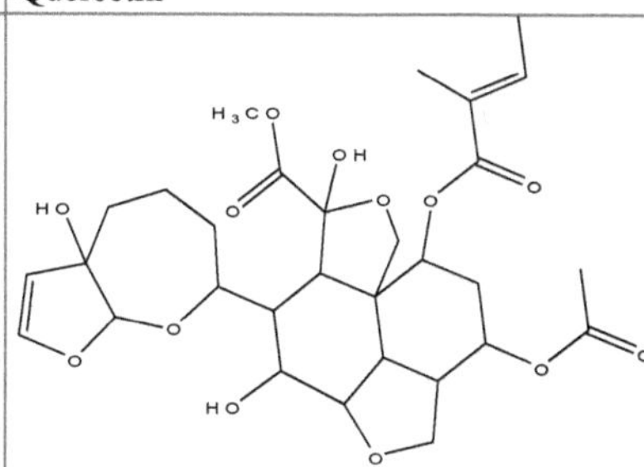 Nimbin
Mangifera indica	Phenolics	Yellow	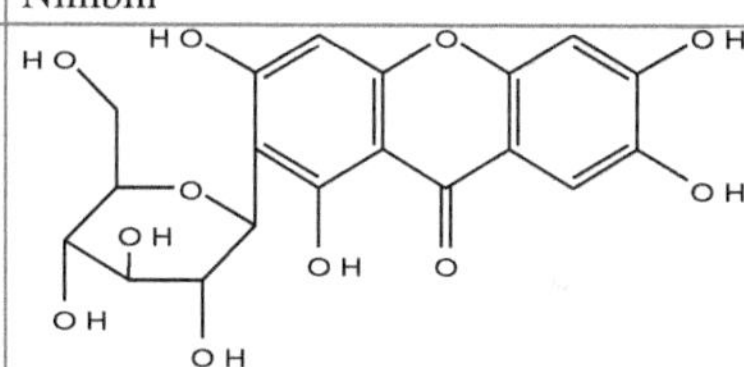 Mangiferin
Morinda lucida	Flavonoids	Light yellow	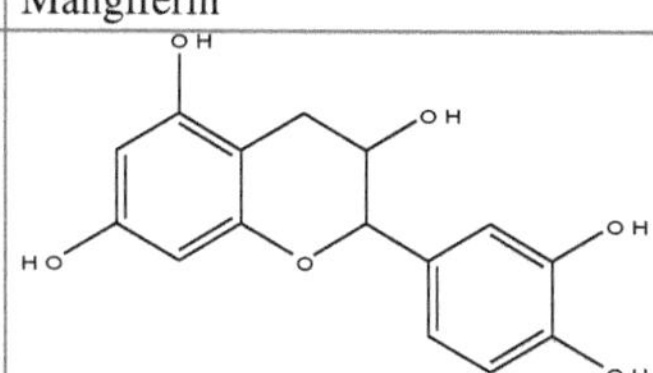 Morindin
Acacia nilotica	Flavonoids	Light yellow	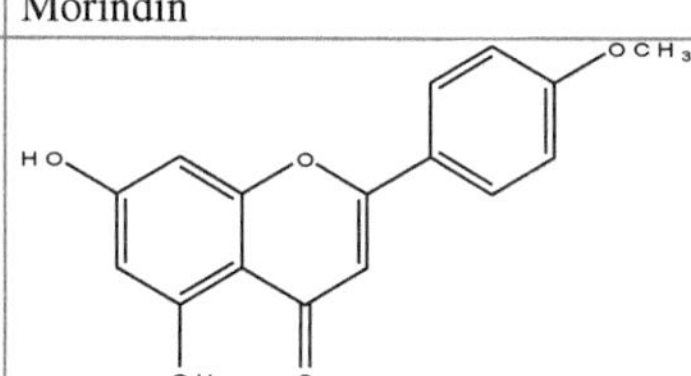 Acacetin

Fig. 3 Natural dye-bearing (**a**) sorghum, (**b**) Indian spinach, and (**c**) bloodleaf stem

3.3.7 Plant Leaves

Plant leaves (Fig. 7) are blade-like flattened organs that emanate from the plant stem. They are the organs responsible for supplying food to the plant through a process called photosynthesis. They are the agents of plant transpiration; as a result, they assist in water conservation in the plant kingdom. For instance, plants like cocoa and almond shed off their leaves during the dry season for minimizing the rate of transpiration, whereas by making water absorbed during the rainy season available for the plants throughout the dry season. Therefore, leaves are essential plant parts. They are often referred to as sustainable plant parts because they are obtainable from the stem with little or no harm to the plant. They are used for different purposes, ranging from ornamental to biological and medicinal.

Although plant leaves are of different colors, they frequently appear as green-colored substances. Several colored dyes have been extracted from plant leaves by many researchers [6]. Some of the natural dye-yielding leaves, dye colors, chemical classes, and structures are listed in Table 10.

4 Chemical Structure-Based Classification

The color chemists (colorists) use chemical structure to group the natural dyes according to their molecular chromophoric system. Natural dyes have been identified as unsaturated compounds, usually of aromatic rings called chromogens containing other groups called chromophores whose role is to impart color and auxochromes that moderate dye solubility and intensify and enhance shade substantivity on the substrate. This classification accurately identifies a particular dye for a specific chemical group with unique properties. Therefore, natural dyes are appropriately classified into indigoids, quinonoids, terpenoids, and phenolics [9].

Table 6 Plant stem natural dyes

Source	Class	Color	Chemical structure and name [4]
Pterocapus erinaceus	Flavonoids	Yellow to red	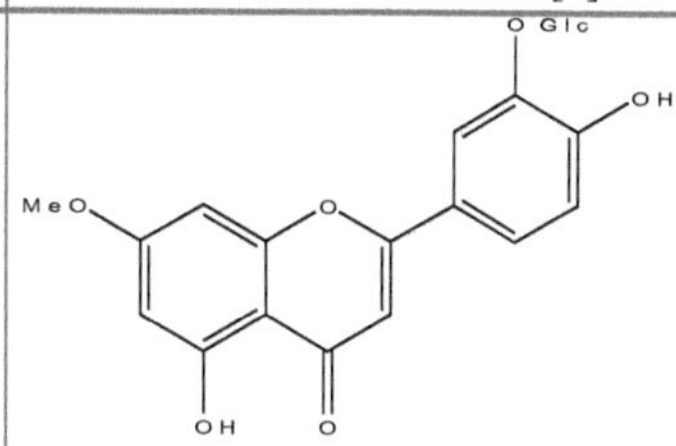 Retusin glycosides
Haematoxylon campechianum	Phenolics	Purple	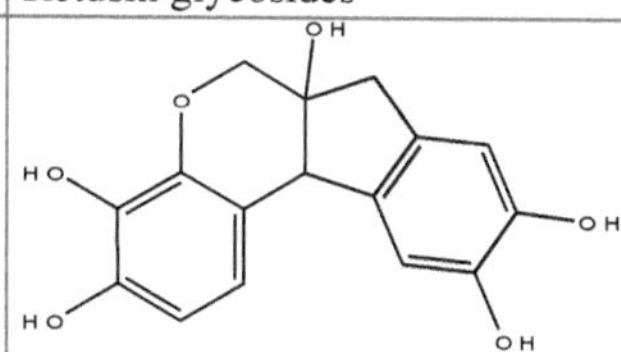 Haematoxylin
Sorghum vulgare	Flavonoids	Wine-red	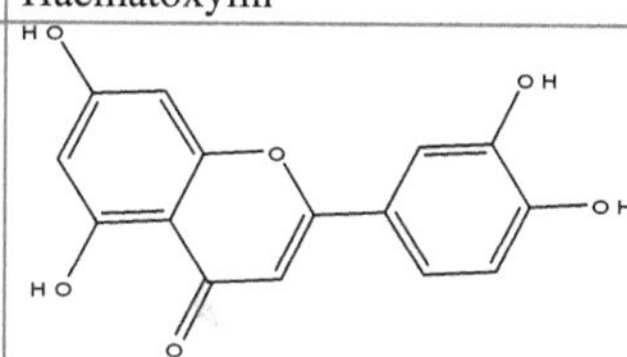 Luteolinidin
Corylus avellana	Phenolics	Dark yellow	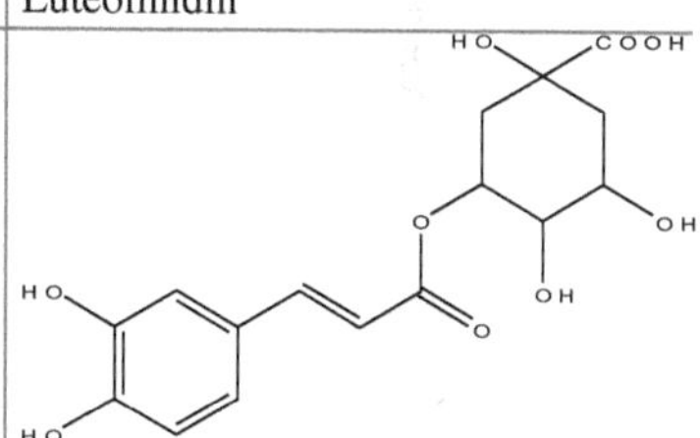 Caffeoylquinic acid
Caulis spatholobi	Phenolics	Brown	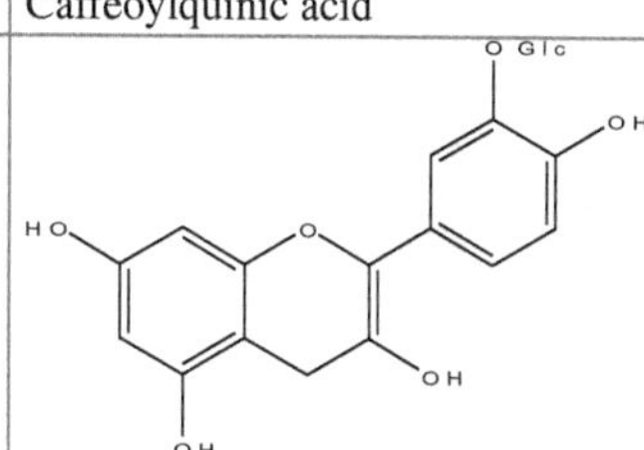 Cyanidin

4.1 *Indigoids*

Indigoids are a group of important and primitive natural dyes used by ancient craftsmen for coloring substrates. The first known member of this group, named indigo, with the chemical structure shown in Fig. 8a was first extracted from the leaves of

Fig. 4 Natural dye-bearing (**a**) hibiscus, (**b**) sun, and (**c**) cotton flower

an indigofera plant called *Indigofera tinctoria*. The *Indigofera tinctoria* was initially discovered in India and later cultivated in Java, China, and Central and West Africa. Indigo is well known for its intense blue shade and excellent colorfastness properties [18]. Tyrian purple (Fig. 8b), which is another member of the indigoid group of natural dyes, is an extract from some species of mollusks listed in Sect. 3.2; it is a derivative of indigo dye containing bromine as a substituent at positions 6 and 6′ on the original indigo chemical structure. Unlike indigo dye, Tyrian purple was a highly priced purple dye often referred to as imperial or royal purple, because it was used for coloring royal clothes. Other indigoids dyes are indirubin (Fig. 8c) obtained from *Isatis tinctoria*, which originated from Europe, and isatan (Fig. 8d), which is obtained from *Calotropis procera*, which is a native of North Africa, Pakistan, Indonesia, and West and South Asia [17].

4.2 Quinonoids

Quinonoids have the most diverse chemical structures than any other class of natural dyes with shades often varying from yellow to red, violet, blue, and green. They are obtainable from a lot of plants and animals. Quinonoids can be broadly grouped into anthraquinones, naphthoquinones, and benzoquinones [19].

4.2.1 Anthraquinones

Anthraquinones with a general chemical structure shown in Fig. 9 are dyes containing anthracene as chromogene, a two-electron withdrawing carbonyl functional groups as chromophore, and electron-releasing substituents (such as –NHPh, NH_2, –OH, etc.) as axochrome [4].

Where R’ to R”“‘ are axochromes, which can be the same or different from one another.

Table 7 Natural dyes from plant flowers and petals

Source	Class	Color	Chemical structure and name [17]
Crocus sativus	Terpenoids	Yellow	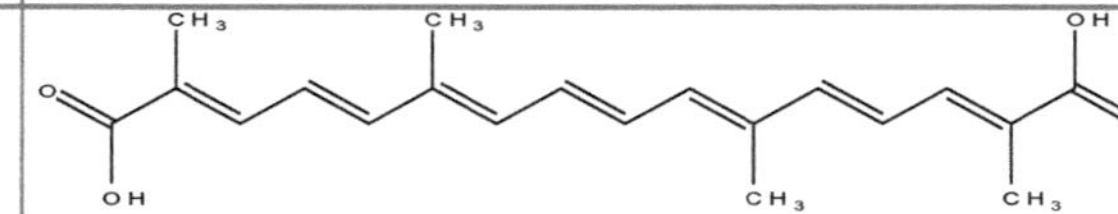 Crocetin
Chrysanthemum segetum	Flavonoids	Bright yellow	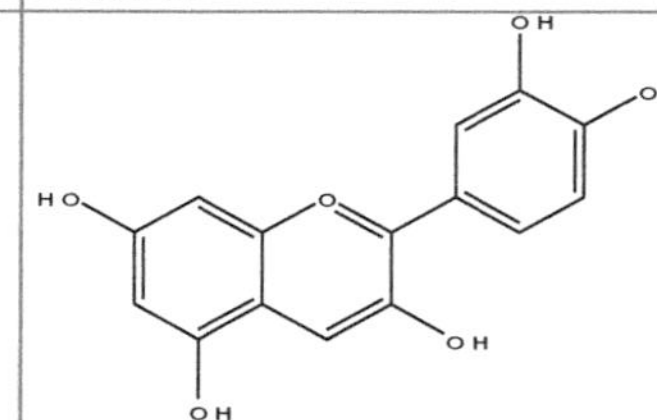 Anthocyanin
Helianthus annuus	Phenolics	Greenish-yellow	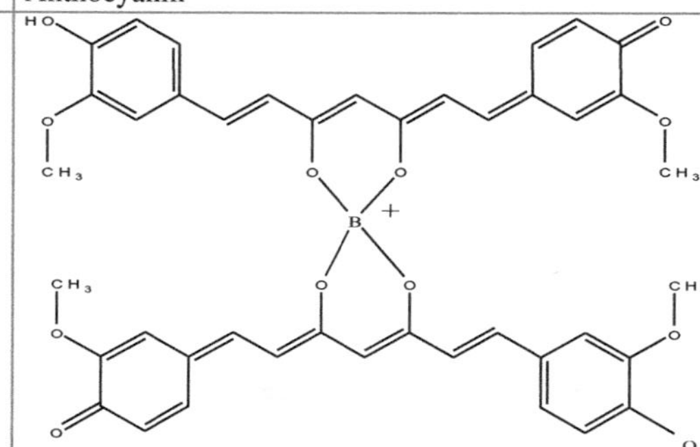 Rosocyanine

(continued)

Table 7 (continued)

Source	Class	Color	Chemical structure and name [17]
Hibiscus rosa-sinensis	Flavonoids	Red	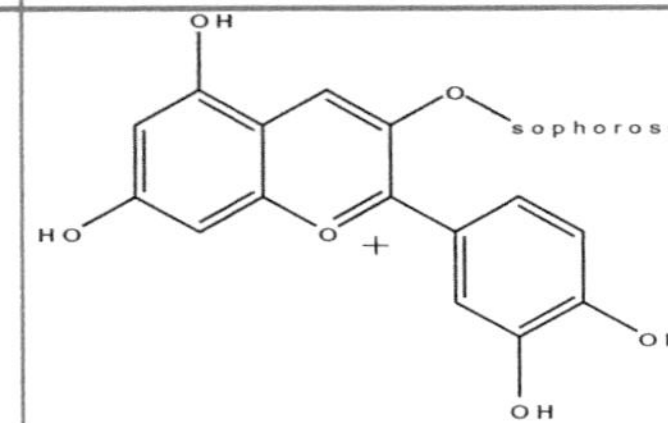 Cyanidin-3-sophoroside
Coreopsis tinctoria	Flavonoid	Gold, nasty-red	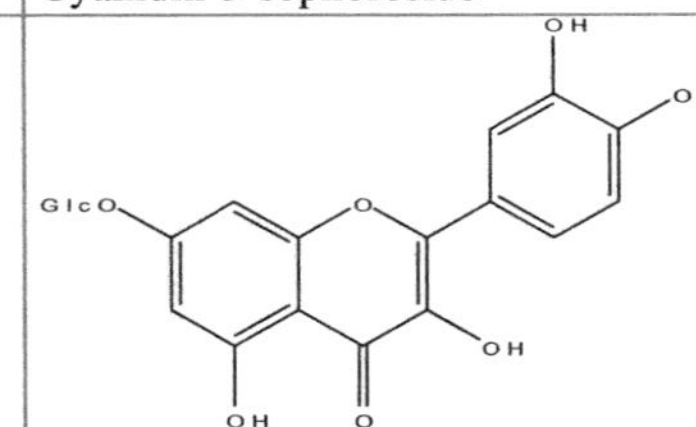 Okanin glycoside

Fig. 5 Natural dye-yielding (**a**) rotten tomatoes, (**b**) banana peel, and (**c**) orange rind

Insect-based natural dyes from *Kermes ilicis* to *Coccus ilicis* and *Kerria lacca* (Fig. 10) belong to this group of dyes.

In the plant kingdom, dye extracts known as alizarin from *Rubia tinctoria* root, rhein from *Cassia fistula* fruit, and tectoleafquinone from *Tectona grandis* leaf (Fig. 11) belong to this class of dyes.

4.2.2 Naphthoquinones

Naphthoquinones (Fig. 12) are dyes whose chromogenes contain naphthalene and that have two-electron withdrawing carbonyl functional groups as chromophores and –NHPh, NH_2, and/or –OH as axochromes.

The dyes found in these sub-groups are extractable from the plant. For instance, alkannin and lawsone are naphthoquinones isolated from the roots and leaves of *Alkanna tinctoria* and *Lawsonia inermis*, respectively (Fig. 13a, b). Juglone (Fig. 13c) is extracted from the seed shell of the *Juglans regia*.

4.2.3 Benzoquinonoids

Benzoquinones contain the smallest chromophoric unit of the quinonoids, which is a C6 cyclic ring containing two carbonyl functional groups at positions 1 and 4 as shown in Fig. 14.

Benzoquinone is the active biochemical moiety for the biosynthesis of other quinonoid dyes and related compounds in the plant kingdom. For instance, compounds formed by attaching one benzene ring to benzoquinone at positions 2-C and 3-C are called naphthoquinones (Fig. 13), while those formed by attaching one benzene ring at positions 2-C and 3-C and another benzene ring at positions 5-C and 6-C to benzoquinone are known as anthraquinones (Fig. 11). A yellow-colored natural dye named carthamin and a red-colored natural dye named safflower carmine with chemical structures shown in Fig. 15 are members of the benzoquinonoids group of dyes obtained from the safflower's petals and flowers, respectively.

Table 8 Natural dyes from plant fruits

Source	Class	Color	Chemical structure and name [15]
Musa species' peels	Flavonoids	Yellow	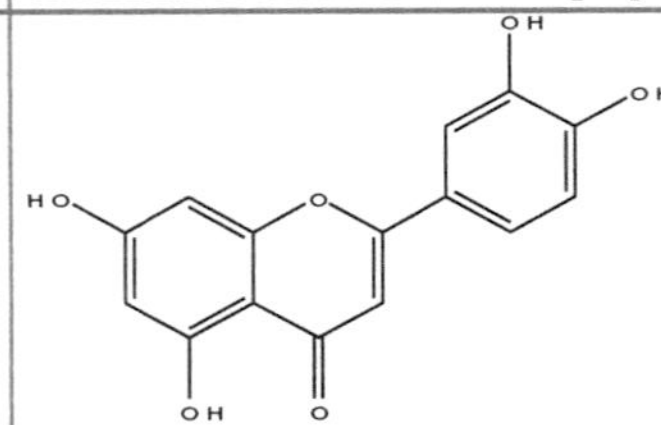 Flavone
Lycopersicon esculentum	Carotenoids	Yellow to red	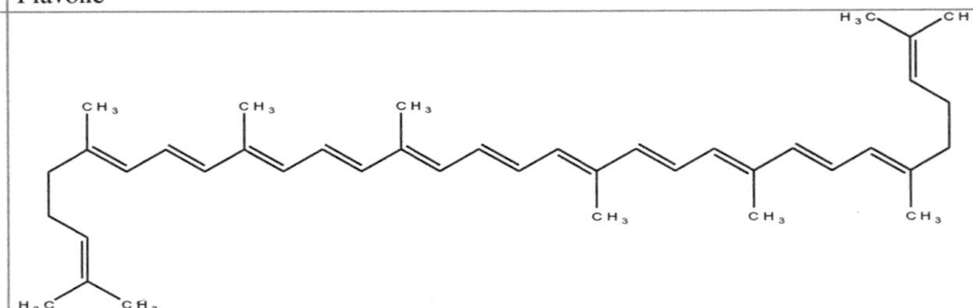 Lycopene
Thespesia populnea	Flavonoids	Yellow	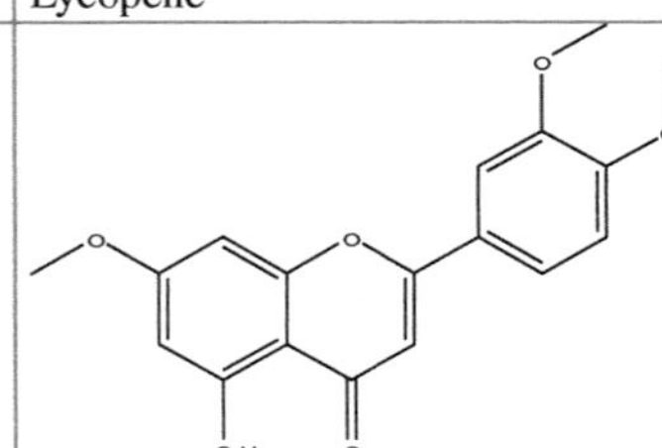 Trimethoxy quercetin

Theobroma cacao	Flavonoids	Brownish red	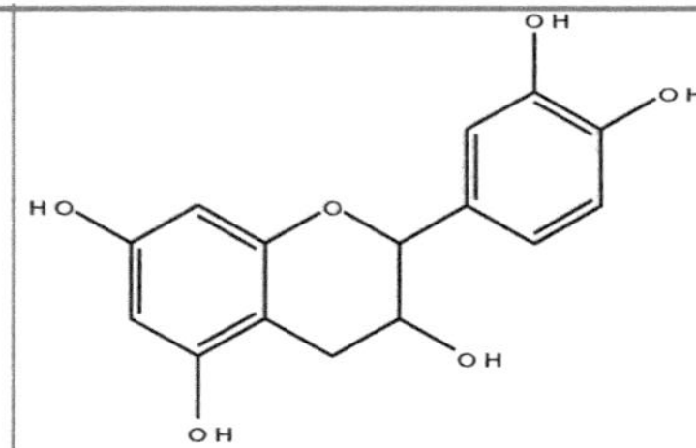 Catechin
Cassia fistula	Anthraquinonoids	Yellow	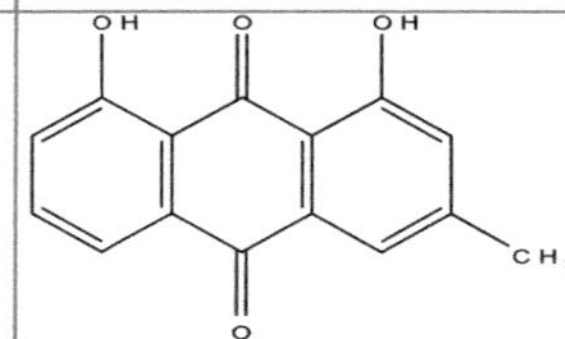 Rhein

Fig. 6 Natural dye-yielding (**a**) cocoa pods, (**b**) coconut coir, and (**c**) bixa seeds

4.3 Terpenoids

Terpenoids are brightly colored natural dyes originating from chromoplsat and/or chloroplast in virtually all members of the plant kingdom. Their bright colors (ranging from yellow to orange and red) can be traced to the absorption of visible light (between wavelengths of 400 and 500 nm) by the conjugated carbon-carbon double bonds in their chemical structures. Terpenoids are otherwise called isoprenoids, because they are built up from a number of derivatives of isoprene units. The hemiterpenoids (C5), monoterpenoids (10), sesquiterpenoids (C15), diterpenoids (20), sesterterpenoids (C25), triterpenoids (C30), and tetraterpenoids (C40) are members of terpenoids group of dyes having one to eight isoprene units, respectively, in their chemical structures [18]. Many of the plant-based natural dyes belonging to the terpenoids group are tetraterpenoids that are otherwise known as carotenoids. The chemical structures of some of the plant-based carotenoids like lycopene, crocetin, bixin, curcumin, and β-carotene from *Lycopersicon esculentum*, *Crocus sativus*, *Bixa orellana*, *Curcuma tinctoria*, and *Daucus carota* respectively, are shown in Fig. 16.

4.4 Phenolics

Colorants derived from all parts of the plant, containing benzenoid ring(s) with hydroxyl functional groups as substituents in their chemical structures, are called phenolics. They can be categorized into (i) tannins and (ii) flavonoids depending on their molecular weights and the quantity of the hydroxyl functional group they contain.

Table 9 Natural dyes from plant seeds

Source	Class	Color	Chemical structure and name [16]
Bixa orellana	Carotenoids	Orange	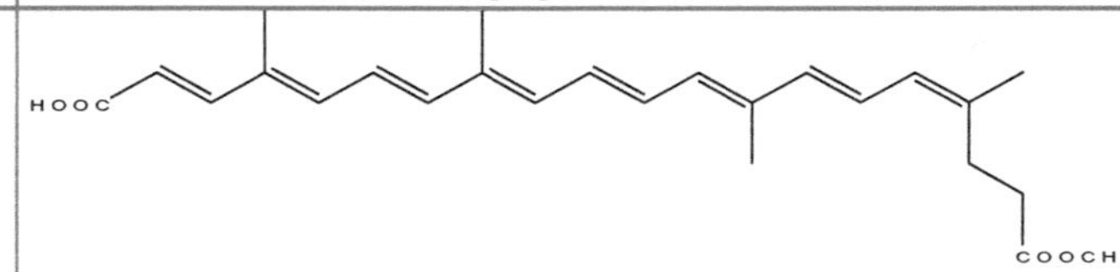 Bixin
Zizania aquatica	Flavonoids	Blood-red	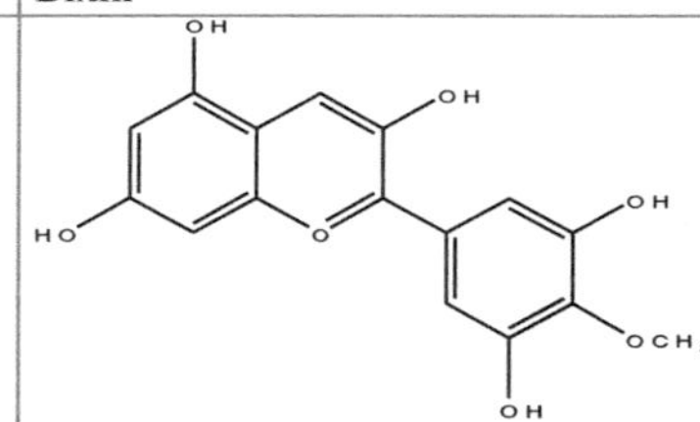 Peonidin
Cola nitida	Flavonoids	Brown	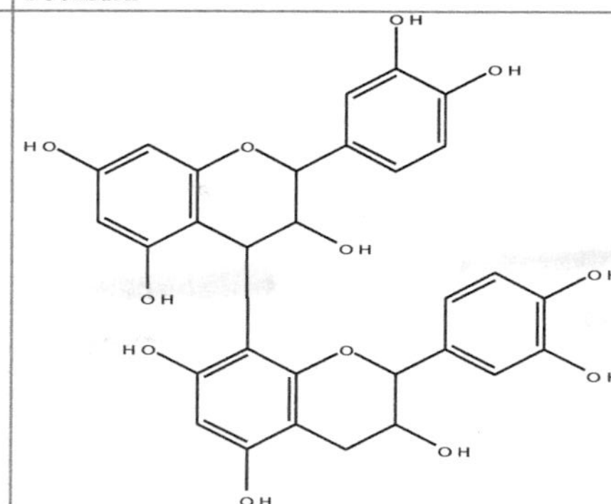 Procyanidin

(continued)

Table 9 (continued)

Source	Class	Color	Chemical structure and name [16]
Acalypha wilkesiana	Flavonoids	Orange	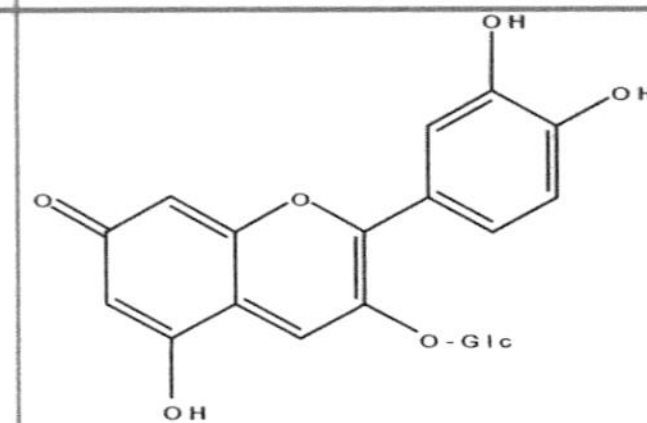 Cyaniding-3-glycoside
Juglans regia	Naphthoquinonoids	Brown	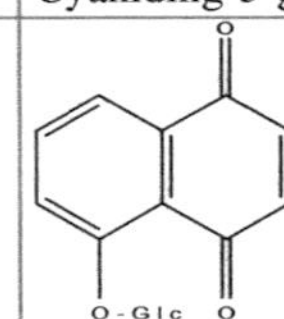 Juglone

Fig. 7 Natural dye-yielding (**a**) teak, (**b**) sweet potato, and (**c**) almond leaves

4.4.1 Tannins

Tannins are high molecular weight colorants, containing a number of benzenenoid rings as chromogens, at least one carbonyl functional group as chromophore, and a number of hydroxyl functional group substituents as axochromes. They are otherwise referred to as polyphenolics, probably due to their molecular weight and/or the availability of multiple hydroxyl functional groups in their structures. Tannins reside majorly in the cell wall of the epidermis layer of the plant, in the roots, stems, leaves, seed pods, shell, or husk. They are usually classified as hydrolyzable and condensed tannins. The hydrolyzable ones are named ellagitannins or gallotannins, while the condensed tannins are epicatechin and catechin, which are monomeric units of 15C high molecular weight polyhydroxyflavan-3-ol [20]. The structures of typical tannins are given in Fig. 17.

4.4.2 Flavonoids

Flavonoids are plant-based natural dyes with a general chemical structure of the 15C flavone (6C-3C-6C) skeleton of a heterocyclic ring and two benzyl rings (Fig. 18).

Flavonoids are the most abundant natural dyes of multiple colors, such as red, orange, violet, and pink, which are obtainable from all parts of the plant, ranging from roots to leaves, fruits, flowers, and seeds. They are sub-grouped into flavones, anthocyanins, flavan-3-ols, flavonols, flavanones, and isoflavones [20]. The chemical structure of a member of each of the sub-groups is presented in Fig. 19.

Table 10 Plant leaf-based natural dyes

Source	Class	Color	Chemical structure and name
Vernonia amygdalina	Terpenoids	Dark green	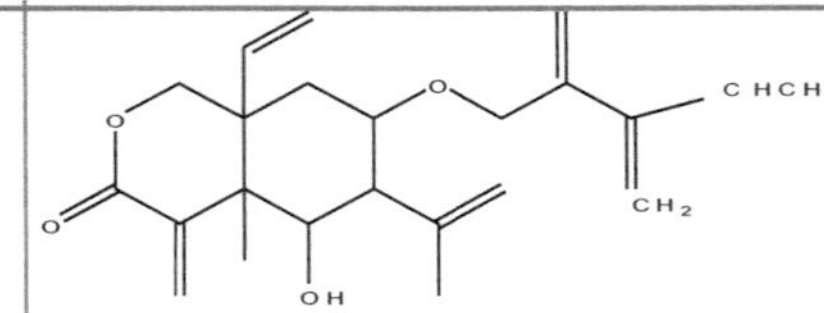 Vernodalol
Corylus avellana	Flavonoids	Yellow	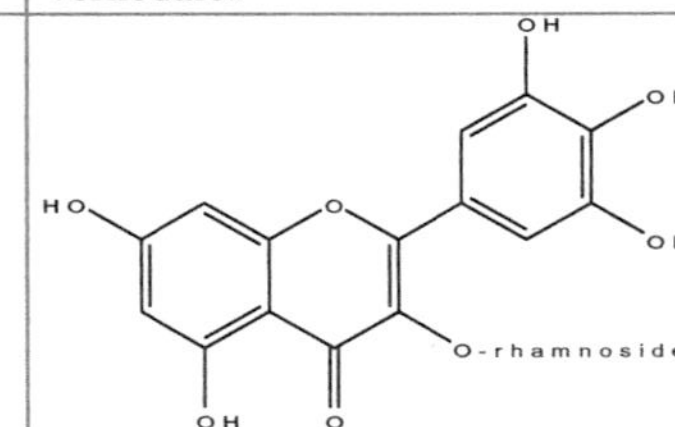 Myricetin 3-rhamnoside
Lawsonia inermis	Naphthoquinonoids	Brown	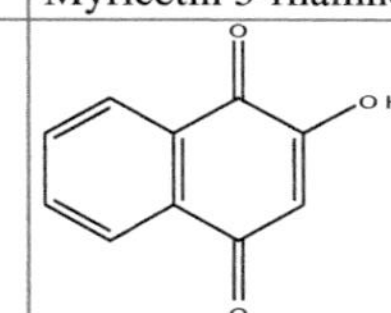 Lawsone

Ipomoea batatas	Flavonoids	Purple	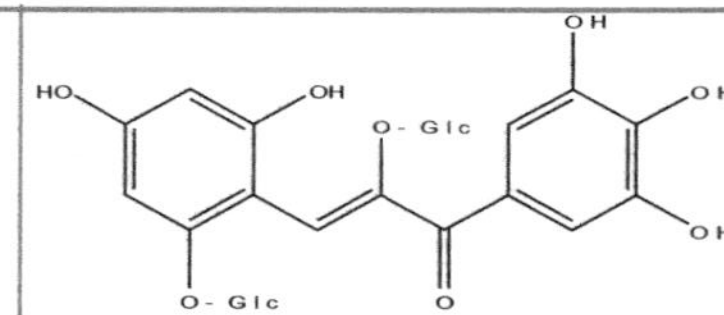 Chalcone
Tectona grandis	Anthraquinonoids	Pink	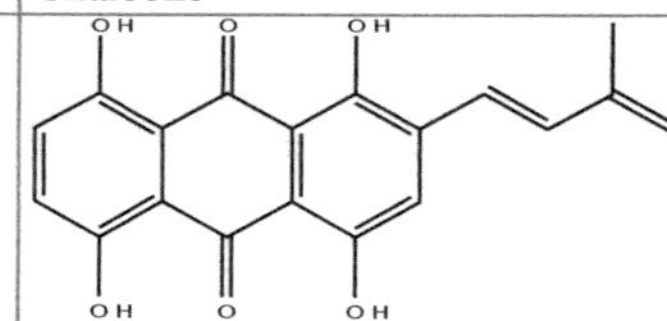 Tectoleafquinone

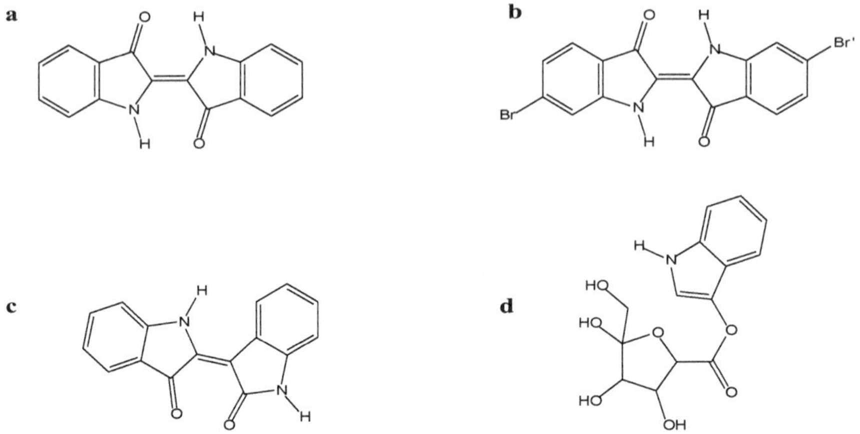

Fig. 8 Chemical structures of (a) indigo, (b) Tyrian purple, (c) indirubin, and (d) isatan dye

Fig. 9 Anthraquinone's general chemical structure

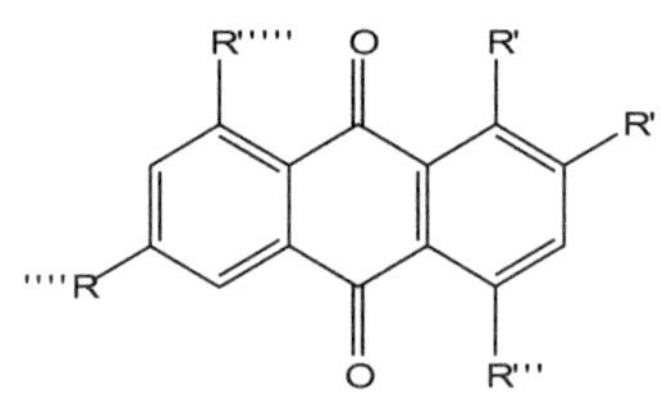

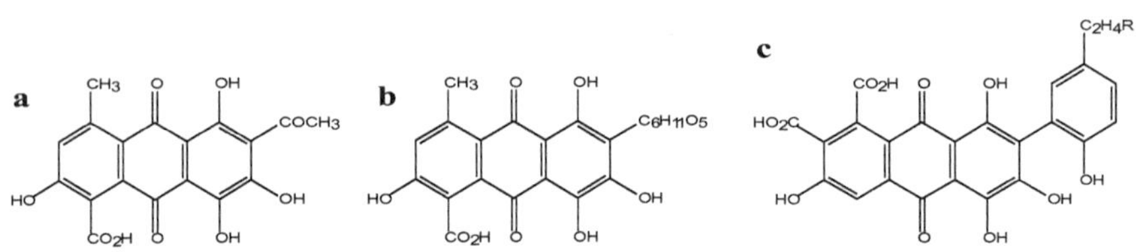

Fig. 10 Dyes obtainable from (a) *Kermes ilicis*, (b) *Coccus ilicis*, and (c) *Kerria lacca*

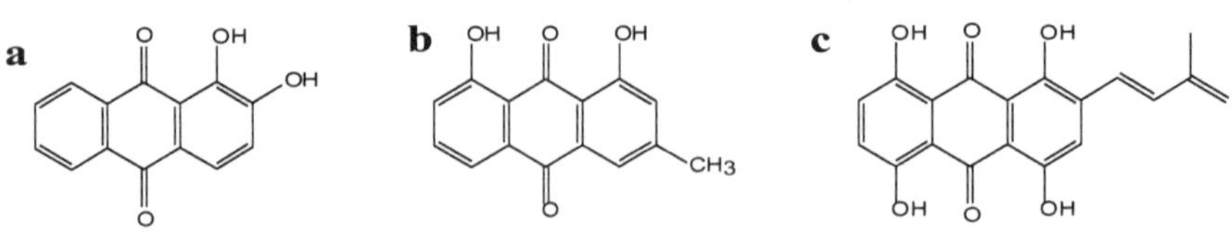

Fig. 11 Chemical structures of dyes obtainable from (a) *Rubia tinctoria* root, (b) *Cassia fistula* fruit, and (c) *Tectona grandis* leaf

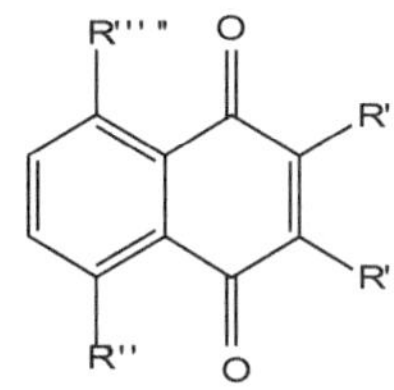

Fig. 12 General chemical structure of naphthoquinone

a **b** **c**

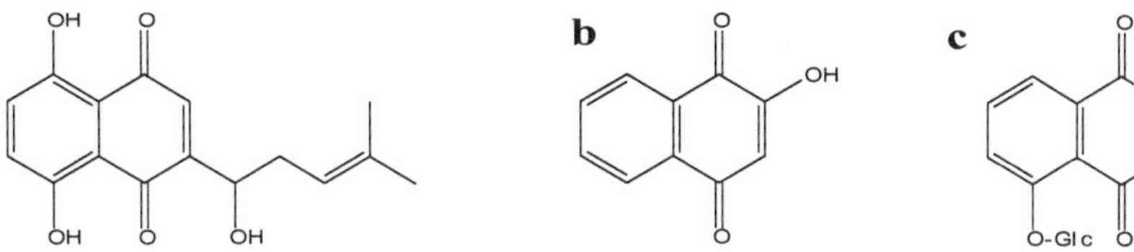

Fig. 13 (**a**) Alkannin, (**b**) lawsone, and (**c**) juglone dyes obtainable from the plant

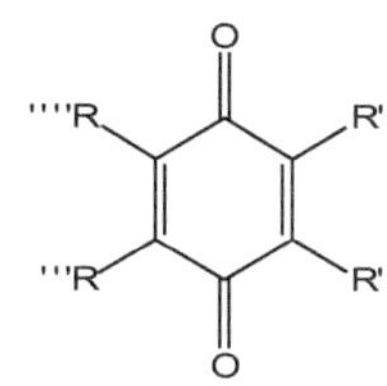

Fig. 14 General chemical structure of benzoquinonoids

5 Application Methods for Natural Dyes

This classification was adopted by the dyers for easy grouping of natural dyes. The main interest of dyers is to produce consumers' desired colored garments via the selection and valorization of dyes whose affinity for the chosen substrates is high. Therefore, the dyers classify natural dyes into different groups regardless of their sources, chemical constituents, or structures, but based on their capability to color a specific group of substrates. Right from pre-historical days, they have been utilized for coloring only natural fibers like silk, wool, and cotton [21]. A recent literature survey did not only affirm their suitability in coloring natural fibers but also for coloring synthetic fibers like polyamides, polyesters, acetates, and regenerated cellulose as well [20–24]. However, the following highlighted groups have been identified for natural dyes based on methods of application:

5.1 *Vat Dyes*

Natural dyes in this group are applied onto cellulosic fibers through the fermentation vatting process, in which they are temporarily reduced to a water-soluble (leuco) form in an alkaline medium (such as sodium hydrosulfite or sodium

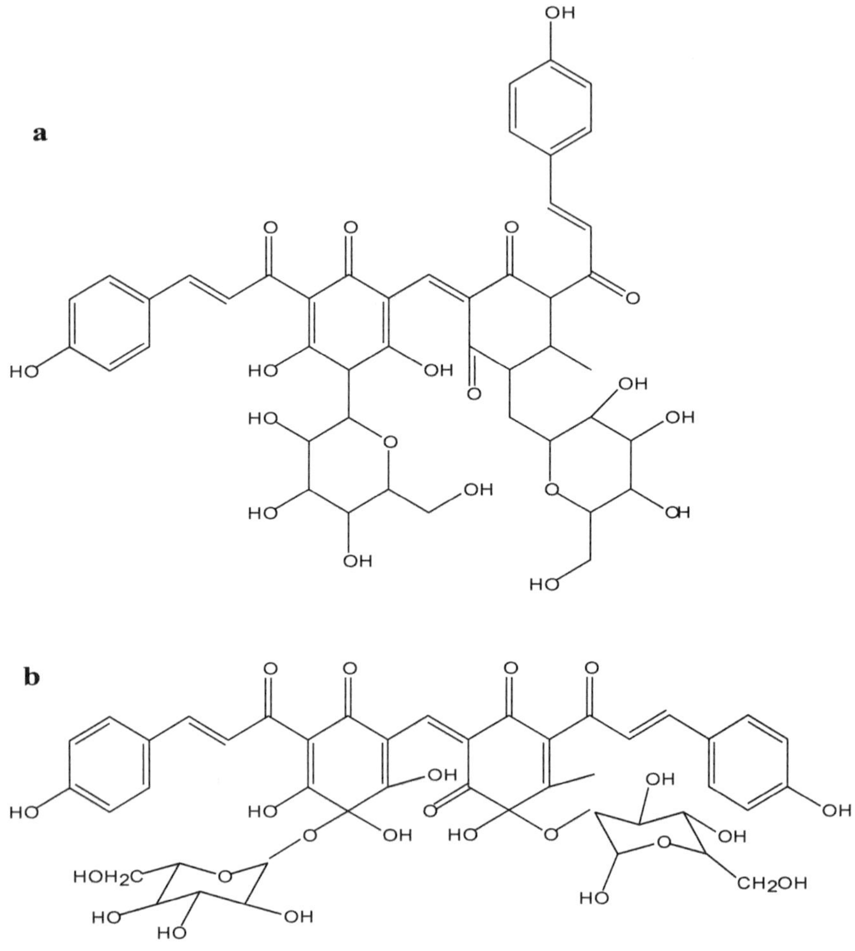

Fig. 15 Chemical structures of (**a**) carthamin and (**b**) safflower carmine from safflower

dithionite in diluted sodium hydroxide) before dyeing, due to their water insolubility trait. After the dyeing process, re-oxidation of the water-soluble (leuco) dye molecules to the vat (water-insoluble) form in the fibers will be achieved through treatment with an oxidizing agent like $KMnO_4$, $K_2Cr_2O_7$, or H_2O_2 or exposure of dyed fibers to atmospheric air [17]. Indigo dye, Tyrian purple (dibromo-indigo), indirubin, isatan dye (shown in Fig. 8), and haematoxylin (Fig. 20) are the most common natural dyes in this class [25].

Overall, the vat dyes contain one or more carbonyl groups (quinonoid) as chromophore, which is reduced temporarily to alcoholic group(s) (leuco form) in alkaline medium for easy dissolution and adsorption onto the substrates (usually cellulosic fibers) [4].

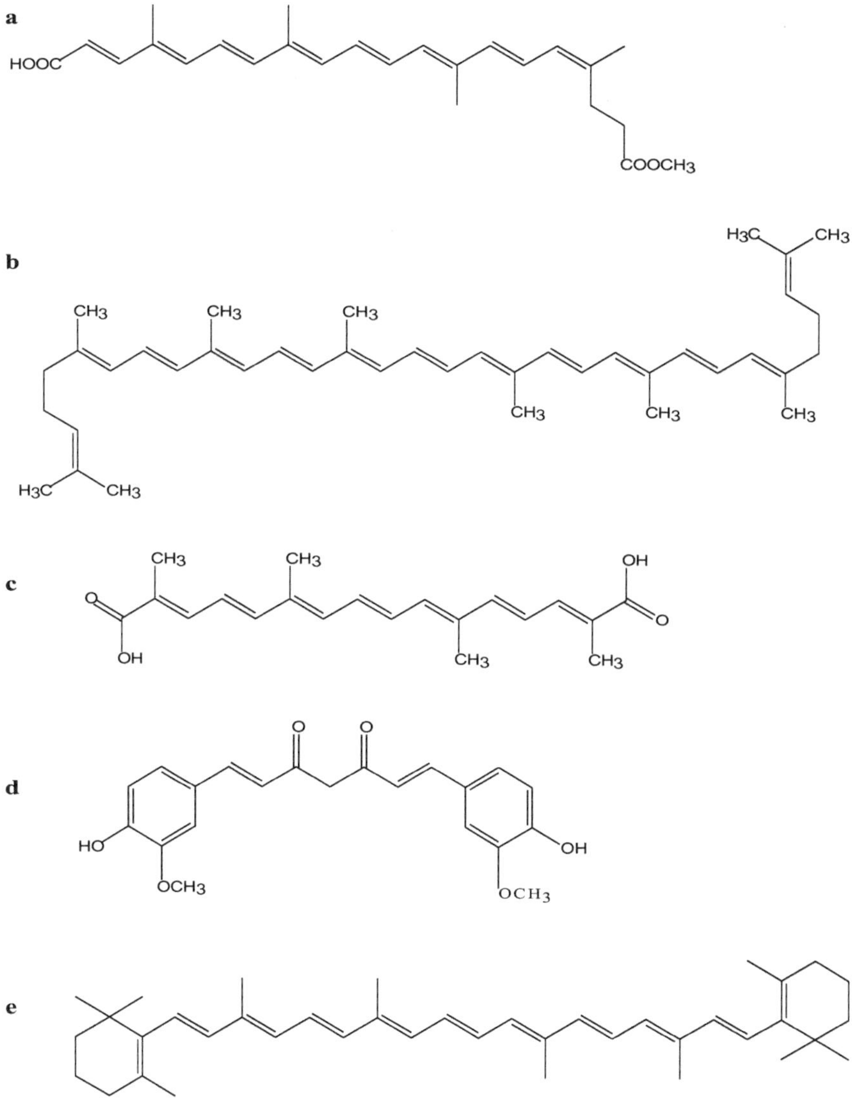

Fig. 16 Structural units of (**a**) lycopene, (**b**) crocetin, (**c**) bixin, (**d**) curcumin, and (**e**) β-carotene

Thereafter, the leuco dye is re-oxidized to a water-insoluble form in the substrate by simple exposure to atmospheric air (Scheme 1). The fixing process of vat dyes conferred excellent colorfastness to light, perspiration, rub, and wash on dyed substrates [25].

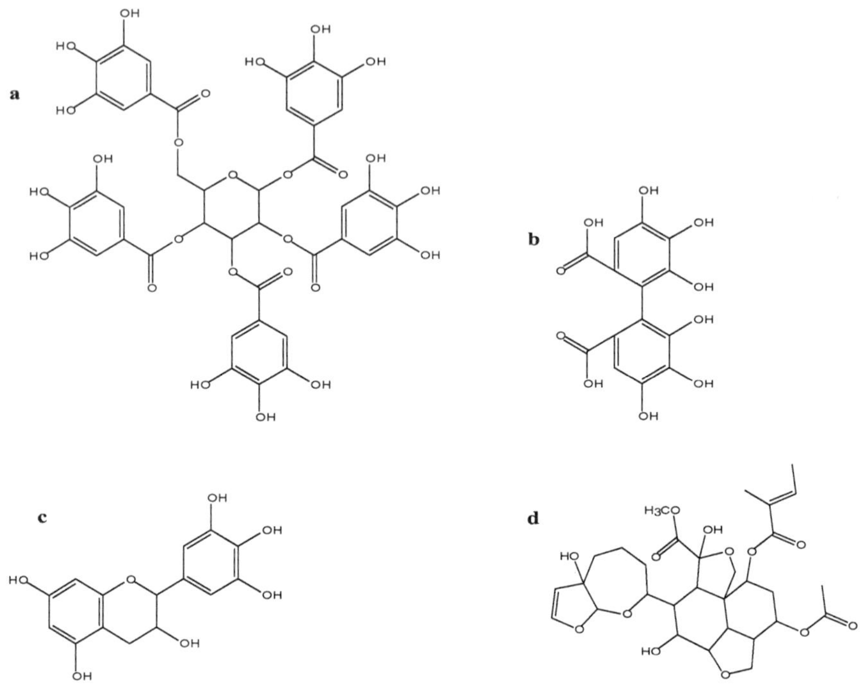

Fig. 17 Structures of (**a**) penta-galloyl-glucose, (**b**) hexahydroxydiphenic acid, (**c**) gallocatechin, and (**d**) nimbin

Fig. 18 General chemical structure of flavonoids

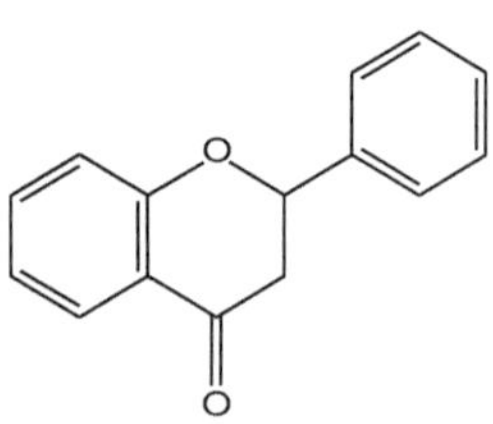

5.2 *Direct Dyes*

Natural dyes under this sub-classification are easily applied on cellulosic fibers, such as cotton and viscose rayon, due to their high affinity for these fibers with bright color formation. Their potential capability to directly apply to the textile fibers like hemp, bast, jute, linen, cotton, etc., in the absence of mordants qualified them to be termed "direct dyes" [18].

Generally, direct dyes possess large molecular weight; examples of such are bixin, penta-galloyl-glucose, and carthamin. The large size (that is responsible for

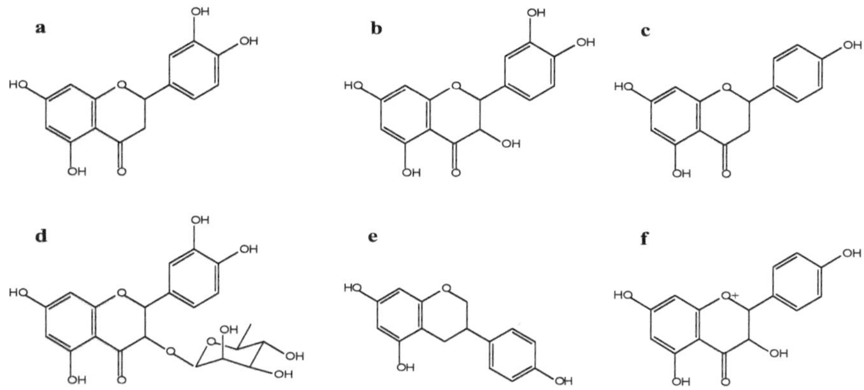

Fig. 19 Structures of (**a**) luteolin (flavones), (**b**) quercetin (flavonol), (**c**) naringenin (flavanone), (**d**) astilbin (flavan-3-ol), (**e**) genistein (isoflavone), and (**f**) pelargonidin (anthocyanin)

Fig. 20 Haematoxylin black dye from *Haematoxylum camphecianum* [4]

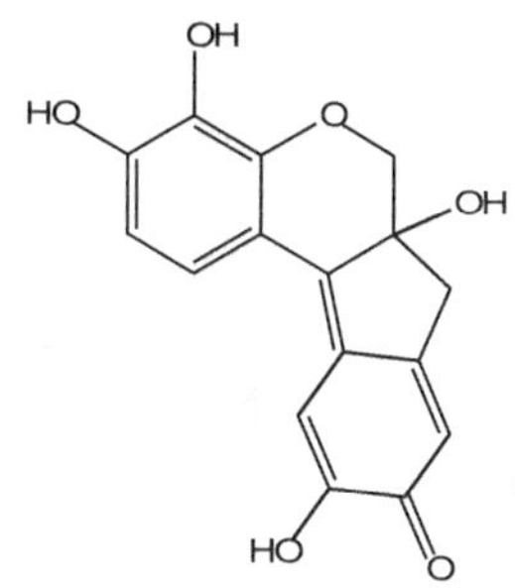

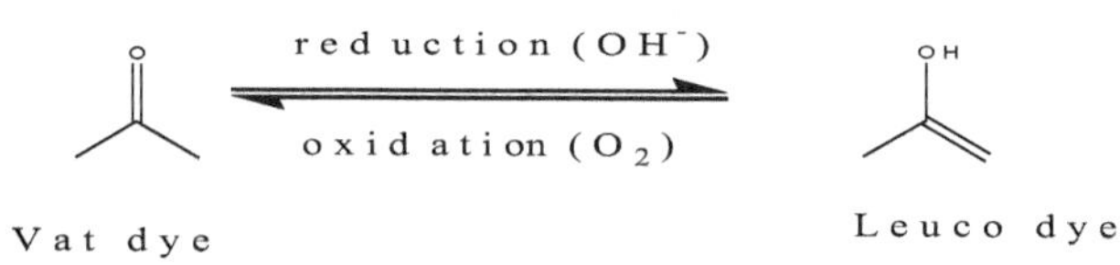

Scheme 1 Fermentation vatting process

the mechanical and van der Waals force of adhesion of the dye molecules to the fiber matrix) and possession of hydroxyl groups(as axochromes that form H-bonds with cellulosic fibers) are the unique features of these dyes. The poor colorfastness to washing of the directly dyed cotton fiber can be linked to weak fiber-dye interaction (Scheme 2).

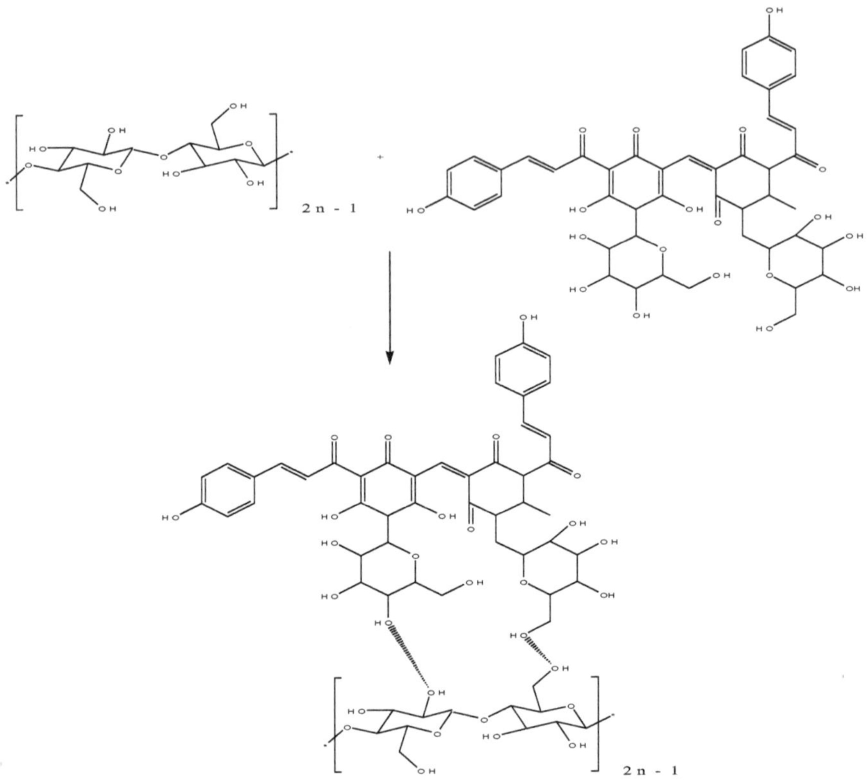

Scheme 2 Proposed direct dye-cotton fiber interaction

5.3 Disperse Dyes

Disperse dyes are low molecular weight organic compounds containing slight water solubilizing groups as auxochromes. As a result, their solubility in aqueous medium is low and they are often used in dyeing hydrophobic fibers like cellulosic man-made and synthetic fibers [25]. The adsorption of these dyes onto cellulosic man-made (regenerated cellulose, viscous rayon, and acetate) and synthetic (polyamide, polyester, and acrylates) fibers is done in neutral or mild acidic medium. These dyes are rarely used for coloring protein fibers (silk and wool) that share these same functional groups with synthetic fibers. Quinones, such as lawsone and juglone, and a few flavones belong to this class of dyes. The forces of attachment of these dyes to the fibers are van der Waals forces and H-bond since the dyes lack ionic groups.

5.4 Acid Dyes

They are habitually exploited for coloring wool, regenerated protein, silk, and acrylate fibers. These dyes are called acid dyes because they are used to color the above-listed fibers in a mild acidic medium. The unique features of dyes in this group are the possession of carboxylic or sulfonic acid group in their chemical structure and the formation of strong ionic bonds between the dye molecules and fibers upon which they are applied [4]. The mechanism of acid dyeing and the proposed fiber-dye interaction is presented in Schemes 3 and 4, respectively. The acid dyes impart bright shade and good colorfastness to wash onto the fibers. Female insect-based red dyes, such as kermesic, carminic, and laccaic acid, belong to this group.

5.5 Basic Dyes

Natural dye that ionizes when dissolved in water to acquire a cationic charge and intense color is known as a basic dye. Basic dyes are often used for coloring fibers carrying carboxylic groups like wool and silk (protein fibers). When applied on protein fibers in neutral to mild acidic medium, they are fixed to the fibers by ionic bonds to form bright-colored dyed fibers with poor colorfastness to light [25]. Amazingly, when these dyes are applied on acrylates, they form a bright color with an improved color fastness to light. Anthocyanide dyes are members of this class of dyes. The proposed chemical interaction between protein fibers and basic dyes is presented in Scheme 5.

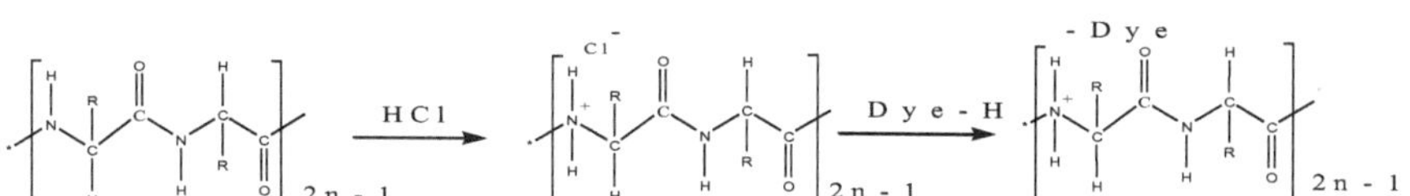

Scheme 3 Mechanism of acid dyeing of protein fiber

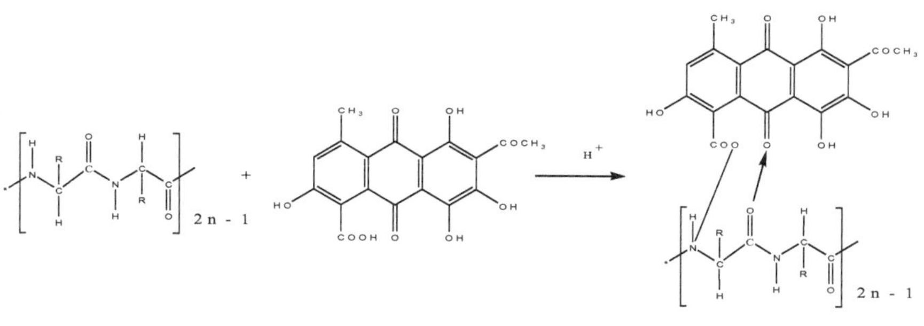

Scheme 4 Proposed acid dye-wool fiber interaction

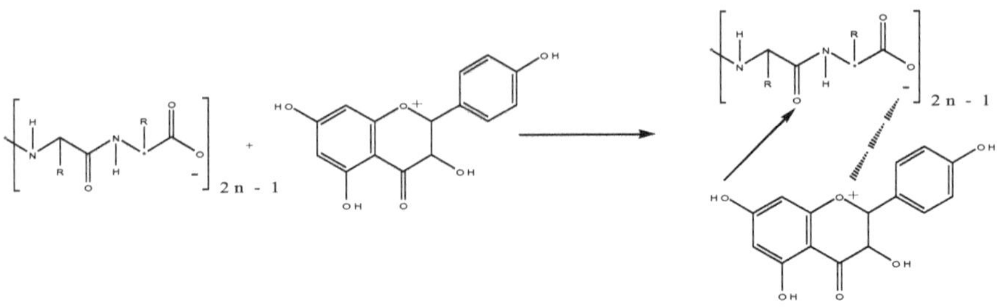

Scheme 5 Proposed chemical interaction between the basic dye and protein fiber [1]

5.6 Mordant Dyes

These are natural dyes that possess a slight affinity for the substrates. The affinity or chemical interaction of this group of dyes with the substrates is assisted by the incorporation of auxiliary chemicals called mordants [25]. The mordants often used for this purpose are metal salts (alum, salts of iron, copper, cobalt, tin, etc.) that chelate the natural dyes with the substrates to form a stable complex (Scheme 6) with the development of varieties of new shade (in most times) depending on the metal salt used [26]. The toxic nature of most of the chemical mordants encourages the entire world to search for bio-anchors which are less hazardous or toxic-free alternatives with nearly equal or better performance as mordanting agents for boosting the colorfastness of naturally dyed substrates [27]. Most of the available natural dyes can be linked to this group; the specific exceptions are indigo and dibromo-indigo dyes. The chemical interaction among the dye, mordant, and substrate is shown in Schemes 6 and 7.

6 Classification Based on Dye Color

This classification interests the consumers more than any of the other classifications because they are often more concerned about the shade on the substrates. As a matter of fact, color has a significant influence on human socioreligious cultural practices [6]. For instance, in Nigeria and some other places on the whole, a particular religious group is well recognized for wearing white garments, while another religious set can be linked to red-colored attire. During the yearly lovers' day celebration (February 14), the social people put on red and black attire as a symbol of belonging and/or remembrance for Valentine. Equally, from the middle toward the end of December, the world visitor called Father Christmas from Rome puts on a red and white–/cream-colored dress as a symbol of the Christmas festival. Naturally, putting on a dress of a particular color comforts some groups of people, and in most cases, identification of a particular race can be associated with skin color, while hair color has a direct relationship with human age. Based on the above-given reasons, the following are the colors of some important natural dyes:

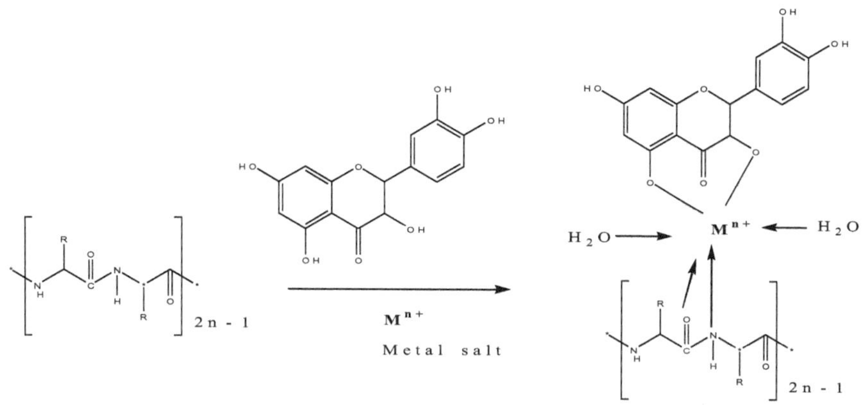

Scheme 6 Proposed mordant dye-electrolyte-protein fiber interaction [5]

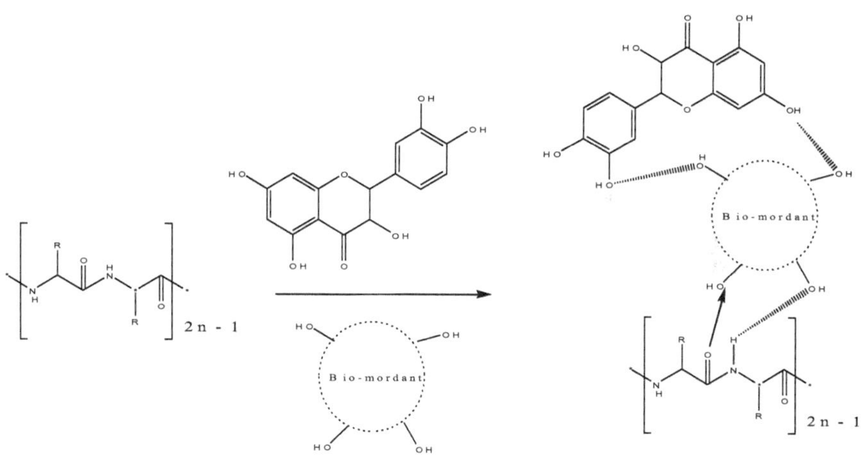

Scheme 7 Proposed mordant dye-biomordant-protein fiber interaction [21]

6.1 Yellow

A large number of natural dyes are yellow in color. This color is inferior technically due to its low tint strength and colorfastness to light in particular, when compared with other shades like green, red, black, and blue. Most of the flavonoids and terpenoids natural dyes like lutein, curcumin, acacetin, rosocyanin, and leucopene from *Moringa oleifera, Curcuma tinctoria, Acacia nilotica, Helianthus annus,* and *Lycopersicon esculentum*, respectively, are forms of yellow shade [28].

6.2 Orange

Orange is a color with a visible-spectrum wavelength appearing between those of yellow and red. Some of the natural dyes that display this color on dyed substrates are bixin, catechin quercetin, and nimbin from *Bixa orellana* seeds, *Theobroma cacao* leaf and fruit pod, *Bridelia ferruginea* bark, and *Azadirachta indica* bark, respectively. Other forms of orange natural dyes are pelargonidin and cyaniding-3-glycoside from leaves of *Phaseolus vulgaris* and *Acalypha wikesiana* [25].

6.3 Purple

Natural dyes having this color were highly priced in the Roman Empire prior to the nineteenth century, where it was regarded as a royal dye because it was only kings and priests that could afford to buy the attire dyed with this dye. The first kind of these dyes named Tyrian purple (6,6- dibromo-indigo) was extracted from the glands of mollusk species [29]. Afterward, dyes of this color are obtainable from plants, like chalcone from the leaves of *Ipomoea batatas* and haematoxylin from the heartwood of *Haematoxylum camphecianum.*

6.4 Brown

Brown-colored dyes obtained from different parts of many plants are used for coloring natural fibers like cotton, wool, and silk with good tint strength. For instance, cotton fibers were reportedly dyed brown with cutch from stems of *Acacia catechu* in the literature; other brown dyes suitable for coloring natural fibers are naphthoquinones like lawsone and juglone obtainable from leaves and seed pods of *Lawsonia imermis* and *Juglans regia*, respectively [30, 31].

6.5 Red

Red-colored natural dyes possess unique attributes like very good colorfastness and high tint strength when applied onto both natural (such as silk, wool, and cotton) and man-made (such as polyamides, polyesters, rayon, acrylates, and acetates) fibers. Most natural dyes with this color are of anthraquinoid skeletal frame, like kermesic acid, carminic acid, alizarin, alkannin, and purpurin from dried body of *Kermes ilicis, Dactylopius coccus,* root of *Rubia tinctorum*, bark of *Alkannin tinctoria*, and *Galium verum*, respectively. Among others, which are non-anthraquinone structural frameworks, are retusin glycoside and luteolinidin from *Pterocapus erinaceus* bark and *Sorghum vulgare* stem, respectively [32].

6.6 *Blue*

Blue dyes are usually of desired properties, like excellent colorfastness and high color strength. The ancient blue dye named indigo, which is still being used, belongs to this group. Among others in this group are sulfonated indigo extracted from *Indigofera tinctoria*, just like indigo itself, indica, and tsuykuss from Japan [33].

6.7 *Green*

Natural dye with a green color is rare. What most dyers do to obtain green shade on substrates is a combination of two dyes of different colors, usually blue and yellow, or mordanting of yellow dyed substrates with salt or iron. However, some plant-based green dyes for coloring substrates are vernodalol, flavonoid, and chlorophyll from *Vernonia amygdalina*, *Convallaria majalis*, and *Urtica dioica*, respectively [34].

6.8 *Black*

Just like the green color, natural black dye is not common. The dyers obtain black shade on substrates by combining two or more dyes or through metal salt mordanting of haematoxylin and laccaic acid. Some of the few black dyes obtained from plants are delphinidin and mangiferin from the roots of *Iris germanica*, tannin from the leaves and bark of *Vachellia nilotica* and *Anogeissus latifolia*, and terpenoid from *Cedrela toona* bark. Black dyes generally have a high affinity for natural and man-made fibers with enhanced colorfastness and high color strength [30].

7 Conclusions

Natural dyes are sustainable, abundant, biodegradable, and environmentally safe water-soluble chemical matters used in all manufacturing sectors for beneficiating man and his surroundings. Coloring matter can generally be in two forms, namely non-extractable (schemochromes) and extractable (biochromes). Biochromes are referred to as natural dyes because they are extractable/obtainable from nature. The bulk supply of natural dyes is plant-based. These dyes can be easily identified by their unique chemical structures, which belong to indigoids, quinonoids, terpenoids, or phenolic groups. Among these structural groups, phenolics are the most available with a large collection of colors and a high affinity for textiles. Fermentation vatting, direct, disperse, basic, acid, and mordant dyeing techniques used for coloring substrates by the dyers depend largely on the chemical constituents and structure of natural dyes. Nature is gifted with dyes of a set of colors, varying from yellow to orange, purple, and brown. Other frequently seen colors of natural dyes are red, blue, green, and black. These colors significantly influence the socioreligious cultural activities of human beings.

References

1. Jabar, J. M., Alabi, K. A., & Lawal, A. K. (2020). Synthesis, characterization and application of novel 1, 3-bis[(furan-2-l)methylene] thiourea functional dye on wool and cotton fabrics. *SN Applied Science, 2*, 1850.
2. Bukhari, M. N., Rather, L. J., Shabbir, M., Islam, S., Singh, U., Khan, M. A., & Mohammad, F. (2019). Dyeing of wool with anthraquinone based natural colorants from *Cassia fistula* fruit. *Journal of Natural Fibers, 16*(6), 855–865.
3. Li, J., Liang, Q., Hong, J., Yan, J., Dolgov, L., Meng, Y., Xu, Y., Shi, J., & Wu, M. (2018). White light emission and enhanced color stability in a single-component host. *ACS Applied Materials and Interfaces, 10*(21), 18066–18072.
4. Popoola, A. V. (2015). *Chemistry of colours in dyes and pignment*. Wits Publishing Ltd.
5. Jabar, J. M., Ogunmokun, A. I., & Taleat, T. A. A. (2020). Color and fastness properties of mordanted *Bridelia ferruginea* B dyed cellulosic fabric. *Fashion and Textiles, 7*, 1.
6. Jabar, J. M., Adedayo, T. E., & Odusote, Y. A. (2021). Green, eco-friendly and sustainable alternative in dyeing cotton fabric using aqueous extract *Mucuna slonaei* F dye: Effects of metal salts pre-mordanting on color strength and fastness properties. *Current Research in Green and Sustainable Chemistry, 4*, 100151.
7. Jablonsky, N. (2006). *Skin: A natural history*. University of Califonia Press.
8. Fang, J., Meng, C., & Zhang, G. (2022). Agricultural waste of *Ipomea batatas* leaves as a source of natural dye for green coloration and bio-functional finishing for textile fabrics. *Industrial Crops and Products, 177*, 114440.
9. Yusuf, M., Shabbir, M., & Mohammed, F. (2017). Natural colorants: Historical and sustainable prospects. *Natural Products and Bioprospecting, 7*, 123–145.
10. Kumar, A., Dixit, U., Singh, K., Gupta, S. P., & Beg, M. S. J. (2021). *Structure and properties of dyes and pigments*. IntechOpen.
11. Adeel, S., Rehman, F., Pervaiz, M., Hussaan, M., Amin, N., Majeed, A., & Rehman, H. (2021). Microwave assisted green isolation of lacaic acid from lac insect (Kerria lacca) for wool dyeing. *Progress in Color, Colorants and Coatings, 14*, 293–299.
12. Hosseinnezhaid, M., Gharanjig, K., Adeel, S., Rouhani, S., Imani, H., & Razani, N. (2022). The effect of ultrasonic on environmentally extraction and dyeing of wool yarns. *Journal of Engineered Fibers and Fabric, 17*, 1–10.
13. Yılmaz, F., & Bahtiyari, M. İ. (2020). Antibacterial finishing of cotton fabrics by dyeing with olive tree leaves fallen during olive harvesting. *Journal of Cleaner Production, 270*, 122068.
14. Jiang, H., Hu, X., Meena, B. I., Khan, A., Hussain, M. T., Yao, J., & Wang, J. (2019). Extraction of natural dye from the stem of caulis spatholobi and their application on wool. *Textile Research Journal, 0*(00), 1–9.
15. Sinha, K., Saha, P. D., & Data, S. (2012). Extraction of natural dye from petals of flame of forest (Butea monosperma) flowe: Process optimization using response surface methodology (RSM). *Dyes and Pigments, 94*(2), 212–216.
16. Haque, M. A., Mia, R., Mahmud, S. T., Bakar, M. A., Ahmed, T. A., Farsee, M. S., & Hossain, M. I. (2022). Sustainable dyeing and functionalization of wool fabrics with black rice extract. *Resources, Environment and Sustainability, 7*, 100045.

17. Gupta, V. K. (2019). *Fundamentals of natural dyes and its applications on textile substrates*. IntechOpen.
18. Jabar, J. M. (2014). Effect of rate of dyeing on the fastness properties of indigo dyed cotton fabrics. *International Refereed Multidisciplinary Journal of Contemporary Research, 2*(1), 43–47.
19. Fraga-Corral, M., Garcia-Oliveira, P., Pereira, A. G., Lourenco-Lopes, C., Jimenez-Lopez, C., Prieto, M. A., & Simal-Gandara, J. (2020). Technological application of tannin-based extracts. *Molecules, 25*, 614.
20. Dulo, B., Pham, K., Githaiga, J., Raes, K., & Meester, S. D. (2021). Natural quinone dyes: A review on structure, extraction techniques, analysis and application potential. *Waste and Biomass Valorization, 12*, 6339–6374.
21. Jabar, J. M., Owokotomo, I. A., & Ogunsade, A. F. (2022). Sustainable dyeing of cotton fabric with mangiferin: Roles of microwave-rays and bio-mordants on fabric colorimetric and fastness properties. *Sustainable Chemistry and Pharmacy, 29*, 100822.
22. Mia, R., Islam, M. M., Ahmed, T., Waqar, M. A., Khanam, N. J., Sultana, S., Bhuiyan, M. S. K., & Uddin, M. N. (2022). Natural dye extracted from Triadica sebifera in aqueous medium for sustainable dyeing and functionalizing of viscous fabric. *Cleaner Engineering and Technology, 8*, 100471.
23. Saleem, M. A., Pie, L., Saleem, M. F., Sahid, S., & Wang, J. (2021). Sustainable dyeing of nylon fabric with acid dyes in decamethylcyclopentasiloxane (D5) solvent for improving dye uptake and reducing raw material consumption. *Journal of Cleaner Production, 279*, 123480.
24. Rehman, F. U., Adeel, S., Haddar, W., Bibi, R., Azeem, M., Mia, R., & Ahmed, B. (2022). Microwave-assisted exploration of yellow natural dyes for nylon fabric. *Sustainability, 14*, 5599.
25. Mansour, R. (2018). Natural dyes and pigments: Extraction and application. In *Handbook of renewable materials for coloration and finishing* (pp. 75–102). Scrivener Publishing LLC.
26. Jabar, J. M., & Abayomi, T. G. (2015). Effect of temperature on affinity of natural dye from coconut coir fibre for cotton fabric. *International Journal for Research in Applied Chemistry, 1*(1), 1–9.
27. Jabar, J. M., Ogunsade, A. F., Odusote, Y. A., & Yılmaz, M. (2023). Utilization of Nigerian mango (*Mangifera indica* L) leaves dye extract for silk fabric coloration: Influence of extraction technique, mordant and mordanting type on the fabric color attributes. *Industrial Crops and Products, 193*, 116235.
28. Gulrajani, M. L., Gupta, D. B., Agarwal, V., & Jain, M. (1992). Some studies on natural yellow dyes, part 1: CI Natural Yellow 3 (tumeric). *Indian Textile Journal, 102*(4), 50–56.
29. Paul, R., Jayseh, M., & Naik, S. R. (1996). Natural dyes: Classification, extraction and fastness properties. *Textile Dyer and Printer, 29*(22), 16–23.
30. Hao, S., Wu, J., Huang, Y., & Lin, J. (2006). Natural dyes as photosynthesizers for dye-sensitizerd solar cell. *Solar Energy, 80*(2), 209–214.
31. Bhuiyan, M. A. R., Islam, A., Islam, S., Hossain, A., & Nahar, K. (2017). Improving dyeability and antibacterial activity of Lawsonia inermis L on jute fabrics by chitosan pretreatment. *Textiles and Clothing Sustainability, 3*, 1.
32. Santis, D., & Moresi, M. (2017). Production of alizarin extracts from Rubia tinctorium and assessment of their dyeing properties. *Industrial Crops and Products, 26*(2), 151–162.
33. Sinha, K., Saha, P. D., & Datta, S. (2012). Response surface optimization and artificial neural network modelling of microwave assisted natural dye extraction pomegranate ring. *Industrial Crops and Products, 37*(1), 408–412.
34. Singh, R., & Srivastava, S. (2017). A critical review on extraction of natural dyes from leaves. *International Journal of Home Science, 3*(2), 100–103.

Natural Indigo Dyes: A Potential Dye for Sustainability

R. Surjit, H. Shanruthi, V. Sruthi, and K. S. Tarunvalavan

Abstract Recently, the use of denim garments has increased in India due to their excellent color fastness. The indigo plant was a very important dye historically, but it was later replaced by synthetic indigo dyes. Many synthetic dyes exert carcinogenic and allergic effects to humans. Most vat dyes are not biodegradable; they accumulate on lands and water bodies, causing ecological problems. In recent years, natural indigo has been recognized for its artistic value and reduced environmental impact. Although production of this dye presents several disadvantages, the use of exogenous enzymes in its production could lead to a new and sustainable methodology. Knowledge about indigo culture thrived greatly among peasant traditions in India. There are various ways through which indigo dyes could be sustainably extracted compared with vat indigo dyes. *Isatis tinctoria* (medieval indigo dye) and *Indigofera tinctoria* are few plants from which these dyes are extracted and are renewable. Valorization of waste generated from the extraction process of *Indigofera tinctoria* is possible. This dye is mainly used in denim jeans; smaller quantities of this dye are also used in dyeing wool and silk. Natural indigo dyes are highly sustainable with minimal adverse effects on the environment. The scope of usage of indigo dye is going to be very large as most of the people now are aware of sustainability to make a better future to live in and more works are being carried out using these dyes. This chapter will discuss in detail about natural indigo dyes and their effect on sustainability along with their applications.

Keywords Natural dyes · Traditional textiles · Indigo · Denim · Renewable

R. Surjit (✉) · H. Shanruthi · V. Sruthi · K. S. Tarunvalavan
Department of Fashion Technology, PSG College of Technology,
Coimbatore, Tamil Nadu, India
e-mail: rst.fashion@psgtech.ac.in

S. S. Muthu (ed.), *Natural Dyes and Sustainability*, Sustainable Textiles: Production, Processing, Manufacturing & Chemistry,
https://doi.org/10.1007/978-3-031-47471-2_9

1 Introduction

People have been using natural dyes as a coloring medium since ages. Among those dyes, indigo plays a significant role in history. Indigo is a well-known natural dye that is extracted naturally from different plants belonging to the *Indigofera* genus. However various developments of this dye were brought about using different synthetic methods. This dye was highly demanded in the ancient times due its exclusive blue color. With its increasing demands, the number of methods for its synthesis also increased. Various plant species were used to extract indigo. Moreover there were distinct methods for extracting these dyes. As a natural dye, indigo can be used to dye fabric in several ways. It possesses properties such as durability, fastness, etc. It is one of the vat dyes used for fabrics such as denim. This dye, being associated with vat dye and its natural origin, has a remarkable place in the aspect of sustainability. Along with these, several studies have been conducted based on natural indigo dye. Nowadays, synthetic vat dyes are increasingly used. Normally, health and the environment are affected by the harmful effects of synthetic fibers. The major reason for the increasing use of synthetic dye is the fast fashion of the current era. Environmental hazards posed by the use of synthetic vat include air, water, and soil pollution. These hazards are harmful to both the factory workers and consumers, even animals and plants.

Synthetic indigo dye affects the surrounding environment in the form of water, air, and soil pollution. In the case of water pollution, the production of synthetic vat dyes in dyeing industries causes degradation of water bodies through acidic substances, such as lubricants, and dissolved salts, such as chlorides and bicarbonates. These substances are dumped into natural water and manmade bodies such as oceans, rivers, lakes, and dams. They are very corrosive and explosive. Furthermore, the production of synthetic vat dyes results in the emission of glycol ethers as well as harmful combustion gases and reactive components. Emission of volatile organic compounds (VOCs) is also a major disadvantage of synthetic vat dyeing. This could lead to acid rain and depletion of ozone layer, thereby leading to air pollution. The residue of these dyes causes irregular photosynthetic cycles. These dyes are also nonbiodegradable and take a long time to decompose, thus inducing long-term pollution. Furthermore, they can destroy natural habitats in the soil and thus cause soil pollution. VOCs are among the carcinogenic, toxic, and mutative compounds in the world. They are released in small concentrations during the production of garments using synthetic vat dyes. Furthermore, they could be very harmful to factory workers and people residing near the factory. Nevertheless, these problems can be eliminated by using natural indigo dyes. Compared with synthetic vat dyes, natural indigo dyes are eco-friendly and have no carcinogenic or allergic effects on humans, particularly those working in dyeing factories. Natural indigo dyes are affordable and are increasingly used for dyeing traditional fabrics. Indians are considered as the masters in the art of using natural indigo dyes. On the other hand, natural vat dyes are very good in eliminating pollution caused by synthetic vat dyes.

In this chapter, the history of natural indigo dyeing is discussed. Natural indigo dye has been used since ancient times. The extraction of indigo dye from different plant sources is also feasible using various methods, which are discussed in this chapter in detail. Most of the significant properties such as color fastness and efficiency of this dye to have ideal exhaustion on the fabric provides an understanding that this natural indigo dye has the caliber to be one of the most efficient and sustainable dyes in the present time.

2 History of Dyeing

In the past, when talking about a fabric, people always focused on the fabric's weave structure, embroidery, and design despite the fact that color is the very first thing that is observed by the viewer. Color is a very definable and dominating aspect that adds beauty to a textile. However, the use of these colors was not justifiable enough. Only in the late twentieth century did people start looking forward to choosing fabrics with different colors and prints. The process of dyeing involves imparting a form of interaction between a chemical called dye and a fiber or fabric where the dye spreads throughout the entire material and colorizes it. It works on the mechanism of combination or a combined effect of adsorption and diffusion. Textile dyeing could be traced back to the Neolithic period or the New Stone Age, almost around 10,200 BC. Evidences from China highlight that dyes from natural sources like insects, barks, and plants were used about 5000 years ago. Much earlier evidences indicate the origin of textile dyeing in Sindh province, Pakistan. Initially cotton was dyed using vegetable dyes in a more basic process of natural dyeing.

2.1 History of Indigo Dyes

About more than 5000 years ago, countries like India, East Asia, and Egypt were well known for their blue dye. This dye was said to have been extracted from a plant, *Indigofera tinctoria*. The history of natural indigo dye dates back to the pre-Vedic period in India where people perceive *Indigofera tinctoria* L. as a primary source of blue dyestuffs. However, the production of indigo was closely related to the utilization of slaves in European colonies of West Indies and North and South America for extraction. A similar method of production took place in Asia; thus, the people in this continent called indigo "devil's dye." Indigo used to be one of the rare commodities that were expensive. It was a product that only the royalties and aristocracies could afford. This was because of the various difficulties that came along with its import from other countries. Almost 150 years ago, the first jean was patented by a Bavarian immigrant. Indigo is said to have gotten its name from "The Indian" or "from India."

The leaves of the *Indigofera* plant were harvested and soaked in water, which led to the release of glucose by the amino acid-containing indican. The released indicans were fermented along with other plant enzymes that cause the water to turn yellow. As a result, indoxyl was obtained, which was air-dried, resulting in its oxidization and coagulation, thus turning into blue indigo. It was later mixed with an alkaline solution, such as bicarbonate, turning into a sludge-like byproduct. These byproducts were pressed into cakes and allowed to dry and then ground to fine powder with dark-blue pigments. Indigo cultivation was not possible in Europe due to cold climate, which poses unfavorable condition for its growth. Since the twelfth century, *Isatis tinctoria* plant, commonly known as woad, yielded results similar to those of *Indigofera*. However, it yielded 1/30th of the amount of the dye that was obtained from the same quantity of the *Indigofera* plant. Hence it necessitated the search for an alternative. During the seventeenth century, indigo enthusiasts found themselves being dependent on American colonies. Hence, they began cultivating direct relatives of the *Indigofera* plant. In South Carolina, there was an attempt to cultivate *Amorpha fruticosa*, which was termed as "false indigo" or "bastard indigo." The beginning of the eighteenth century was marked by the trade republic of Genoa becoming the largest buyer of "false indigo." Genoa was known for its abundance in deep blue cotton cloth, which was an ideal material for tough work clothes preferred by fishermen, seamen, dockworkers, and miners. People in Europe spoke more French. They called these blue cotton clothes "Blue de Genes," which meant Blue of Genoa. "Jeans" is an English derivative of it. These were mostly exported to Nimes, a French town from European Metropolis of weavers. However, they began producing their own blue cloths that were comparatively more durable.

During this period, "Twill" was developed from the combination of raw silk and wool that had parallel ribs. These ribs were made out of two threads, a warp and a weft. Indigo was used as warp thread and white as weft thread. This twill had two sides, the outward-facing that had more warp side with a dark blue color and the back-facing that had more weft side with light blue color. Due to its special weave technique, it is highly durable. Twill was found to be no less than Italian Blue de Genes and hence got its new name "Temie de Nimes" or "de Nimes," which meant "cloth from Nimes." "Denim" was the English and American derivative of it. By this time, the demand for indigo started growing. However, due to its expensive value, there was serious pressure to look for cheap alternatives. In America, as people started using workwear, they started demanding more of it. This demand for workwear in turn increased the demand for indigo.

2.1.1 Development of Synthetic Alternatives

In the 1860s, Adolf von Baeyer, a German chemist, conducted his first experiment of producing artificial Indigo. He made use of a synthetic production method in which he utilized amino acid aniline. The plant ash oxindole was produced and then later reduced to indol using zinc dust. It was further reduced to isatin and isolated. Hence, Adolf von Baeyer produced the first synthetic substitute to natural indigo that had significant dye properties. In 1873, the first-ever pair of workwears was patented. Workwears were in high demand as hardworking men began liking the durability of denim and the reinforced pockets. A significant transition with respect to the use of indigo blue was observed. Indigo, which was once a luxury, meant for royal people trickled down to the use of working class people.

By the end of the nineteenth century, BASF and Hoechst, modern chemistry giants, became known due to their remarkable investment on the development of synthetic alternatives. In 1904, an indigo agreement was signed by both of them. Along with that, a simplified isatin production process was related to a more consistent and stronger dye. Within a year, natural indigo was completely replaced with synthetic dyes [2].

2.1.2 Indigo in India

A fertile country like India, located in a tropical area with nutrient-rich soil and abundant rainfall, favors farming and agriculture. India's soil has been capable of growing various crops, including rice, wheat, pulses, cotton, and peanuts, since the ancient times. Apart from these crops, cultivation of indigo holds a significant place in the history of India. The introduction of indigo plantation in India and the subsequent effects it had on the locals in the country, including how they managed to turn against the British rule and fight for their justice, made indigo a part of the country's struggle for freedom. It was in 1777 when indigo plantation was first introduced in India, particularly in Bengal, by Frenchmen. Louis Bonnaud, a Frenchman, was the first indigo planter in India. This was followed by the plantation of indigo on a large scale, which led to the popularity of indigo as a commercially profitable crop. The market value of indigo grew due to the high demand for blue dye in Europe. This called for much expanded indigo plantation and production. Almost a thousand indigo factories boomed in and around Bengal by 1830. Peasants living across Bengal were forced to grow indigo in place of food crops by indigo planters, even in their own land. These indigo planters took various steps to keep the peasants completely under their control. They even provided loans with very high interest rates, which was called the Dadon. It was actually a one-way trap for the farmers. Once they took up loans from those planters, they would be in debt and crisis for their entire life. At times, this economic burden was passed down to the succeeding generations. Aside from forcing them to grow indigo, the planters paid only 2.5% of the original market price they would have gained if they had sold it in the market. This made it difficult for the farmers to live a normal life. Moreover, farmers during these times saw literally zero profit by growing indigo. In 1859, a movement against the indigo planters took place, which was led by the peasants growing indigo in Bengal. This movement lasted for almost a year [13].

When they were given the right to own the land, the planters managed to suppress and force the peasants to grow indigo, and along with it was the deprivation among the peasants of the due price they got from selling their goods in the market. Furthermore, during this time, the value of indigo in the international market significantly decreased. Upon learning this, the farmers refused to further grow indigo; however, the planters managed to dominate and force them to continue growing indigo. At first, most of the farmer protested and movements by the locals were nonviolent and passive. However, after continuous suppression, they resorted to violence. They started burning down indigo storage areas that were owned by the planters. As a result, the government introduced the Indigo Commission in 1860. This commission mainly focused on the repression faced by the common people and farmers because of the inhumane behaviors of the planters. It stated how the peasants were deprived of growing food crops in their own land to cultivate indigo. It also highlighted the high interest rates of loans the planters offered to the farmers to keep them in debt forever. The fact that planters signed contracts with these poor farmers, forcing them to grow only indigo, was also included in the commission report. Despite highlighting the oppression among peasants by the planters, the report actually focused on the need for British rule in India. Moreover, the commission did not propose any new laws to protect these peasants. It just requested the government to stop interfering with the issues revolving around the cultivation of indigo. This commission had a yet another remarkable impact on India and farmers who grew indigo during that time [1].

3 Indigo Plant Sources

Indigo plant sources include *Indigofera*, *Isatis tinctoria*, *Polygonum tinctorium*, *Strobilanthes flaccidifolius*, and *Lonchocarpus cyanescens*.

Indigofera
This genus belongs to the Leguminosae family and includes almost 800 species. It is the third largest in this family. However, only very few of this has a significant amount of indican, which is a precursor for the production of indigo dye. *Indigofera tinctoria* is a notable species of this genus.

Isatis Tinctoria
This is a particular plant species belonging to the genus *Isatis*. This plant is commonly called "woad." This plant is abundant in the Mediterranean and Western Asia. The huge seed pods that are dark are described to be "like little blackish toongs." The Turkish species of this plant is said to contain indirubin, "indigo red."

Polygonum tinctorium
It is also known as "Chinese or Japanese indigo." This plant grows up to 50 cm on an average. The fleshy stem of this plant bears large leaves that are dark bluish green in color. The color of these leaves ranges from green to red. It also has white to dark pink flowers. However, it is proven to be efficient in producing more quantity of indigo dye when compared with those produced from the leaves of woad.

Strobilanthes flaccidifolius
It is also called "Assam indigo." It is one of the most important dyes extracted from nature in minority areas even today. It is a perennial subshrub. It falls under the family Acanthaceae. It is seen in the mountainous and remote parts of Asia.

Lonchocarpus cyanescens
It is said to grow up to a height of 3 m. It has huge leaves and flowers that are purple in color. It grows in parts of West Africa. Later it was grown in various parts of Malaysia.

3.1 Dyeing of Natural Indigo Dyes

Natural indigo dyes have the potential to dye all the natural fibers and give them magical shades of blue. They have the ability to give the shades from summer sky to purple black where the exact shade may depend on the dyeing procedure and influencing factors. When dyeing using natural indigo dyes, the three main factors that are needed are the naturally extracted indigo dye powder, a reducing agent, and finally the alkaline medium/base.

3.1.1 Dye

The natural indigo dye molecule has a planar structure with the chemical formula $C_{16}H_{10}N_2O_2$. A study stated that indigo dye has two isomers, namely, cis and trans, with the trans isomer considered to be more stable than the cis isomer [6] as shown in Fig. 1.

The indigotin is the color-giving substance that is present in the natural indigo dye molecule. Flavonoids, terpenoids, alkaloids, glycosides, indigotin, indirubin, and rotenoids are some of the compounds that are present in the dye molecule [28]. Normally the dye molecule contains 15–55% of indigotin in it, in which 15 g of dye powder can dye 450 g of fabric in the shades of dark blue/900 g of fabric in the shades of medium blue. Carbonyl and amino are the functional groups contained in the indigo dye molecule. The adjacent carbonyl and amino groups are bound together by the hydrogen bonds.

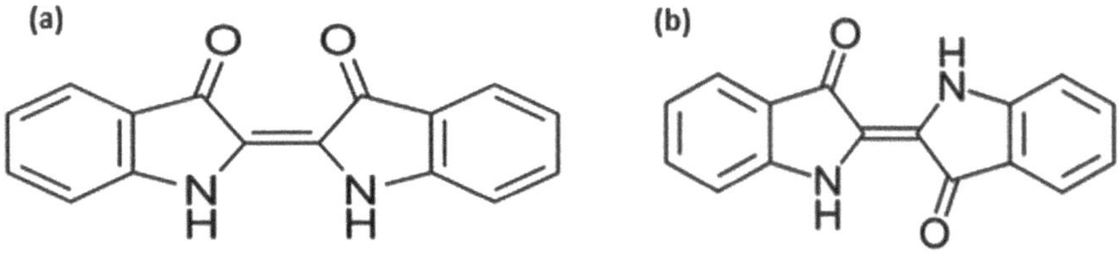

Fig. 1 Isomers (**a**) cis and (**b**) trans [8]

3.1.2 Reducing Agents

Basically, the indigo dye is insoluble in water, acid, or alkaline medium. So, reducing agents are used to reduce the indigo dye molecule and to make it soluble or leuco form. In conventional methods, sodium dithionite ($Na_2S_2O_4$) is used as a reducing agent. The main disadvantage of using sodium dithionite is that it will cause the dye to have low stability in the fabric and that the decomposition of sodium dithionite releases sulfate and sulfite to the environment, which causes environmental issues. In the natural dyeing process, the reducing agents are also extracted from nature itself. There are many studies that have been made to analyze the effectiveness of many natural reducing agents. It is recommended to use reducing agents that are rich in fructose and also fruits that are rich in sugar. Glucose is the commonly used reducing agent. Glucose oxidizes to gluconate and helps in reducing the indigo dye and dyeing [5]. Below are some of the natural reducing agents that have been experimented:

- Mao berries are like blue berries extracted from nature that give higher color strength than the conventionally used sodium dithionite. So, it is preferred as the best green alternative [24].
- *Rubus coreanus* (bokbunja), a Korean black raspberry plant that acts as a very good reducing agent in the reduction process of the synthetic as well as the natural indigo dyes [26].
- Banana, apple, and date palm are used for the reduction processes of natural indigo dyes in an experiment. It was found that among the three, date palm extract has good performance in terms of dyeing, better rubbing and wash fastness with soap and color strength [10].
- French botanist and chemist Michel Garcia had done a series of experiments on the basis of the dyeing ratio of 1:2:3, where 1 represents the dye proportion, 2 represents the alkaline base, and the 3 represents the reducing agent proportion. The first series was the citrus series, which includes extracts of orange rinds, lemon rinds, and commercially available pectins. The sample in which the orange rinds were used showed good reduction and color strength. Another series was named vegetable series in which the extracts of zucchini, turnip, carrot, sweet potato, and white potato were used for reduction. Among these, sweet potato reduced first and gave a good result [5].
- Banana peel, green apple, and figs are some of the other green reducing agents that can be used.

3.1.3 Base

For the reduction process of indigo dye molecule to take place, a basic or alkaline medium is required. Calcium hydroxide (calx) is the base that we use in the dyeing of natural indigo dyes. Calx is used to reduce oxygen in the dye vat. Sometimes soda ash and iye are also used. The reason behind the preference of using calcium hydroxide in natural indigo dyeing is that it does not have any harmful effects on the environment when disposed of after use. Calcium hydroxide combines with carbon dioxide and forms calcium carbonate, so that is very easy to dispose of without causing any adverse effects [22].

3.1.4 Mechanism

The mechanisms behind the dyeing of natural indigo dyes are the reduction and oxidation processes. As said earlier, indigo dye molecule is insoluble and has less affinity toward fiber; so first, it is made into soluble or leuco form, which helps the dye molecule disperse and penetrate into the pores of the fiber and give the desired color to the fabric. The dye vat will be blue in color when the dye powder is mixed with water or other solutions (in insoluble form), so the reducing agent and alkaline medium are added to reduce the indigo vat by removing the oxygen in the vat and giving the green yellow color with the brown bubbles sometimes. Now the fabric can be added into the vat for dyeing. After that, the fabric is taken out and subjected to oxidation so that the blue color will appear.

3.1.5 Process Flow

The process of natural indigo dyeing is very simple compared with that of any other dyes. First, to prepare the vat, boiling water is added in the vessel or vat, and then calcium hydroxide (calx) is added. Subsequently, the temperature of the vat is reduced to 45 °C. Then, the natural indigo dye that was extracted using natural methods is added, and the pH for the dye bath is checked. If the pH is less than 11, the required amount of calx is added to maintain a pH of around 11 in which the reduction and dyeing processes take place effectively. Then the vat is left to stand for 30 minutes. After 30 minutes, vat is stirred very slowly to check the color change; completion of the reduction process is indicated by the color change from blue to green yellow. If the required color is not achieved, then the vat is left to stand for another 30 minutes, and the color change is observed. Sometimes it may take up to 2 hours for the reduction process to take place. Once the green yellow color is achieved, the yarn or fabric is dipped in the dye bath. After some time, the yarn or fabric is taken from the bath and subjected to air oxidation to achieve the desired blue color.

3.1.6 Factors Influencing Dyeing

Various factors that influence the dyeability, intensity, and color strength of the dyed yarn or fabric are as follows:

- *Dye concentration:* The amount or quantity of the dye used has a great impact on the dyeing process. As the concentration of the dye bath increases, the intensity of the color of the dyed yarn or fabric will also increase. At the same time, when the dye concentration continuously increases, the rubbing fastness of the fabric reduces rapidly.
- *pH of the bath:* pH is the most important factor in the dyeing process. It facilitates fixing of the dye in the fabric. The pH in the dye bath or vat should be maintained at the optimum level of around 11 to achieve proper reduction and dyeing for cotton. Bayer's yeast is used as a reducing agent, and the effect of pH on it was studied. When the pH is around 11.0, it adjusts itself to 11 and induces better reduction, whereas in other pH values, it does not.
- *Number of dip*: In natural indigo dyeing, the fabric or yarn is dipped in the vat to achieve the darker shades and taken out many times. The number of dips in the dye bath influences the dye intensity.
- *Time*: The number of hours that we allow the fabric to be in the dye bath matters a lot in the dyeing process. As the time of the fabric in the vat increases, the penetration of the dye into the fabric also increases [29].

3.1.7 Ring Dyeing

As we all know, the wefts in the denim are not dyed, and only the warp yarns are dyed. The warps are dyed only on its surface, and the formation of the ringlike structure of the dye stops the dye molecules from diffusing further into the fabric. So, the ring-dyed fabrics have very low rubbing and wash fastness, which is very helpful in achieving the faded effect in the garments. This is also because of the undyed white weft yarns in it. It can be said that ring dyeing is a distinctive process that can make the denims even more special than any other garments and dyes (colors). When there is a need to achieve a quicker and better stonewashing effect, it is preferred to use the whiter weft yarns and the thinner ring-dyed warp yarns, so that less abrasion is enough to achieve the desired effect.

4 Properties of Natural Indigo Dyes

Natural indigo dyes have a superlative property. When it comes to dyeing, the dyes should fulfill some of the basic requirements. Some of the important properties that a dye should have are the suitable color, fixation to the fabric, and color fastness property to the fabric. When it comes to dyes, solubility is also an area of concern. Improperly solubilized dyes cannot dye the fabric evenly. For any dye to be used for dyeing, it must dissolve or be solubilized in the liquid medium, only then the dispersion will take place in the fabric. Here the dye is basically insoluble in nature. As we discussed earlier, we need a reducing agent to reduce the oxygen in the indigo dye molecule and to get it in the leuco form.

When talking about colors in the fabric, the first thing that comes into our mind is fastness. Fastness refers to how long the color withstands in the fabric even after abrasion or strain or something like that. There are numerous dyes that are available and can be extracted from nature, for example, turmeric, jack fruit, onion peel, tea powder, beetroot, henna, etc. Blue color is induced by two major plants, *Indigofera tinctoria* and *Isatis tinctoria*; here our major concern is indigo. Indigo always gives deeper shades and good fastness properties than woad or any other dyes because it is processed via vatting. Looking into the fastness properties of some other natural dyes, henna is said to have a good fastness property when compared with turmeric, tea leaf, and pomegranate [25]. However, the fastness properties of synthetic vat dyes is always higher than any other natural or synthetic dyes, but the environmental side effects are very much dangerous in case of the synthetic vat dyes. Fastness can be assessed by the gray scale. After subjecting the fabric to rubbing or washing or any abrasion, the color fade can be identified using the gray scale.

The fastness properties of natural indigo are amazingly good when compared with other natural dyes. Exploring the after-effects of the residue in the indigo dye vat, it was found that it can be reused as an agriculture fertilizer; thus it does not exert any unfavorable effects on the environment. In a study conducted by Zarkogianni et al. [30], it has been found that the natural indigo-dyed fabric without mordant exhibited excellent wash fastness properties. It has been understood that by nature, the natural indigo does not need any mordant for its dyeing process. In another study conducted by Comlekcioglu et al. [4], when the extraction was done by hot water treatments with a pH of exactly 11.0, the natural indigo gave a dark shade of blue and the washing fastness ranged from 2 to 4/5. It possesses moderate rubbing-fastness and poor light-fastness properties. The indigo silk-dyed fabric with and without mordant exhibited the same level of fastness, ranging from 4 to 5. In a study conducted by Glowacki et al. [7], it has been said that natural indigo dyes can also be used as a semiconducting material.

The properties of natural indigo-dyed fabric can be enhanced in many ways. Many studies have been conducted on the basis of ultraviolet properties and in the area of absorbance. On dyeing of silk fabric with natural indigo dye, it gives the UPF (Ultraviolet Protection Factor) value of 4.4–6.32 [16]. The post mordanting with ferrous sulfate as mordant and the higher concentration of the natural indigo gives greater values of UPF leading to better UV protection. The cotton fabrics dyed with natural indigo and coated with nano chitosan and nano zinc oxide exhibited increased surface area, which helps in the absorption of the dye molecule in the fabric or yarn and also in the improvement of fastness properties. Another experiment has been conducted to test the UV protection properties of plain cotton knitted white fabric and naturally dyed fabrics. Among the naturally dyed fabrics, the indigo-dyed fabric showed better UV protection values and color fastness ratings [20].

5 Application of Indigo Dyes

Even though natural indigo and synthetic indigo give the same colors, the effects they exert after use are concerning. In this sustainable fashion era, we are promoting the use of naturally available resources for dyeing and also other textile processes. The residues of synthetic vat dyes have unwholesome effects on human and nature. As natural indigo dyes are extracted from nature and has no environmental side effects, it is widely applied in the aspects of medicine, food coloring, and dyeing because it has good antimicrobial, anti-allergy, anti-septic, and anti-inflammatory properties. It helps in curing many diseases such as diabetes, skin diseases, etc.

When talking about denim or jeans, the focus is only on the shades of blue. The blue color in denim garments has a great impact on people. Even though we currently have jeans in many colors, blue always has a special attraction and affection. The blue color can be achieved using natural indigo dyes. The era of fast fashion has paved the way for the love of garments among normal people. People were keen on filling their wardrobes with upcoming trends and favorite dresses. Denim garments are mostly used in the era of fast fashion. Indigo-dyed denims are used not only for making garments but also many accessories such as bags, hairbands, shoes, and even earrings. A report says that denim garments have a separate and the largest place in the wardrobes of many people. When people's love for denim increases, the production also increases, but with the use of synthetic dyes. Synthetic dyes exert lots of harmful effects on the environment such as water, soil, and air pollution. Nowadays people started to be concerned about environmental sustainability a lot. So, the use of greener methods that cause very less or no negative effects on the environment for the production of any products has increased. Similarly, natural indigo dye production and application of the dye to the denim increased. The cotton denims are mostly dyed with natural indigo when compared to all other natural and synthetic denims. The dye can also be applied better on most of all the natural fibers such as silk, wool, linen, hemp, and synthetics.

6 Natural Indigo and Sustainability

As said earlier, natural indigo dye is extracted from the *Indigofera tinctoria* L. plant. It grows mainly in India. Based on sustainability factors, *Indigofera* is one of the most credited crops. It is a very hardy crop and grows easily in all types of soil without the aid of any chemical fertilizers. It is a legume crop, so it has the ability to extract nitrogen from air and transfuse into the soil. It can be used as a source of natural fertilizer and as a rotational crop to vitalize the soil. The stems of the plant are used as firewood after harvesting in some parts of India and Bangladesh such that the whole plant is used.

At present, sustainability has become the main goal for textile industries and all brands, and factories are trying very hard to achieve sustainable development. From fiber to garment processing, every factor contributing to sustainability is analyzed, and dyeing using natural dyes is key to a sustainable processing. It can be a major remedy for the harmful effects caused by the synthetic dyes. Natural indigo dye could be considered as a potential dye for sustainability and could eliminate the synthetic vat, which is widely used for the production of denim garments. As discussed before, garments made from synthetic dyes lead to non-sustainable environment as serious pollution problems are envisaged, whereas for sustainable way of clothing, natural vat dyes can be used. Natural indigo dyes are biodegradable and exhibit antibiotic, antiallergic, noncarcinogenic, and nontoxic properties. They are easily available. They are a renewable source of natural and bright colors. The application of synthetic vat dyes releases large amounts of waste to water. It is found that none of the synthetic vat dyes have 100% exhaustion. So they produce tons of unfixed dyes and cause light transpose and lead to improper photosynthesis. Synthetic vat dyes cause health hazards, pollution, and disturbance to eco-balance. So the use of natural indigo for the production of denim and other indigo-dyed garments is drawing attention.

When synthetic dyes are used to dye denim jeans with blue color, 50–100 l of water is required, and after dyeing, the effluent contains toxic reducing agents and dissolved salts [21]. Due to advancements in textile technology, the fixation rate of the indigo dye has increased. Nanocellulose–indigo dyeing of textiles is very environmentally friendly. Nanofibrillated cellulose (NFC) was generated and it enclosed indigo particles, and fixation to the fabrics was done by physical crosslinking using compounds such as chitosan [21]. For denim fabric dyeing, the NFC-natural indigo dyeing technology can be a very essential practice for sustainable processing in dyeing industries as conventional vat dyeing using pure indigo powder and prereduced indigo utilizes high quantities of reducing agent, alkali, and water, whereas the NFC-indigo dyeing uses no reducing agent and alkali. Its water consumption is also very low compared with other methods. It is an eco-friendly method where over 90% fixations could be driven by the addition of chitosan and NFC. It has also no ill effects on comfort properties such as air permeability, bending stiffness, etc. This method helps reduce chemical concentration and water consumption, which avoids multiple dipping and hence enhances sustainable dyeing.

Although indigo dye is very sustainable and purely nonartificial, its production process is complicated. Obviously, the production process of natural indigo (from leaf extract to the liquefied form of dye) is harmful and damages the environment. This would reverse the whole concept of sustainability. Therefore, they have found that a biochemical protecting group can be used as a stabilizer for indigo precursor indoxyl to form indican [11]. Here the glucose moiety is used as the stabilizer. If the stabilization occurs properly, a sustainable indigo dyeing strategy can be successfully introduced. Then reducing agent and toxic reagents would not be needed for the production process as they are the main causes of pollution during the process.

Water is highly used in the textile sector, especially in dyeing, printing, and wet processing. It is significantly more than all of the other processes combined. The annual consumption of fresh water only in the dyeing industry at the continent level is 600 billion liters, which is very huge compared with annual rainfall and other environmental considerations [27]. Wastewater should be treated properly and reused. In the indigo dyeing of denim, indigo dye used in the process is directly discharged into the natural water bodies such as rivers, oceans, and lakes or they are discharged into the wastewater. To prevent this hazardous condition, a mechanism has been proposed based on wastewater management where the effluents containing the indigo dyes were treated through four different ultrafiltration membranes. The chances of removal of color from the wastewater were studied. The result indicated that up to 99% of the concentrated dye was removed, leaving toxic dye-free water. It was also observed that 80% of chemical oxygen demand (COD) has been decreased. And the recovered dye is used again for the dyeing process. So this prevents light transpose and helps in proper photosynthesis cycles, thereby coordinating healthy wastewater management. So this helps in the sustainable dyeing of denim using indigo dye [3].

As natural dyes have become important in recent years, textile researchers are keen about dyeing using natural indigo, and natural dyes have been successfully used in laboratories. As they could not be produced in a sustainable manner, weld dyeing recipes were created using jet dyeing machines, and their properties such as color fastness were investigated [12]. They were produced in such a way that they are sustainable and therefore help in environmental friendly mass production of cotton dyeing with natural indigo dyes. During the extraction of natural indigo dye from *Indigofera tinctoria* L. plant biomass, the yield of pure indigo powder is just 1%, and the remaining 99% of the biomass is unused and discarded. This plant biomass can be recycled and reused to produce several value-added products. Here, the unused biomass can be used as biofuel obtained from sources other than food. The feasibility of this valorization of waste from natural indigo dyes depends on further study of the characteristics of natural indigo [18]. To achieve an ecofriendly and sustainable process of dyeing, utilization of waste biomass is very necessary. Other researches on therapeutic and green manure potential of the plant had been conducted. But this utilization of waste biomass could be more interesting and economical. Based on this, a detailed study of after dyeing biomass (ADB) and before dyeing biomass (BDB) is needed to understand the capability for the production of bioethanol, compost, and biogas.

There has always been a fascination for colors among humans; thus, we see various colorful dyes in our garments. As the population and demand for fashion and the textile industry increase day by day, the use of various dyes is also increasing rapidly. Among all dyes, indigo dye is the most credited one for its unique deep-blue appearance in garments, especially denim [11]. As the demand increases, the production for such garments also increases. The main problem lies in the fact that only synthetic vat dyes are used nowadays. Their production has doubled, and the demand for the synthetic vat dyes in the past century was satisfied by petroleum byproducts [23]. As discussed in the Introduction section, synthetic vat dyes affect the environment in several ways, particularly the denim industry; natural dyes should be preferred over vat dyes for the reason of sustainable development [17]. A great revolutionary change is needed for the denim industry to sustain in the world of apparel technology and sustainable practices should be done in order to reduce water footprint of denim industry [17]. Bio-indigo can be used as another alternative for chemically synthesized indigo. It is produced using organic microbes. Tryptophan is also utilized as a substrate for bio-indigo production using *Escherichia coli* [9]. However, no large-scale production was attempted based on this technique using recombinant microorganisms. Bio-indigo is a type of natural indigo dye that is extracted from the *Indigofera tinctorium* L. plant. It has several health benefits and medicinal properties. It is a legume that acts as a nitrogen source for plants and trees near its periphery. Its contribution to the fertility of the soil is huge and appreciable, and it acts as compost. The environmental impact of dyeing 1 kg of cotton using natural indigo dye is more than 10% lesser than that of dyeing using synthetic vat dyes. Fresh water consumption is an exception. Denim brands and companies can utilize bio-indigo to acquire a permanent sustainable solution for the toxic effects exerted by synthetic vat dyes in factories.

In cotton yarn dyeing in the textile sector, the widely used dyes are undoubtedly the dark-blue ones. However, dye water consumption and the amount of wastewater discharge in indigo dyeing are very enormous under the present conditions. So a cationic surfactant is added at the end of fixation; this type of dyeing technology can reduce alleviate serious wastewater pollution during indigo dye production [15]. A study based on a survey shows the differences between a sustainable design and indigo dyeing colors. The main aim is to show that the dark-colored dyes are less sustainable than light-colored dyes [14]. Microorganisms can decompose indigo-dyed fabrics. Light-colored fabrics take a short time to decompose, but the dark-colored ones take a long time. Light-colored indigo dyeing is eco-friendly as it saves dye, water, and electricity. The textile industry is a labor-intensive industry. Natural indigo dyeing provides job opportunities for people who do cultivation of crops, extraction and application of these dyes on textile. This can be a source of income for weaker sections in the society in rural and suburban areas both for dyeing and nonfood farming. So this generates sustainable employment. Natural indigo dyes have never failed to produce special aesthetic qualities in the dyed fabric. The use of natural dyes has given added value to apparel production. In the textile sector, if appropriate dyestuffs are used without exploiting the natural resources, it could help make the future better for upcoming generations, thereby achieving sustainable goals.

7 Future Scope of Natural Indigo Dyes

Denim garments, mostly jeans, are popularly used by people from the nineteenth century due to their durability and color fastness. If we check a random person's wardrobe, jeans rank high in terms of average number of garments. Denim garments are mostly made of indigo dyes. Levi Strauss & Co. entered into an agreement with Stony Creek Colours to use IndiGold™. It is a natural source indigo reduced by 20%. It is prereduced and the dyes are about to be commercialized in early 2023. The prereduction will help achieve consistent dye results and require less water and energy. Stony Creek Colors have also tied up with Archroma (a pigment maker) to use plant-based indigo dye in an industrial scale. Archroma is planning to use natural indigo in the dye and then test the properties of the denim produced. Adriano Goldschmied, a premium denim designer, incorporated natural indigo during the production of Tencel-Modal cellulosic fiber. This fiber is produced by Lenzing AG (an Austria-based company). The fiber is blue in color, and the yarns produced from these fibers are already colored in blue. These yarns are then made into t-shirts with motifs on awareness about sustainability printed on them. The market of global indigo dyes is expected to reach USD 1639 million by 2028 at a growth rate of 5% [19]. Indigo dyes are also used predominantly in hair coloring, heat removal in Chinese medicine for treating diseases, printing ink, antiseptic properties, mosquito repellency, burn healing, tea, etc. Many companies such as Candiani, BUAISOU, and Blue Print Amsterdam see immense potential in using natural indigo and are also determined to use natural indigo in the future. Cone Denim and Bossa Denim also work with natural dye manufacturers to create natural indigo-dyed denim collection. It is expected that workers in indigo plantation, farming, and dyeing are well compensated so that indigo will thrive more in the market. Sustainability is now a major concern due to the occurrence of natural disasters such as global warming, acid rain, etc. These harmful effects are mainly due to water, air, and soil pollution, which is also caused by the use of artificial indigo dyes. Considering both people's enthusiasm for denim garments and the environmental factors, natural indigo dye could be used as they resemble synthetic dyes, which have good color fastness and durability. The future scope for natural indigo dyes is immense and its usage is expected to be very high compared to other natural dyes. It could also help peasants cultivating indigo plants, therefore promoting agriculture. Although the cost of production of natural indigo is higher, increased production will lead to economies of scale and will be accessible at different price ranges for the customers on a sustainable front.

8 Conclusion

This chapter is an attempt to elucidate that natural indigo dye plays a vital role in the achievement of sustainable development. Therefore it is a potential dye for achieving sustainability. For more than 3000 years, indigo plant cultivation has been part of our culture. This indicates the significance of knowing the history of indigo dyes. Indigo dyes are extracted from the leaves of *Indigofera tinctorium* L., ensuring its renewability and usage towards the well-being of our environment. It also highlights the harmful effects of synthetic vat dyes in comparison with the eco-friendly organic vat dye. Denim garments are among the most used garments nowadays. So, natural indigo dye can have diverse applications in the production of sustainable clothing. Valorization of waste generated from the natural indigo dye extraction process is possible using a sustainable method, and sustainable designs using natural indigo are created. Hence using natural indigo dye on garments, we can derive many eco-friendly processes to make our future better. With the well-known fact that sustainability is closely connected with the availability of resources and cost of living of the future, a study like this that co-relates usage of a natural dye which has significant properties would widen the thoughts while selecting and using a dye in the textile field.

References

1. Bhattacharya, S. (1977). The indigo revolt of Bengal. *Social Scientist*, 13–23.
2. Bojer, T. S. (2017). *The history of indigo dyeing and how it changed the world* [Online]. Available from shorturl.at/exAEV. Accessed on 25 Jan 2023.
3. Buscio, V., Crespi, M., & Gutiérrez-Bouzán, C. (2015). Sustainable dyeing of denim using indigo dye recovered with polyvinylidene difluoride ultrafiltration membranes. *Journal of Cleaner Production, 91*, 201–207.
4. Comlekcioglu, N., Efe, L., & Karaman, S. (2015). Extraction of indigo from some Isatis species and dyeing standardization using low-technology methods. *Brazilian Archives of Biology and Technology, 58*, 96–102.
5. Ellis, C. (2016). Applications of cross dyeing with natural dyes. *Crosscurrents: Land, Labor, and the Port.* Textile Society of America's15th Biennial Symposium.
6. El-Mansy, M. A. M., Yahia, I. S., & Alfaify, S. (2015). Conformational and vibrational properties of indigo dye: DFT approach. *Org. Opto-Electronic, 1*, 39–45.
7. Głowacki, E. D., Voss, G., Leonat, L., Irimia-Vladu, M., Bauer, S., & Sariciftci, N. S. (2012). Indigo and Tyrian purple–from ancient natural dyes to modern organic semiconductors. *Israel Journal of Chemistry, 52*(6), 540–551.
8. Gupta, V. K. (2019). Fundamentals of natural dyes and its application on textile substrates. In *Chemistry and technology of natural and synthetic dyes and pigments*. IntechOpen.
9. Han, G. H., Bang, S. E., Babu, B. K., Chang, M., Shin, H. J., & Kim, S. W. (2011). Bio-indigo production in two different fermentation systems using recombinant Escherichia coli cells harboring a flavin-containing monooxygenase gene (fmo). *Process Biochemistry, 46*(3), 788–791.
10. Hossain, M. D., Khan, M. M. R., & Uddin, M. Z. (2017). Fastness properties and color analysis of natural indigo dye and compatibility study of different natural reducing agents. *Journal of Polymers and the Environment, 25*, 1219–1230.
11. Hsu, T. M., Welner, D. H., Russ, Z. N., Cervantes, B., Prathuri, R. L., Adams, P. D., & Dueber, J. E. (2018). Employing a biochemical protecting group for a sustainable indigo dyeing strategy. *Nature Chemical Biology, 14*(3), 256–261.
12. Karadag, R. (2022). Sustainable and mass production of cotton dyeing with natural dye (weld) in the textile industry. *Journal of Natural Fibers, 19*(15), 10935–10945.
13. Kling, B. B. (2016). *The blue mutiny: The indigo disturbances in Bengal, 1859–1862*. University of Pennsylvania Press.
14. Lai, C. C., & Chang, C. E. (2021). A study on sustainable design for indigo dyeing color in the visual aspect of clothing. *Sustainability, 13*(7), 3686.
15. Luo, Y., Pei, L., & Wang, J. (2020). Sustainable indigo dyeing and improvement of rubbing fastness of dyed cotton fiber using different fixing agents for obtaining eco-friendly cowboy products. *Journal of Cleaner Production, 251*, 119728.
16. Mongkholrattanasit, R., Saiwan, C., Klaichoi, C., Rungruangkitkrai, N., & Sasivatchutikool, P. (2015). UV-protection property of silk fabric dyed with natural indigo by using padding technique: A focus on effect of ferrous sulfate mordant. In *Applied mechanics and materials* (Vol. 804, pp. 217–220). Trans Tech Publications Ltd..
17. Pal, H., Chatterjee, K. N., & Sharma, D. (2017). Water footprint of denim industry. In *Sustainability in denim* (pp. 111–123). Woodhead Publishing.

18. Pattanaik, L., Padhi, S. K., Hariprasad, P., & Naik, S. N. (2020). Life cycle cost analysis of natural indigo dye production from Indigofera tinctoria L. plant biomass: A case study of India. *Clean Technologies and Environmental Policy, 22*, 1639–1654.
19. Periyasamy, A. P., & Periyasami, S. (2023). Critical review on sustainability in denim: A step toward sustainable production and consumption of denim. *ACS Omega, 8*(5), 4472–4490.
20. Pupatana, O. (2019). *Cotton knit fabric dyed with natural indigo and UV protection*. [Online]. Available from: https://ph02.tci-thaijo.org/index.php/past/article/view/242998. Accessed on 25 Jan 2023.
21. Rai, S., Saremi, R., Sharma, S., & Minko, S. (2021). Environment-friendly nanocellulose-indigo dyeing of textiles. *Green Chemistry, 23*(20), 7937–7944.
22. Ramesh, T. N., Kirana, D. V., Ashwini, A., & Manasa, T. R. (2017). Calcium hydroxide as low cost adsorbent for the effective removal of indigo carmine dye in water. *Journal of Saudi Chemical Society, 21*(2), 165–171.
23. Rebelo, S. L., Linhares, M., Simões, M. M., Silva, A. M., Neves, M. G. P., Cavaleiro, J. A., & Freire, C. (2014). Indigo dye production by enzymatic mimicking based on an iron (III) porphyrin. *Journal of Catalysis, 315*, 33–40.
24. Saikhao, L., Setthayanond, J., Karpkird, T., & Suwanruji, P. (2022). Mao berry as a green reducing agent in natural indigo dyeing on cotton fabrics. *The Journal of The Textile Institute, 113*(8), 1538–1544.
25. Shariful Islam, S. M., Alam, M., & Akter, S. (2020). Investigation of the color fastness properties of natural dyes on cotton fabrics. *Fibers and Textiles, 27*(1), 1–6.
26. Shin, Y., Choi, M., & Yoo, D. I. (2014). Eco-friendly indigo reduction using bokbunja (Rubus coreanusMiq.) sludge. *Fashion and Textiles, 1*, 1–8.
27. Vajnhandl, S., & Valh, J. V. (2014). The status of water reuse in European textile sector. *Journal of Environmental Management, 141*, 29–35.
28. Wahyuningsih, S., Ramelan, A. H., Wardani, D. K., Aini, F. N., Sari, P. L., Tamtama, B. P. N., & Kristiawan, Y. R. (2017, April). Indigo dye derived from Indigofera tinctoria as natural food colorant. *IOP Conference Series: Materials Science and Engineering, 193*(1), 012048. IOP Publishing.
29. Xin, J. H., Chong, C. L., & Tu, T. (2000). Colour variation in the dyeing of denim yarn with indigo. *Coloration Technology, 116*(9), 260–265.
30. Zarkogianni, M., Mikropoulou, E., Varella, E., & Tsatsaroni, E. (2011). Colour and fastness of natural dyes: Revival of traditional dyeing techniques. *Coloration Technology, 127*(1), 18–27.

Natural Dyes: Coloring Tool for Textiles

Shanthi Radhakrishnan, Yamunadevi Ayyaswamy, and Jasirabanu Ahamed Basheer

Abstract Natural dyes have been used since ancient times yet their importance has been low. Interest toward natural dyes has been revived with stringent environmental standards and health consciousness. Though many merits have been attributed to natural dyeing, there are a few limitations in terms of colorfastness and cost. To enhance the colorfastness, there is a need for mordants which may not be eco-friendly. Hence, there is a need for research to find ways to bond the natural dye to different natural fibers to make the process completely environment friendly and sustainable. Fundamentals about natural dyes and their chemistry should be understood for the conduct of different research approaches. An attempt has been made in this chapter to explain the basics of natural dyes and its application on textile materials.

Keywords Natural dyes · Mordants · Natural dyeing process · Characterization

1 Introduction

Natural dyes are coloring agents available in nature. The sources may be plant, animal, or mineral, but the characteristics need to be analyzed and well utilized to be termed as an eco-friendly alternative to chemical counterparts. The craft of natural dyeing dates back to the Neolithic times and has been mentioned in the early

S. Radhakrishnan (✉)
Department of Costume & Apparel Design, PSGR Krishnammal College for Women, Coimbatore, Tamil Nadu, India
e-mail: deancad@psgrkcw.ac.in

Y. Ayyaswamy · J. Ahamed Basheer
Department of Garment Designing, PSGR Krishnammal College for Women, Coimbatore, Tamil Nadu, India
e-mail: yamunadevi@psgrkcw.ac.in; jasirabanu@psgrkcw.ac.in

S. S. Muthu (ed.), *Natural Dyes and Sustainability*, Sustainable Textiles: Production, Processing, Manufacturing & Chemistry,
https://doi.org/10.1007/978-3-031-47471-2_10

civilization cultures. Natural dyeing is a connect between humans and the Earth and a symbol of tradition. In earlier days, natural dyes and pigments were used in painting, dyeing, printing on textiles and walls, food preparation, Ayurveda, and others. Natural dyes have also been used to color different substrates like leather, wood, cane (basketry), pith, and ivory.

In India, this craft is practiced in Gujarat [1]. Natural dyes were used in small quantities by artists and craftsmen and were a localized art. The natural dyed textiles were famous for the color brilliance and right processing techniques; the ethnic output was attributed to the locally available colors, the mordanting process, the auxiliaries, and the quality of the water available in the region. Water may be from rain, well, river, or ocean, but the dyed material has the unusual touch relevant to a particular place, season, or time. Natural dyes bond well with natural fibers like cotton, silk, hemp, jute, ramie, and sisal. They are more complex than synthetic dyes as they are deeper and rustic and take on a different shade or tint according to its use. The same source produces a wide range of colors, and they all seem to blend very well, suitable to the term "living colors" [2].

Commercially natural dyes cannot compete with synthetic dyes but are a renewable resource and a major contributor to rural development. Some natural dyes have excellent fastness to light, water, and wear due to passage of time and use [3]. Quoting the paintings of the Sittanavasal Cave, in Pudukottai, Tamilnadu, India, they were made with Fresco-secco techniques made over rough stone using rough plaster. Fresco-secco, a wall painting technique, uses pigments combined with an organic binder/lime and applied on to dry plaster [4]. Pigments used were from mineral resources and included white color from lime, black from wood charcoal/lamp black, yellow from yellow ocher, red from red ocher, blue from ultramarine/lapis lazuli, and green from terre verte.

Lime (white color) is an inorganic material which occurs as calcium oxide and hydroxide [5]; wood charcoal is the conversion of biomass into charcoal derivatives known as carbon black or pyrogenic and varies from lightly charred matter to graphite rich soot particles [6]; yellow ocher also called yellow earth/gold ocher gives yellow color due to the presence of limonite which is a combination of iron containing minerals of which the main component is goethite – iron oxide-hydroxide [7, 8]; red ocher imparts the red color due to the presence of the mineral hematite-anhydrous iron oxide Fe_2O_3 [9]; the pigment ultramarine is obtained by grinding lapis lazuli stone, which consists of a blue cubic mineral called lazurite and is more expensive than gold [10, 11]; terra verta, also known as green earth or Verona green, is an inorganic pigment obtained from minerals celadonite and glauconite [12]. The paintings on the walls and ceiling of the domes in Sittanavasal is over 1000 years old [13]. Similarly, kalamkari temple paintings were popular in Andhra Pradesh, South India, and were based on plant extracts, namely, blue from indigo and red from madder; alum and mineral salts of iron, tin, and copper were used as mordant. Kalamkari worked fabrics are very colorfast and last a lifetime [14].

As shown in Fig. 1, the ecological cycle of natural dyes is sustainable – growth of plants, harvesting of the raw materials, preparation and processing of the dye and base materials, leftover materials and used water returned to nature. This beautiful

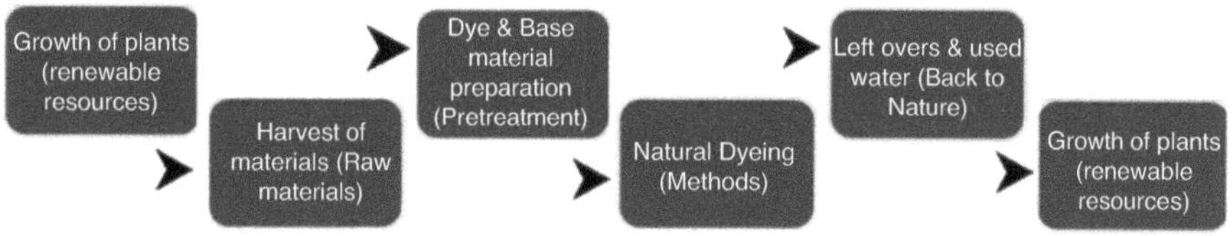

Fig. 1 Ecological cycle of natural dyes

cycle of *soil to soil* is part of Fibershed movement which strives to create local fiber communities which have respect for the soil [15, 16].

The impact of natural dyeing on the human mind is multifaceted. Natural dyes are creative tools for the human imagination as there is no fixed recipe for any hue and different combinations of raw materials can produce the same hue of a color family. The "eureka" or "wow" factor is classically and continuously existent as new colors have been born over centuries and are going to be born in the futuristic centuries. This stimulates the mind and encourages the thirst for creativity in crafting exclusive colors, gathering new raw material sources, experimenting the process, and envisaging the possible outcomes. The doors of research are wide open in the field of natural dyeing as something new and niche can be artistically developed. Working with natural dyes can bring about new life to old pieces and reduce mass consumption which has created havoc to our environment. The possibility of learning and conducting trials with Mother Nature is extremely empowering and enriching. The time spent in identifying materials for dyeing or collecting spent materials from backyards helps to reconnect one with nature. Creating something novel and beautiful with our hands and reconnecting with the environment can produce incredible happiness and satisfaction. Learning traditional skills and revival of ancient art are a meaningful fulfillment in the modern era which is totally technology driven.

2 Natural Dyes and Their Classification

Natural dyes have been in use since time immemorial though the exact date of its origin is still unknown. However, there have been evidences of the use of natural dyes like the El Castillo Cave, Spain, painted 40,000 years back; colored flax fiber from prehistoric cave in Georgia in 34,000 BC; iron oxide dyes obtained during the Neolithic period in Southern Anatolia; data about dyes from natural resources in China dating 5000 years ago; recovered vegetable dyed pieces of a coat from the archeological site of Mohenjo-daro; and so on [17].

Natural dyes are currently becoming a popular trend because they are noncarcinogenic, friendly to the environment, nonallergic to humans, and nontoxic in nature. Due to these characteristics, they are used in food, medicines, handicrafts, and textiles and have become a marketing label in fashion houses and boutiques. In some cases, there is an added bonus apart from coloring when natural dyes are used for

UV protection, wound healing, moisturizing, etc. The ultraviolet protection factor (UPF) has been studied by researchers [18–20], for cotton, flax, hemp, and ramie fabrics dyed with natural dyes like indigo, madder, cochineal, and so on. Metallic mordants and orange peel extracts have the capacity to improve the UPF value in wool, silk, and cotton [21]. The use of UV protected fabrics can prevent tanning, premature sun burns, premature aging, and skin burns. A study [22] on the natural dyeing of wool with cochineal, indigo, and madder (anthraquinone based) was helpful to create an environment free from moth and fungus attack, thereby giving rise to an insect-proof and repellent fabric. This facet of usage of dyes from nature had become a field of research in identifying the coloration modalities and the added medicinal or esthetic benefits to humans.

Natural dyes have been used for ages, but there is no methodologies specified for others to follow, and still a lot of research has to be undertaken. The characteristics and potentials need to be researched to evaluate the other impacts apart from coloration. This provides a huge scope for research and development [23]. Despite this background, the USA and EU-France, Germany, the UK, and Italy are the major importers of natural dyes. Carpets from Turkey are famous for their esthetics and the beauty with natural dyes.

2.1 *Merits and Demerits*

The advantages and limitations of using natural dyes are given in Table 1.

2.2 *Classification of Natural Dyes*

Natural dyes are found in nature and may be classified based on various ways, namely, (1) the source/origin, (2) the type of hue, (3) chemical structure and color compounds, and (4) the application of the dye as given in Fig. 2.

2.2.1 Classification Based on Source or Origin

The various sources may be of plant or animal origin. The different parts of the plant like seeds, root, stem, bark, leaves, and flowers, where the source of color exists, are extracted and applied to the fabric. Some of the popular and commonly used dyes are given in Tables 2 and 3.

While taking the insect sources for natural dye extraction, few are available as given in Table 3. The secretions from the insects or dye containing parts of the insects become the source for the dye. In the case of mineral sources, light colors are obtained along with metallic salts. Mineral sources include red clay, ball clay, and mineral ores.

Table 1 Merits and demerits of natural dyes

Sl. no.	Merits	Demerits
1.	Nontoxic with no health hazards [24]	Highly suited to natural fibers – cotton, linen, wool, and silk [28] [29–33]
2.	Easy to extract making it suitable for use	Colorfastness is medium to low [26]
3.	Renewable sources and biodegradability	Non-reproducible shades of color due to variation in agrocrops
4.	Pleasing shades with special fresh aroma [25]; soft, lustrous, and soothing to the human eyes	Color recipes and methods may vary – no standards available [27]
5.	Mild dyeing environment	Metallic mordants used – which may be non-eco-friendly
6.	No/minimum effluent generation; has potential to earn carbon credit	Lesser availability to meet the existing demands
7.	Encourages creativity, mix and match, small variations to form new colors, instant harmony of colors	High cost due to skilled workmanship, use of more dyes to get the right color
8.	Utilization of waste lands for dye yielding plant cultivation	Time for regeneration of resources
9.	Labour driven, providing job opportunities and startups	Lack of sufficient scientific backup and research data; lack of technical knowledge for dye extraction and application

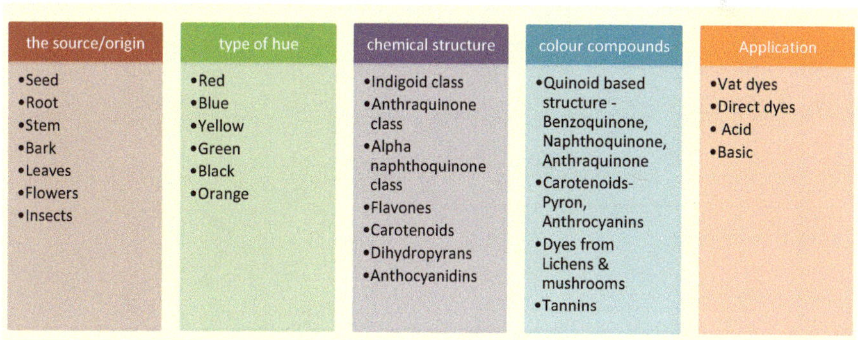

Fig. 2 Classification of natural dyes

2.2.2 Classification Based on Chemical Constitution

The identity, arrangement, and ratio of elements that are present in a compound are known as chemical constitution. Understanding these fundamentals will help the user to make use of the compound in its application and understand the problem, thereby trying to alter the process to make the best use of the dye. The different properties of the classes are given in Table 4.

Table 2 Details of natural dyes from plant sources: jackfruit wood, turmeric root, onion skin, and henna leaf [47]

Particulars	Dye 1	2	3	4
Source	Jackfruit wood	Turmeric root	Onion skin	Henna leaf
Botanical details	*Artocarpus heterophyllus* Lam – family Moraceae	Curcuma botanical group	*Allium cepa*	*Lawsonia inermis* L.
Dye extraction	Chopped wood powder boiled in water	Powdered root boiled with water; mordant improves fastness	Skin boiled with water; with/without mordants	Dried and crushed leaf boiled with water – good with mordants
Color compounds	Morin	Curcumin	Pelargonidin	Hennotannic acid
Color range	Yellow to brown	Yellow shades	Orange to brown	Mustard yellow to brown
Fabrics dyed	Cotton or jute	Cotton, wool, and silk	Cotton fabric	Wool, silk, and polyester
Molecular structure	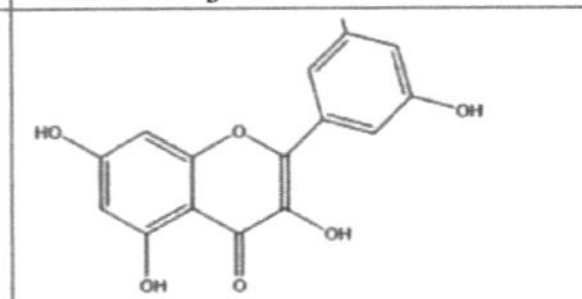	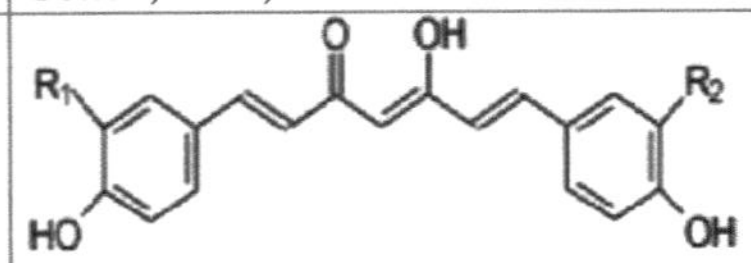	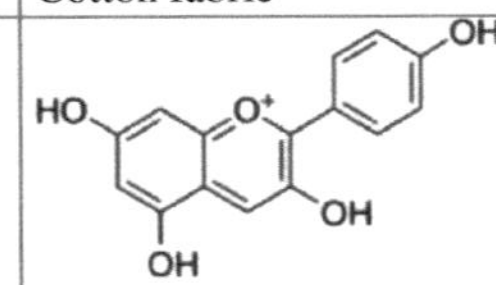	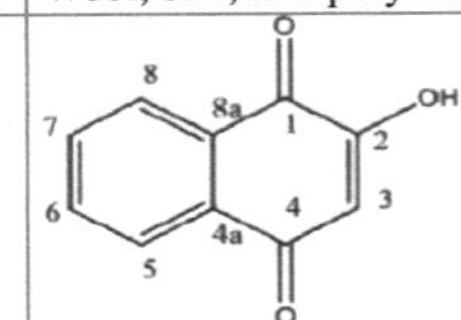
Details of molecular structure	Molecular structure of morin (3,5,7,2′,4′ pentahydroxyflavone)	Molecular structure of curcumin (diarylheptanoid)	Molecular structure of pelargonidin (5,5,7,4 tetrahydroxy antocyanidol)	Molecular structure of lawsone (2-hydroxyl-1-4-naphthoquinone)

Table 3 Details of natural dyes from plant and insect sources: madder root, safflower petals, lac insect, and cochineal insect [47]

Properties	Dye 1	2	3	4
Source	Madder/Manjistha/*Rubia* root	Safflower petals	Lac insect – resinous serum	Cochineal insect
Botanical details	*Rubia tinctorum*	*Carthamus tinctorius*	*Laccifer lacca* Kerr/*Coccus lacca*	*Dactylopius coccus*
Dye extraction	Powdered root boiled in water	Petals soaked in distilled water and boiled in water and filtered; filtrate is vacuum dried	Stick lac + water and sodium carbonate solution Precipitated in lime	Extracted from female cochineal insects
Color compounds	Polyphenolic compounds – 1,3/1,4/1,2-dihydoxyantraquinone and 1,2,4-tri hydoxyantraquinone	Natural pigment carthamine	Laccaic acid A and B	Carminic acid
Color range	Red to violet	Cherry red to yellowish red	Scarlet to crimson red	Crimson/scarlet/pink; with metallic mordant purple to gray
Fabrics dyed	Wool, silk, and cotton	Wool, silk, and cotton	Wool, silk, and cotton	Wool, silk, and cotton
Molecular structure	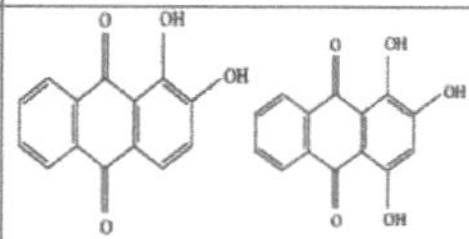	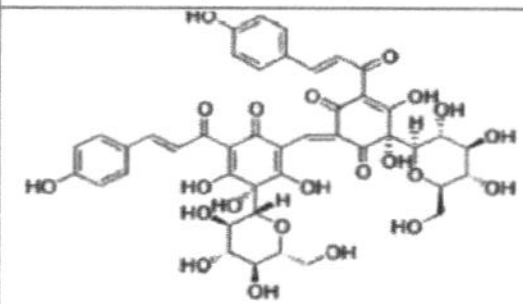	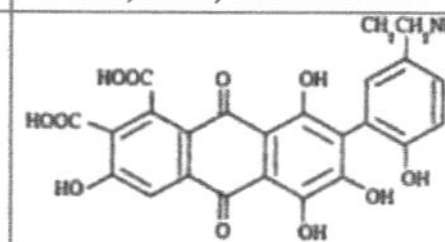	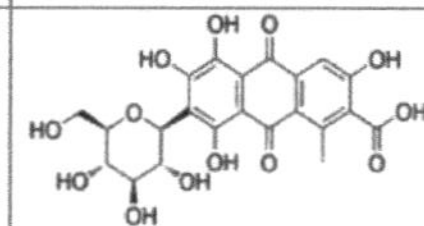
Details of molecular/chemical structure	Molecular structure of alizarin and purpurin	Molecular structure of carthamine	Chemical structure of laccaic acid A	Chemical structure of carminic acid

Table 4 Details of natural dyes based on chemical constitution [34, 35]

Sl. no.	Class	Occurrence	Dyes present	Chemical structure
1.	Indigoid	Glucose indicant in plants	Indigo blue, Tyrian purple, woad	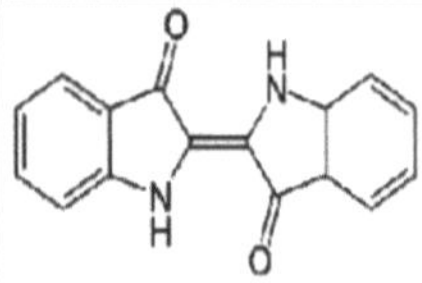 Indigoid structure
2.	Anthraquinone	Insect and plant – lac, kermes, cochineal, and madder	Red shade	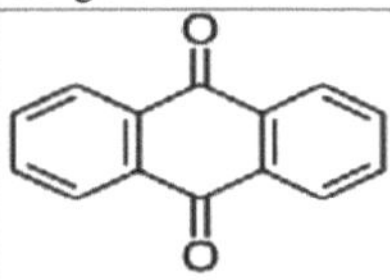 Anthraquinoid structure
3.	Alpha naphthoquinone	Two hydroxy 1–4 naphthoquinone composition, e.g., hina, lawsone, juglone	Greenish brown	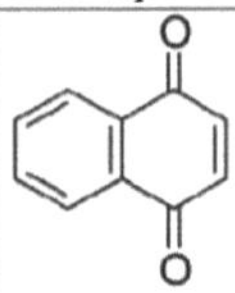 Naphthoquinone structure
4.	Flavones	Derivatives of hydroxyl and methoxy substituted flavones or isoflavones, e.g., weld	Yellow shades	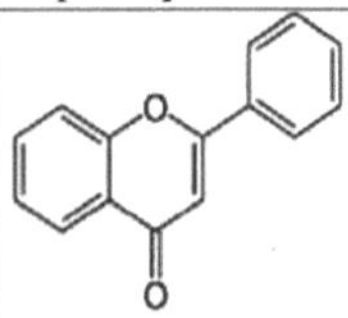 Flavone structure
5.	Carotenoids	Long chain conjugated double bonds, e.g., saffron and annatto	Orange	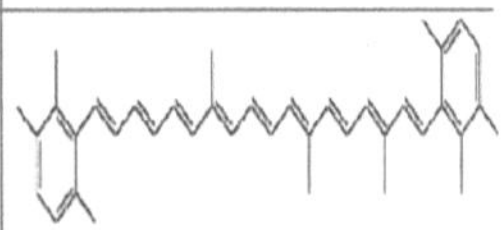 Carotenoid structure
6.	Dihydropyrans	Logwood and sappan wood	Black shades	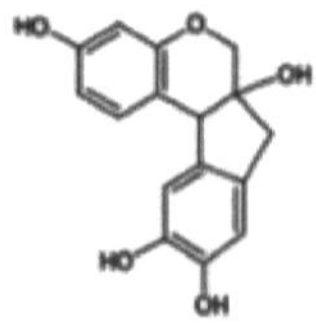 Dihydropyran structure
7.	Anthocyanidins	Carajurin	Blue and orange shades	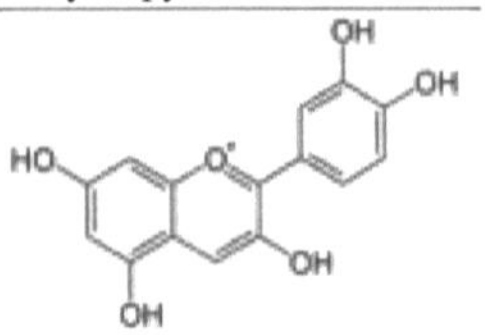 Anthocyanidin structure

2.2.3 Classification Based on Dye Structure

Coloring matter in dyes is present as chromophoric or auxochromic groups. The presence of a particular group in the dye will determine the chemistry of the dye.

- *Quinoid* based dye structure is of three types, namely, benzoquinone (carthamine), naphthoquinone (juglone and lawsone), and anthraquinone (alizarin).
- *Carotenoids* are found in red, yellow, and orange pigments in plants and animals [36]. Carotenoids are made up of two components: hydrocarbon carotenoid and oxygen containing xanthophylls. They possess high UV light resistance and usually exist as β-carotene. Pyron dyes contain flavonoids and anthocyanins. Flavanols, flavones, anthocyanidins, isoflavones, flavon-3,4-diols, and coumarins fall under the flavonoid classification. The most popular flavonoid is yellow quercetin. Anthocyanins are found in fruits and vegetable as 500+ varieties producing colors like red, pink, orange, and violet. Important anthocyanins include peonidin, cyanidin, petunidin, malvidin, delphinidin, and pelargonidin. Besides anthocyanin plants also contain quercetin and chlorophylls, thereby producing different colors with these combinations.
- *Dyes from lichens and mushrooms*
  Lichens are small gray or yellow plant that grows on rocks, walls, and trees; they grow in symbiosis with algae and fungi but do not have flowers. In the Middle Ages, Royal and Tyrian purple were obtained from mollusks [37]. Dyes from lichens had poor light fastness and hence are limited in use. Dyes were also obtained from fungi, namely, *Echinodontium tinctorium* and *Polyporales*. The colors in lichens and fungi are benzoquinone derivatives which are formed by the pre-compounds (lecarnoric acid, atranorin, and gyrophoric acid) of orchil and

Table 5 Details of natural dyes based on dye structure

Benzoquinone	Naphthoquinone	Anthraquinone
# Carthamine [CI Natural red 26] pigment from safflower # Dye extraction – water + Safflor yellow; dilute Na_2CO_3 + Safflorcarmin followed by neutralization # Used for dyeing wool, silk, and cotton	# Present in glycosidic form Hennosid A, B, and C # Lawsone extracted from Hina leaves which contain luteolin, flavonoid colorant # Lawsone forms 1:2 composition with Fe(II) and Mn(II) – wool and silk # Juglone extracted from different parts of nut trees # Insect resistant enhancement for wool dyed with juglone; mordanting improves colorfastness # Lawsone and juglone – used to dye both natural and man-made synthetic materials	# Large groups of colorants # Root of the rhubarb plant gives CI Natural yellow 23 # Alum as mordant produces yellow to orange and also enhances light fastness # Alizarin, from root of madder plant, produces CI Natural red 8 # Madder roots contain di- and trihydroxyl anthraquinone glucosides

Table 6 Types of tannins [41–44]

Hydrolyzable tannins	Proanthocyanidins/condensed tannins
# Found in roots of plants # These tannin synthesis branches directly from the shikimate pathway – direct aromatization of 3-dehydroshikimic acid to produce gallic acid which upon esterification forms polyol # Three types – gallotannins (gallic acids), ellagitannins (ellagic acids), or combination of both tannins # Pentagalloyl glucose is commonly occurring gallotannins # Esters of hexahydroxydiphenic acids are ellagitannins	# Are derived from the flavonoid pathway, one of the branches of the phenylpropanoid pathway, or polyketide pathway # The phenylpropanoid pathway takes the aromatic amino acid (L-phenylalanine) which is deaminated to E-cinnamate by phenylalanine ammonia-lyase # The polyketide pathway takes malonyl moieties for the aromatic ring formation in the biosynthesis of flavonoids

litmus through enzymatic hydrolyzation/decarboxylation/oxidation reactions [38–40]. The pH of the solution has a bearing on the colors obtained for both litmus and orchil. If the solution has a basic pH, the color produced is blue violet anion and in an acidic pH red cation is formed.

Mushroom species like *Sarcodon*, *Phellodon*, and Hydlnellum contain terphenyl-quinone compounds which are benzoquinone derivatives that produce blue color. *Cortinarius* species of mushrooms have antraquinone derivatives that are brown, red, olive green, and violet in color.

- *Tannins* are present in the epidermal tissues of plants like wood, leaves and stems, buds, flowers, and roots [41, 44, 45]. Tannins are classified as hydrolyzable tannins and proanthocyanidins or condensed tannins. The properties of tannins are given in Table 6.

2.2.4 Classification Based on the Hue or Color Produced

All dyes are classified on the basis of a world recognized numbering system that helps easy identification called the Color Index [46]. It describes the colorant of the dye or pigment by application and chemical class. Similarly natural dyes are also classified according to the hue of the color as described in Fig. 3.

2.2.5 Classification Based on Application

This classification is based on the application of the dye to the substrate or materials as given in Table 7.

Thus, dyes from natural resources are varied, and lot of research and development is required to get acquainted to their structure and properties in order to make the best use of the resources in an economical and commercial way.

Fig. 3 Classification of natural dyes based on hue

Table 7 Classification of natural dyes based on application

Vat dyes	Direct dyes	Acid dyes	Basic dyes
# Components: water insoluble dye # Requirements: Solubilization is performed in the presence of reagents, e.g., sodium hydrosulfite and sodium hydroxide for natural indigo # Method: on solubilization applied to the cellulosic fiber, subjected to oxidation with H_2O_2 for color development # Prominent dyes: indigoid class of vat dye, e.g., indigo	# Components: Dye molecules contain amino, hydroxyl, and sulfonic groups # Requirements: Dyes should be water soluble, presence of long and planar molecular structure, occurrence of conjugated bonds (single or double bond) # Method: temperature 100 °C; common salt for better exhaustion of the dye # Prominent dyes: turmeric, harda, pomegranate rind, and annatto	# Components: Dye molecules possess sulfonic or carboxylic groups # Method: dyeing in acidic pH 4.5–5.5. Post mordanting with tannic acid to improve fastness. Common salt added for leveling effect # Prominent dyes: saffron on wool and silk	# Components: Dye molecules to contain NH2 groups and react with –COOH groups # Requirements: Dye molecules produce colored cation after dissolution in water at acidic pH # Method: Dye bath pH is 4–5 by adding acetic acid

3 Natural Dyeing Process

Natural dyes can be prepared in many ways and in many combinations. However, some common steps have been given below as guidelines for beginners. Small tips and nuances are to be adhered to get good results. Some of the general steps are as follows:

- Prepare a checklist of all the supplies and materials required.
- Prepare the fiber/yarn making it free from dirt and chemicals; the fiber may require a wash with cleaning agent (scouring). After cleansing, the fiber should be soaked in water overnight to ensure even coloring.
- Establish the ratio for fiber to mordant and dye ratio. This is usually based on weight of material to be dyed.
- Prepare the mordant and decide on the technique of mordanting (premordanting, simultaneous mordanting, and post mordanting).
- Select the appropriate natural dye extraction method (aqueous extraction using water; nonaqueous and other solvent assisted systems using ethanol/oxalic acid, supercritical fluids (SF), organic solvent water; extraction by acid and alkali assisted system using HCl and Na_2CO_3; fermentation with bio enzymes; other methods using ultrasound, enzymes, etc. for extraction).
- Natural dyeing of the material, the recipe for natural dyeing, should be ready, and the number of trials to be undertaken should be noted.
- Curing of the dyed material has to be defined.
- Rinse the material with fresh clean water at room temperature, and use the neutral pH soap to remove excess dye. Rinse till the water leaves no dye and dry in shade.
- Overdyeing and redyeing can produce different or intense coloration [48].

3.1 Natural Dyeing Process for Different Fibers

3.1.1 Cellulosic Fibers

The conventional methods of natural dyeing of cellulosic fibers like cotton include the following:

- Pretreatment (desizing and scouring), dunging (overnight soaking in cow dung water), washing, drying and steaming, steeping in alkaline lye, rinsing.
- Mordanting with myrobalan or harda till the material assumes a yellow tint. This is followed by dipping in a solution of alum along with gum/tamarind kernel paste or Fuller's earth and is ready for dyeing.
- Dyeing – boiling the cloth in aqueous extract solution till the cloth absorbs the color.
- Drying – rinsing of material in freshwater and air-drying.
- Finishing – starching followed by drying [49].

This process has been shortened to make it more economical and less time-consuming. Acid bath desizing, soap and soda scouring, and hydrogen peroxide bleaching are being adopted as pretreatment. The dyeing conditions include time (30–120 min based on shade), temperature (70–100 °C), material liquor ratio (MLR) (1:20 up to 30), dye concentration (10–50% on weight of material (owm)), salt concentration (5–20 gpl), and pH (10–12). Posttreatment includes repeated washing in hot and cold water for 15 min, soaping with 2 gpl at 60 °C, rinsing with water, and line drying. Improvement in wash fastness can be achieved by cationic treatment with eco-friendly agents.

3.1.2 Protein Fibers

In the case of wool and silk, protein based fibers, care should be taken based on the fiber chemical composition and structure. Damage will occur to wool fiber by alkali and high temperature. Mordanting is done with tannin rich sources like harda, gall nut, or metallic salts like alum, aluminum sulfate, or ferrous sulfate. The parameters include selective single or double mordanting techniques with 5–20% owm for 30–40 min at 80–90 °C with MLR of 1:5 up to 1:20. Dyeing conditions are acidic dyebath at 50–60 °C that can be raised up to 90 °C along the dyeing time of 30–40 min followed by rinsing, soaping, and washing. In the case of post mordanting, after dyeing, the material is dipped in 1–2% owm metallic salt solution followed by rinsing, soaping, and washing.

3.1.3 Lignocellulosic Fibers

The natural dyeing process parameters for jute and other lignocellulosic fibers vary slightly. Double premordanting process is beneficial, e.g., harda and alum treatment at 10–20%, respectively. Dyeing conditions include pH of 10–12 and temperature of 80–90 °C for 60–90 min with MLR of 1:20. Dyeing is followed by rinsing, soaping, and washing; higher colorfastness can be achieved with 2% cationic agent treatment.

Thus, raw material characteristics, mordant and dye selection, mordanting and dyeing conditions, pH, time, and temperature play a key role in natural dyeing of textile substrates. Nonconventional dyeing methods include exhaust dyeing of cationized fabric with nanoscale pigment dispersion [50], ultrasonic energized dyeing conditions [52], and use of supercritical carbon dioxide [51] for natural dyeing.

4 Chemistry Behind Natural Dyeing Process

Natural dyeing, though present from earlier times, was developed in India as a folk art with no data on the scientific fundamentals underlying the process. When synthetic dyes came into the industry, natural dyeing was totally lost into oblivion. The natural dyeing techniques and their application were founded on practices and not on scientific concepts with reasoning. Since they were not documented but only stayed as archaeological remains, lot of research is required to understand the principles and overcome the limitations of natural dyeing. Natural dyed materials are low in light fastness and wash fastness; hence, a mordanting process is required to fasten the dye to the substrate. This term is referred to as "fixation of natural dyeing." Only few natural dyes can be dyed directly to the textile substrate. In most cases, the dye has not much affinity to the fiber to be dyed; hence, a two-step process is required, namely, mordanting and dyeing.

4.1 Mordants

Mordants are substances that have affinity to the dye and the substate, thereby forming a link between the textile fiber and the dyestuff. Lack of dyeing affinity toward the fibers can be reversed by using mordants; when fibers have an affinity to the dyestuff, the use of mordants helps in enhancing the color brilliancy and improving fastness of the natural color. This is facilitated by forming an insoluble complex of dye and mordant within the fiber complex. Protein based fibers like wool and silk hold the color well unlike cellulosics which are difficult to dye leading to duller shades. This can be explained as the absence of amino and carboxyl groups in cellulosic fibers leads to the deficiency of attachment sites for the dye molecules. There are three generic types of mordants – metal salts or metallic mordants, oil mordants, and tannins.

- Metallic mordants are metal salts of Al, Cr, Sn, Cu, and Fe commonly used in natural dyeing of textile materials. Eco regulations have banned the use of chromium, copper comes under the restricted category, and presence of tin in the effluent is not desirable. Alum and iron are considered to be ecologically safe. The same dyestuff can produce different colors when combined with different metallic mordants. For example, alizarin + aluminum produces a red lake or complex while alizarin + iron forms a violet lake. Similarly onion skin dye becomes orange with stannous chloride and turns gray with ferrous sulfate.
  Aluminum: Potash alum is widely used as aluminum mordant in natural dyeing either alone or along with cream of tartar. Generally alum used is 10–20% on the weight of material (owm), and cream of tartar is around 40% owm. Alum is the basic material which is aluminum sulfate. While mordanting cotton with alum, sodium hydroxide/carbonate is added to the alum solution to form a precipitate which is redissolved by stirring. To obtain good results, alumina is fixed to the fiber by precipitation. *Iron*: Ferrous sulfate is used for natural dyeing and printing. Iron salts impart a black or gray color; tannin pretreated cotton has good pickup of iron, while cream of tartar is used in combination with iron for animal fibers. Fermented iron solution is used by traditional dyers, e.g., kalamkari black. *Copper*: To improve the light fastness, copper sulfate is used as mordant.

According to the eco standards, the extent of extractable copper in textiles should be 3–100 ppm and should be used in small amounts. Addition of copper sulfate gives brown color from catechu and black color from logwood. *Tin and chromium*: Stannous or stannic chloride and potassium dichromate are used as mordants.

According to the Global Organic Textiles Standard [GOTS], the amount has been restricted to below 0.2 ppm and therefore it can be avoided.

- Oil mordants: Castor and sesame oils were used as mordants but have been replaced by sulfonated castor oil which is Turkey Red Oil. Alum is used along with the oil mordant and combines with madder to form a red color.
- Tannins: may be as tannic acid or as found in vegetable tannins like myrobalan, pomegranate rind, oak galls, or sumac. Tannin in myrobalan is ellagitannin type found in fruit peels imparting a yellowish color to textiles. It is used as a mordant in dyeing cotton for producing black shades. Leaves and twigs of *Rhus* species or sumac (15–20% tannin) are the gallotannin type producing a green color. Sumac imparts a reddish brown color and is to be avoided for lighter tints and bright colors. When tannins are premordanted in cotton, they absorb all types of metallic mordants.

4.2 Types of Mordanting

Mordanting methods may be as premordanting, post mordanting, and meta or simultaneous mordanting. In premordanting, mordants are treated before the dyeing process. Kalamkari of Andhra Pradesh and Sanganeri of Rajasthan use this method of mordanting. After dyeing, mordanting is being done in a separate bath, known as post mordanting, to develop the color. Iron salts are mordanted in this manner. In simultaneous mordanting, the dye and the mordant are applied to the fabric in the same bath, e.g., cotton fabric.

4.3 Fundamentals of the Mordanting Process

Generally dyes are organic compounds with double bonds; here electrons can be delocalized, e.g., benzene ring or nitro group called chromophores. A delocalized electron, in an atom, ion, or molecule, is an electron that is not connected to a single atom or covalent bond. Such electrons contribute to the conductivity of the atom/ion/molecule [53]. When light falls on the chromophores, the energy of specified wavelengths is absorbed, and the others are reflected which are visible to human eye as the respective color. Dyes also contain charged chemical groups that have the capacity to bind with the substrate. They facilitate the formation of ionic and hydrogen bonds with the charged part of the substrate, e.g., cellulose in cotton and keratin in wool.

The bridges between dye molecules and fabric are metal cations called mordants. Metals like copper, aluminum, and tin have valencies (+2/+3) which allow multiple electrons to bond. The chemical groups like hydroxyl and carboxyl groups of the dyes and substrate form covalent bonds with the intermittent metallic cations. These bonds with the mordant are much stronger than the ionic or hydrogen bonds that would attach the dye directly with the substrate. This results in long-lasting colors with better fastness. Further, the multiple vacancies in the mordant help more dye molecules to attach with the substrate resulting in a deeper color. Choice of the mordant will depend on the chemistry of the dye and the substrate. The nature of bonding will depend on the structure of the dye and the use of metal ions for chelation.

The combining or bonding of the dye to the fiber is intricate and complex and involves direct bonding, H-bonds, and hydrophobic reactions. Mordants form a bridge between the dye molecule and the fiber improving the coloring ability of the dye and thereby the fastness property. The formation of insoluble compounds within the dye is the work of the mordants. The metal ion is linked to the dye by the presence of certain functional groups in suitable locations. Generally one or two hydroxyl groups with a carbonyl/nitroso/azo group in adjacent positions are responsible for the coordination resulting in a wide variety of colorfast shades.

Let's take the example of the chromium-alizarin complex. Here chromium is trivalent and combines with three molecules of alizarin. This combination takes place with covalency and coordinate valency to form chromium complexes called "lake." The hydrogen bonding between the polypeptide linkages and the dye will help the proteinous fibers like silk and wool to retain the dye. The bonding of alizarin + chromium/alizarin + nylon is given [54] 4-1 in Fig. 4.

Color requirement calls for proper selection of dyes or dye resources. It has been reported that tannin containing sources (barks) are used to produce brown/gray colors; yellow color is produced by flavonoids (flowers and leaves); red color is from anthraquinone dyes (animal and plants); indigo is used to produce blue color. Secondary colors are obtained by dye selection, mordanting, and mixing of two dyes. Green color is produced by first dyeing with indigo followed by a secondary color.

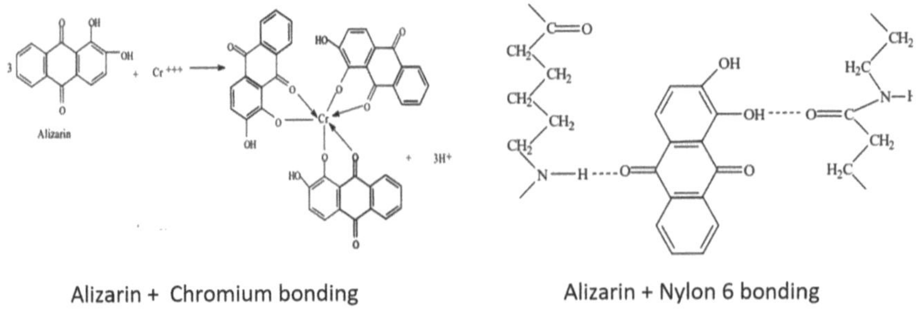

Fig. 4 Bonding of the dye, mordant, and the textile substrate

Thus, to get good results in natural dyeing, a proper understanding of the fundamentals of the dyeing process is required to enable a lot of experimentation into new ventures in color selection, mix and match of different colors, and their impact.

5 Characterization Techniques of Natural Dyes

5.1 *UV-Visible Spectroscopy*

This method helps to identify the color of the dye by taking the maximum absorption wavelength and the most prominent hue. The ability of the dye molecules to absorb UV wavelength and fading characteristics are also studied. Studies show that the absorption bands for Ratanjot dye was 520–525 nm in acidic pH and 610–615 nm in alkaline pH [56]; dye from Mimusops elengi and Terminalia arjun gave fiber absorption results of 21.94–27.46% and 5.18–10.78%, based on the concentration of the dye bath [55, 57–60]. The chemical constituent of the dye is the main factor which affects the maximum absorption wavelength value. The growth environment is responsible for the chemical constituent of the dye which in turn affects the wavelength absorption. The hue of the dye is well understood by means of UV spectroscopy.

5.2 *Chromatographic Technique*

Color components in natural dyes can be identified by thin layer chromatography. The quantitative analysis of natural dye welded by high performance liquid chromatography [HPLC] showed that in methanol + water mixture, there was luteolin (0.448%), luteolin 7-glucoside (0.357%), and luteolin 3′7 diglucoside (0.233%) [62]. It has been reported that HPLC analysis showed the structural change of indigo dye molecules after longer dyeing time [61].

Another important area of research is the physicochemical studies on the dyeing process variables, dyeing kinetics, colorfastness testing, and wearer trials that play an important role in understanding the complexity of natural dyeing. Apart from these techniques, impact analysis methods like toxicity, biotechnological processing, life cycle assessments, and environmental impact studies play an important role to estimate the effect of essential principles/factors that affect the environment. One can become well accomplished in the art of dyeing with natural materials as they experience the taste of experimenting with this color tool. The understanding of the sensitivity of natural materials and the technique of careful handling can make researchers more efficient to the small nuances that can make a great change in the output of this art. Once their expertise enhances, it is the duty of each personnel to highlight the beneficial points of this art by undertaking tests and trials to prove that this wonderful art is also magnanimous to economics, society, and environment. Consumers today would like to contribute to the planet and would definitely go in for eco-friendly purchases that will lead to a greener world.

5.3 Concluding Remarks

Every passing day is showcasing the demand for eco-friendly natural products in all walks of life. The extent to which a consumer can understand the green norm and gains knowledge of characteristics of green product and the right attitude play an important background in influencing the consumer shopping behavior. The traditional craft of natural dyeing is gaining new attention worldwide, and consumers look forward to extend their contribution either directly or indirectly to the society and environment. Textile dyers need to understand that synthetic dyes cannot rule the arena and a shift is required to address the havoc caused in industrial processing. Understanding the basics and the methodologies of natural dyeing can help all the personnel to diversify their interest into the art of natural dyeing. Modern techniques can also help in these processes, but patience and perseverance is required to learn the art and wait for the results. Slowly this craft which is currently at the local level will soon reach new terrain to cater to a majority of consumers who have become highly conscious about green products and marketing. Soon natural dyeing techniques will be taken on an industrial scale with the planning and implementation of sustainable concepts right from agriculture to product development by the designer, manufacturer, apparel brands, retail marketing, and the community.

References

1. AsiaInCH. (1999). *Introduction to natural dyes*. https://asiainch.org/craft/natural-dye-sources/. Accessed 10 May 2023.
2. Vibrant acres. (2023). *Introduction to natural dyes*. https://www.vibrantacres.ca/natural-dyes-introduction/. Accessed 10 May 2023.
3. Kadolph, S. J. (2023). *Natural dyes*. https://www.lovetoknow.com/life/style/natural-dyes. Accessed 20 May 2023.
4. Weyer, A., Roig Picazo, P., Pop, D., Cassar, J. A., Özköse, A., Vallet, J.-M., & Srša, I. (Eds.). (2015). EwaGlos. European illustrated glossary of conservation terms for wall paintings and architectural surfaces (English definitions with translations into Bulgarian, Croatian, French, German, Hungarian, Italian, Polish, Romanian, Spanish and Turkish) (p. 84). Michael Imhof. https://doi.org/10.5165/hawk-hhg/233
5. Anthony, J. W., Bideaux, R. A., Bladh, K. W., & Nichols, M. C. (Eds.). (2017). *Handbook of mineralogy*. Mineralogical Society of America. http://www.handbookofmineralogy.org/
6. Hedges, J. I., Eglinton, G., Hatcher, P. G., Kirchman, D. L., Arnosti, C., Derenne, S., Evershed, R. P., Kögel-Knabner, I., & de Leeuw, J. W. (2000). The molecularly-uncharacterized component of nonliving organic matter in natural environments. *Organic Geochemistry, 31*(10), 945–958. https://doi.org/10.1016/s0146-6380(00)00096-6. ISSN 0146-6380.
7. ColourLex. (2023). *Yellow ochre: Natural inorganic pigment*. https://colourlex.com/project/yellow-ochre/. Accessed 24 May 2023.
8. Collier's New Encyclopedia. (1921). *Ochre-Resorcin* (Vol. 7). P. F. Collier & Son Company.
9. Roelofs, I., & Petillion, F. (2012). *La couleur expliquée aux artistes (Colour explained to artists)* (p. 30). Groupe Eyrolles. ISBN 978-2-212-134865.
10. Roy, A. (1993). Artists' pigments: A handbook of their history and characteristics. *National Gallery of Art, 2*, 39.

11. Plesters, J. (1966). Identification of materials of paintings: II. Ultramarine blue, natural and artificial. *Studies in Conservation, 11*(2), 62–91. International Institute for Conservation of Historic and Artistic Works. Taylor & Francis, Ltd. (34 pages).
12. St. Clair, K. (2016). *The secret lives of colour* (pp. 224–226). John Murray. ISBN 9781473630819. OCLC 936144129.
13. Saxena, S. (2011). *Sittanavasal – The legacy of Chitrasutra*. https://puratattva.in/sittanavasal-the-legacy-of-chitrasutra/. Accessed 20 May 2023.
14. Silk Road Gallery. (2019). *Kalamkari temple paintings from South India*. https://silkroadgallery.co.uk/blogs/news/kalamkari-temple-paintings-from-south-india. Accessed 20 May 2023.
15. Fibershed. (2023). *Borrowed from the soil*. https://fibershed.org/borrowed-from-the-soil-design-challenge/. Accessed 10 May 2023.
16. Fibershed. (2023). *Design challenge resources*. https://fibershed.org/2023-design-challenge-resources/. Accessed 10 May 2023.
17. Hossain, M. H. (2021). *Sources and applications of natural dyes*. https://textilefocus.com/sources-applications-natural-dyes/. Accessed 30 Jan 2023.
18. Sarkar, A. K. (2004). An evaluation of UV protection imparted by cotton fabrics dyed with natural colourants. *BMC Dermatology, 4*(1), 15.
19. Grifani, D., Bacci, L., Zipoli, G., Carreral, G., Baronti, S., & Sabatini, F. (2002). Laboratory and outdoor assessment of UV protection offered by Flex and Hemp fabrics dyed with natural dyes. *Photochemistry and Photobiology, 85*, 313.
20. Grifani, D., Bacci, L., Zipoli, G., Sabatini, F., & Albanete, I. (2011). The role of natural dyes in the UV. Protection of fabrics made of vegetable fibres. *Dyes and Pigments, 91*(3), 279.
21. Gulrajani, M. L., & Gupta, D. (2011). Emerging techniques for functional finishing of textiles. *Indian Journal of Fibre & Textile Research, 36*, 388.
22. Koto, H., Hata, T., & Tsukada, M. (2004). Potentialities of natural dyestuff as antifeedants against varied carpet beetle, Anthrenus verbasci. *Japan Agricultural Research Quarterly, 38*(4), 241.
23. Hill, D. J. (1997). Is there a future for natural dyes? *Review of Progress in Coloration and Related Topics, 27*, 18.
24. Dedhia, E. M. (1998). Natural dyes. *Colourage, 45*(3), 45.
25. Chavan, R. B. (1999). *Chemical processing of handloom yarns and fabric* (p. 6). Department of Textile Technology, IIT.
26. Ghosh, P., Samanta, A. K., & Das, D. (1994). Effect of selective pretreatments and different resin post-treatments on jute-viscose upholstery fabric. *Indian Journal of Fibre & Textile Research, 19*, 298.
27. Gulrajani, M. L., & Deepti, G. (1999). *Natural dyes and their application to textiles* (p. 23). Department of Textile Technology, IIT.
28. Senthil, P., Umasankar, P., & Sujatha, B. (2002). Ultrasonic dyeing of cotton fabric using with neem leaves. *Indian Textile Journal, 112*(6), 14.
29. Samanta, A. K., Preeti, A., & Siddhartha, D. (2007). Dyeing of jute and cotton fabrics using Jackfruit wood extract: Part I—Effects of mordanting and dyeing process variables on colour yield and colour fastness properties. *Indian Journal of Fibre & Textile Research, 32*, 466.
30. Mathur, P., Metha, A., Kanwar, R., & Bhandari, C. S. (2003). Use of neem bark as wool colourant—Optimum conditions of wool dyeing. *Indian Journal of Fibre & Textile Research, 28*, 95.
31. Gulrajani, M. L., Gupta, D. B., Agarwal, V., Jain, V., & Jain, M. (1992). Some studies on natural yellow dyes. *Indian Textile Journal, 102*(4), 50.
32. Mahale, G., Sakshi, & Sunanda, R. K. (2003). Silk dyed with Acalypha (Acalypha wilkesiana) and its fastness. *Indian Journal of Fibre & Textile Research, 28*, 86.
33. Katti, M. R., Kaur, R., & Shrihari, N. (1996). Dyeing of silk with mixture of natural dyes. *Colourage, 43*(12), 37.
34. Kumar, A., & Chowdury, A. K. R. (2018). Eco-friendly dyes and dyeing. *Advanced Materials and Technologies for Environmental Sciences, 2*(1), 145–176.
35. Khoo, H. E., Azlan, A., Tang, S. T., & Lim, S. M. (2017). Anthocyanidins and anthocyanins: Colored pigments as food, pharmaceutical ingredients, and the potential health benefits. *Food & Nutrition Research, 61*(1), 1361779.

36. Krizova, H. (2015). Chapter 18: Natural dyes. In M. Kryštůfek & W. Vik (Eds.), *Textile dyeing—Theory and applications* (1st ed., pp. 317–334). TUL: Vysokoškolskýpodnik Liberec s.r.o., Studentská 2.
37. Cardon, D. (2007). *Natural dyes, tradition, technology and science*. Archetype Publications.
38. Casselman, K. D. (2001). *Lichen dyes and dyeing: The new source book*. Dover Publications.
39. Grierson, S., Duff, D. G., & Sinclair, R. S. (1985). Natural dyes of the Scottish highlands. *Textile History, 16*, 23.
40. Gill, M., & Steglich, W. (1987). In W. Herz, H. Grisebach, G. W. Kirby, & T. Chadd (Eds.), *Progress in the chemistry of organic natural products* (Vol. 51, p. 125). Springer. Pigments from Fungi (Macromycetes)
41. Ayres, M. P., Class, J. P., Jr., Macclean, S. F., Redman, A. M., & Reichart, P. B. (1989). Diversity of structure and antiherbivore activity in condensed tannins. *Ecology, 78*, 1696.
42. Porter, C. J. (1989). Tannins. In J. B. Harborne (Ed.), *Methods in plant biochemistry* (Vol. 1, p. 389). Academic Press.
43. Seigler, D. S. (1998). *Plant secondary metabolism*. Kluwer Academic Publications.
44. Waterman, P. G., & Mole, S. (1994). *Analysis of phenolic plant metabolites*. Blackwell Scientific Publications.
45. Julkunen-Tiitto, R., & Haggman, H. (2009). Tannins and tanning agents. In T. Bechtold & R. Mussak (Eds.), *Handbook of natural colourants* (p. 2003). Wiley.
46. Saxena, S., & Raja, A. S. M. (2014). *Natural dyes: Sources, chemistry, application and sustainability issues*. http://admin.umt.edu.pk/Media/Site/STD/FileManager/OsamaArticle/febarticles/Natural%20Dyes.pdf. Accessed 20 May 2023.
47. Gupta, V. K. (2019). *Fundamentals of natural dyes and its application on textile substrates*. https://doi.org/10.5772/intechopen.89964., https://www.intechopen.com/chapters/70564
48. Fibercurious. (2023). *Natural dyeing process*. https://fibercurious.com/natural-dyeing/#dyeing-equipment-and-supplies
49. Mohanty, B. C., Chandramouli, K. V., & Naik, H. D. (1987). *Natural dyeing processes of India*. Calico Museum of Textiles.
50. Fang, K., Wang, C., Zhang, X., & Xu, Y. (2005). Dyeing of cationised cotton using nanoscale pigment dispersions. *Coloration Technology, 121*(6), 325–328.
51. Kraan, V. D. (2007). PhD Thesis. http://repository.tudelft.nl/file/82909/027985. Aug 2007.
52. Ghorpade, B., Darrekar, M., & Vankar, P. S. (2000). Ecofriendly cotton dyeing with Sappan wood dye using ultrasound energy. *Colourage, 47*(1), 27–30.
53. BYJUS. (2023). *What is a delocalised electron?* https://byjus.com/chemistry/delocalized-electrons/. Accessed 25 May 2023.
54. Vankar, P. S. (2012). Chemistry of natural dyes. *Resonance, 5*(10), 73–80. https://doi.org/10.1007/BF02836844
55. Sankar, R., & Vankar, P. S. (2005). Dyeing wool with Gomphrena globosa flower. *Colourage, 52*(4), 35.
56. Gulrajani, M. L., Gupta, D., & Maulik, S. R. (1994). Bio polishing of tasar silk. *Indian Journal of Fibre & Textile Research, 24*, 294.
57. Bhuyan, R., Sai Kai, C. N., & Das, K. K. (2004). Extraction and identification of colour components from the barks of Mimusops elengi and Terminalia arjuna and evaluation of their dyeing characteristics on wool. *Indian Journal of Fibre & Textile Research, 29*(12), 470.
58. Samanta, A. K., & Priti, A. (2009). Application of natural dyes on textiles. *Indian Journal of Fibre & Textile Research, 34*, 384.
59. Kharbade, B. V., & Agarwal, O. P. (1985). Identification of natural red dyes in old Indian textiles. *Journal of Chromatography A, 347*, 447.
60. Zvic, K. (1994). HPLC analysis of the natural scale insect, madder and indigoid dyes. *Journal of the Society of Dyers and Colourists, 110*(9), 273.
61. Son, A.-Y., Heng, P. J., & Kim, K. T. (2004) An approach to the dying of polyester fiber using indigo and its extended wash fastness properties. Dyes and Pigments *61*(3):263–272 https://doi.org/10.1016/j.dyepig.2003.11.001.
62. Szostek, S., Grwrys, J. O., Surowiec, I., & Trojanowicz, M. (2003). Investigation of natural dyes occurring in historical coptic textiles by high performance liquid chromatography with UV–Vis and mass spectrometric detection. *Journal of Chromatography A, A1012*, 179.

Extraction of Natural Pigments Using Supercritical Carbon Dioxide and Its Application on Different Fabrics

Tarek Abou Elmaaty and Heba Sorour

Abstract Textile dyeing in a sustainable manner can be accomplished by utilizing natural dyes and using environmentally friendly techniques. Natural dyes are becoming increasingly popular in textile dyeing around the world, because of increased awareness of the perilous, poisonous, and unfavorably susceptible reactions associated with synthetic dyes. There have been traditional methods of extracting natural colors from various types of natural substances, including aqueous extraction and solvent extraction. These methods are simple, affordable, and manageable, but these methods have several downsides, such as hazardous organic solvents used, longer time required for extraction, and additional steps required for solvent removal. Inefficient chemical extraction and pigment decomposition may occur as a result of the lengthy performance of conventional techniques. Therefore, it is necessary to develop better extraction techniques to extract dyes from natural sources and apply them to textile dyeing. A green supercritical fluid extraction (SFE) process with zero wastewater emissions, a rapid extraction process, and less solvent usage have recently gained attention. The SFE of natural pigments has enormous prospects in the industry of textiles among many other applications. The focus of this chapter is on the latest developments in recent years, including the extraction of high-quality natural dyes from a variety of natural sources, such as plants, insects, fungi, or minerals, in addition to recent reports on dyeing fabrics with natural dyes.

Keywords Supercritical fluids · Extraction · Natural dyes · Pigments · Extraction techniques · Sustainable

Tarek Abou Elmaaty died prior the publication of this chapter. This is one of the last works of him.

T. A. Elmaaty (Deceased) · H. Sorour (✉)
Department of Textile Printing, Dyeing and Finishing, Faculty of Applied Arts, Damietta University, Damietta, Egypt

S. S. Muthu (ed.), *Natural Dyes and Sustainability*, Sustainable Textiles: Production, Processing, Manufacturing & Chemistry,
https://doi.org/10.1007/978-3-031-47471-2_11

1 Introduction

Sustainability is the attempt to avoid the exhaustion of natural resources to maintain ecological balance and preserve the standard of living. (A) Environment, (B) economy, and (C) society are the three pillars of sustainability. Therefore, if we talk about the relationship between using natural dyes and sustainability, we highlight the three key concepts of sustainable development: profit, humans, and the environment. Natural dyes have some valuable benefits for the economy, society, and environment for dealing with environmental threats and restoring sustainability [1].

(a) Environment: The most causing pollution industry in the world is the textile manufacturing sector. A significant portion of this pollution is caused by synthetic dyes, which are responsible for nearly 20% of the world's water pollution. Natural dyes are generally renewable and made from natural sources.
(b) Economy: The most catastrophic situation for the clothing industry is harmful manufacturing and excessive consumption. Therefore, using natural dyes can help the economy because natural dyes are abundant in nature and derived from plant and animal resources. They will give farmers in countryside an additional source of income [1].
(c) Society: Being a responsible member of society is essential to society's sustainability. The textile industry needs to be more accountable due to its widespread use of synthetic, toxic dyes. Natural dyes are considered environmentally friendly products, which are the best alternative to synthetic dyes [1].

The supercritical carbon dioxide ($scCO_2$) extraction technique is widely used. In addition to having a low viscosity, a high diffusivity, and a liquid-like density, CO_2 is nontoxic, nonflammable, cheap, and easy to use as a solvent [2, 3]. Additionally, the critical temperature and pressure of CO_2 are both low (31 °C and 74 bar, respectively). And therefore, it is the most suitable solvent for extracting molecules with high thermal sensitivity, like carotenoids [4]. The pigment extracted from SFE, according to Mendes et al. [5], is free of any toxicity or hazardous effects associated with some chemical solvents. In addition, $scCO_2$ is a technique that uses CO_2 under supercritical conditions and is generally preferred for textile dyeing due to its high efficiency. It is suitable for extracting a wide range of compounds [6]. $scCO_2$ is categorized as an (SFE), which has a high diffusivity and low viscosity and can disperse easily through solid materials. Crucial variables, such as pressure, temperature, and time, must be investigated during the $scCO_2$ extraction process.

This chapter overviews the different kinds of extraction and the most recent technological developments in the extraction of high-quality natural dyes from various natural sources and the application of natural dyes to different fabrics.

2 Renewable and Sustainable Natural Dyes

Consumers' understanding and awareness of environmental responsibility have increased, and they are asking for products that are sustainable. The term "sustainable" describes products that are produced with a view to reducing their adverse environmental effects. Sustainability in the dyeing and finishing of textiles promotes efficient use of natural resources, environmentally friendly techniques, lower production costs, and less waste production. In simple terms, a sustainable development approach concentrates on the social, economic, and environmental aspects of textile dyeing [7–9]. There has recently been a growing effort to replace hazardous, polluting, and expensive chemical reagents and catalysts with more eco-friendly ones [10–14]. Natural dyes were widely used to color products in a variety of industries, including textiles, leather, pulp, paper, and plastics, from antiquity until the 1880s [15–17]. Natural dyes are substances that are plentiful in nature and have environmentally friendly features. They are nontoxic and can be recycled, biodegradable. Due to their sustainability, they can be easily decomposed in soil after use [18–23]. These dyes are considered a preferable alternative to synthetic colors [24–31]. Natural dyes have gained growing interest around the world [32]. Because certain synthetic dyes are toxic, carcinogenic, and polluting, they are prohibited in Western countries [33, 34]. As a result, until the end of the nineteenth century, natural dyes were the main colorants for textiles [35]. More research is required to identify natural dye sources and to determine the optimum yield of dye extracted from natural sources. This yield is suitable for coloring a wide range of textiles. Natural dyes are the ideal option to replace synthetic dyes because they are considered environmentally friendly products [36]. A lot of studies on natural dyes derived from plants and microorganisms have been conducted in recent years [37, 38]. Natural dyes are derived from plant, animal, mineral, and microbial resources [37]. These dyes can improve a fabric's resistance to microbes [38–40]. In nature, pigments such as carotenoids and chlorophylls exhibit exceptional antimicrobial activity against a wide range of pathogens, including various fungi and bacteria. The possibility for these natural colorants to be used in a variety of industries is very high, especially in the textile, food, and pharmaceutical industries [41–45].

2.1 Classification Systems for Natural Dyes

The history of dyeing is unknown, but it is believed that colored fabrics found in Egyptian mummies date before 2500 BC. The ancient art of dyeing may have first spread south and west from India and may have been prompted by unintentional staining caused by berries and fruit juices [46, 47]. In addition to their chemical composition, dyes can also be classified according to how they are used or applied.

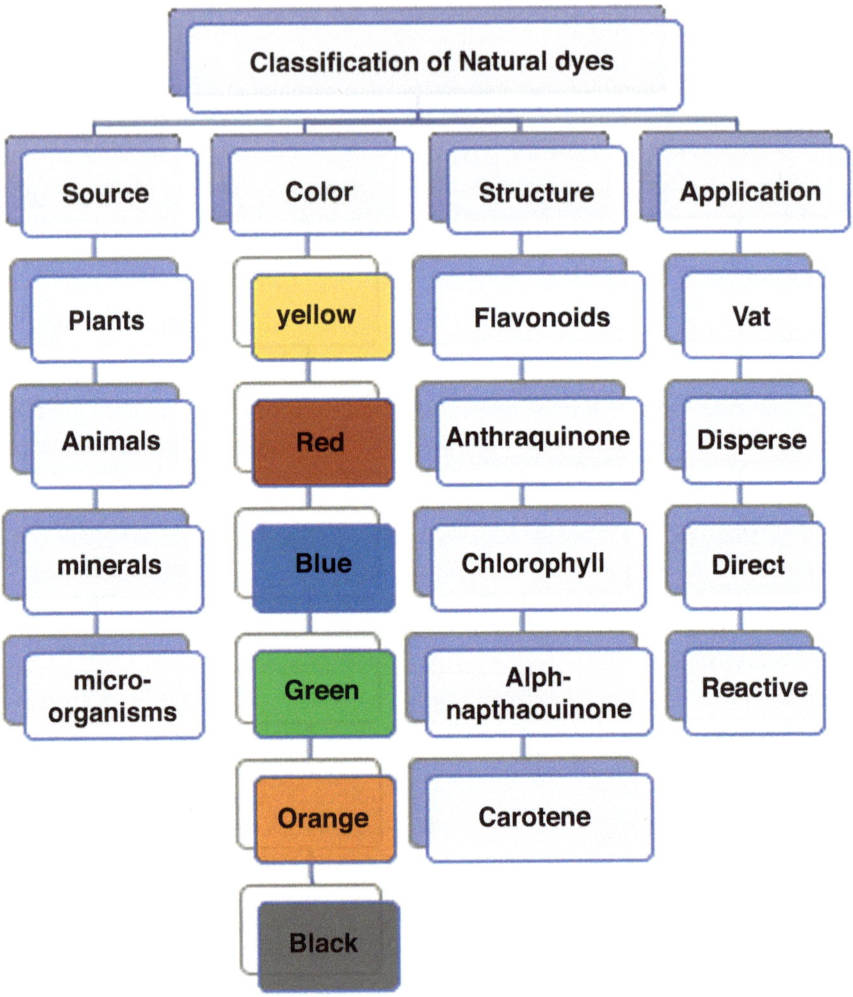

Fig. 1 Classification of natural dyes [48]

Based on their application, these dyes can be classified into four categories: (1) vat dyes; (2) disperse dyes; (3) direct dyes; and (4) reactive dyes. In nature, there are many different shades of colors, including blue, orange, yellow, green, red, purple, and black. Based on the source of dye, natural dyes can be divided into four groups – plant dyes, animal dyes, mineral dyes, and microorganism dyes – as shown in Fig. 1 [48].

2.2 Advantages of Natural Dyes

There are several benefits associated with natural dyes [49–51].

- Shades obtained from natural dyes are soft, shiny, and calming to the eye.
- It is possible to create a variety of colors by mixing natural dyestuffs. Using natural dyes of the polygenetic type, it is possible to change colors and even create entirely distinct hues, which is not always the case with synthetic dyes.
- As they are made from natural sources and are biodegradable, natural dyes are generally renewable, as opposed to synthetic dyes, which are made from nonrenewable basic raw materials.
- Since the production of these dyes for textiles, food, leather, and other products requires a significant amount of labor, there are employment opportunities for everyone involved in their cultivation, extraction, and use.
- Natural dyes can earn carbon credits by reducing the use of synthetic dyes derived from fossil fuels (petroleum).
- Natural dyes contain anti-allergens, are safe for skin contact, and are primarily nonhazardous to humans.

2.3 Limitation or Disadvantages of Natural Dyes

Natural dyes have some disadvantages as well [49–51].

- Since natural dyes and colorants differ from one crop season to the next, and from location to location, it can be challenging to repeat shades.
- In order to define a recipe for the use of natural dyes, it is difficult. This is because the process of natural dyeing and the color that results depend on both the materials and the color components.
- A significant portion of natural dyeing science still requires further investigation.
- In the presence of sunlight, perspiration, and air, the color of the dyed fabric may change.
- Almost all-natural dyes require mordants to be fixed to textile substrates, with a few notable exceptions.

3 Extraction of Natural Dyes

When using natural dyes, the first step is to extract the coloring pigment from a dry source. The extraction of coloring from dye was considered to be a significant process. There are various scientific dye extraction procedures that can be used, depending on the source, concentration, and color of the dye, as well as the cost and equipment requirements. Environmentally friendly and affordable extraction methods are essential for protecting biomaterials without causing them to spoil. Natural dye extraction techniques are becoming increasingly popular due to their cost-effectiveness and satisfaction of consumer needs. A list of the various extraction techniques used for natural dyeing is provided below [52].

3.1 Conventional Extraction of Natural Dyes

Extraction of the desired molecules from complex vegetable matrixes requires the use of multiple techniques. Because they require significant amounts of energy and solvents for their application, they are typically not environmentally friendly [53]. This raises concerns about their safety as well as the lack of sustainable and green extraction protocols [54], as well as the safety of consumers and workers involved. Due to the lengthy nature of some of these processes, excessive energy consumption becomes more prevalent, since the equipment must be continuously run for extended periods of time. This represents one of the largest costs associated with heating and cooling. Additionally, the yields obtained using these methods are generally lower than those obtained using more advanced extraction methods, which are faster and more efficient than those previously used.

3.1.1 Aqueous Extraction

As one of the earliest techniques for removing color from natural dye sources, it is one of the oldest. Due to its environmentally friendly nature, the boiling process is still popular despite the heat required. In this process, enough water and naturally occurring pigment-producing sources are boiled for an extended period of time, probably an hour. In addition to its benefits, this approach also has some disadvantages, such as the long and complex processing time, the high water consumption, and the high temperature [55].

3.1.2 Solvent Extraction

Depending on their nature, organic solvents can also be used to extract natural pigment from natural sources. Many solvents are used for extraction, including acetone, chloroform, alcohol, or a combination of solvents, such as ethanol and methanol, water and alcohol, etc. A variety of factors influence extraction, including the amount of grinding, the duration of extraction, the temperature, the equipment, and the choice of solvent. Degradation is less likely to occur during low-temperature processing. Its main drawbacks are toxic residues, greenhouse gases, and processing challenges caused by chlorophyll and waxy substances [56–76].

3.2 Green Extraction

Extraction relies on techniques that use less energy, permit the use of substitute chemicals and sustainable natural resources, and provide a superior and safe extract. Green extraction tenets are compatible with SFE procedures. In particular, when using supercritical carbon dioxide ($scCO_2$) as an environmentally friendly solvent, it is regarded as the most efficient substitute for conventional solvent extraction techniques. The following are the six guidelines for natural product extraction that belong to sustainability [77–79]:

1. *Utilizing Innovative, Environmentally Friendly Plant Resources*
   Due to the growing demand for organic products and extracts, natural plant resources are being overutilized. Green extraction should emphasize the use of fully renewable resources either through intensive cultivation or in vitro growth of plant cells or organisms.
2. *Utilization of Alternative Solvents, Specifically Water or Agro-Solvents Produced from Agricultural Resources*
   Using the SFE technique, it is possible to obtain pure fragrances, active ingredients, and perfumes without leaving behind any solvent residue. To replace organic solvents, the SFE method utilizes between 30 and 40 °C with a pressure of as much as 300 MPa.
3. *Limiting the Use of Energy Consumption by Energy Recovery and Using Innovative Strategy*
   As compared to conventional extraction methods, the SFE method provides better and faster extraction. Time and energy can be conserved without the need for additional steps.
4. *Utilizing the Bio- and Agro-Refining Manufacturing to Produce Coproducts Rather than Waste*
   As a result of extraction operations, a variety of additional materials are produced, including coproducts, by-products, and waste. Numerous refined compounds can be found in plants. There are many products that can be made from the components of a plant. By using the SFE procedure at low temperatures, novel compounds can be discovered that can be functionalized to enhance the extract's value.
5. *Utilizing Technological Innovation to Decrease Unit Operations While Favoring Secure, Reliable, and Controlled Processes*
   There can be fewer stages in a production chain, which reduces costs and improves energy efficiency. SFE offers several advantages, including the use of a clean solvent and the creation of an extract through the use of a technology that requires a limited number of discrete operations.
6. *Extracts That Are Not Denatured, Biodegradable, and Free of Contaminants*
   Ensure that the extract complies with all applicable laws, rules, quality requirements, and market requirements. Furthermore, the extract is guaranteed not to harm people or the environment. Due to this concept, the SFE extract is extremely pure and free from contaminants.

3.2.1 Green Extraction Techniques

Due to the growing demand for environmentally friendly products in recent years, the use of green extraction methods, such as ultrasound-assisted extraction, high pressure assisted extraction, microwave-assisted extraction, enzyme assisted extraction, and SFE, is becoming increasingly popular [80–87].

3.2.1.1 Ultrasound-Assisted Extraction (UAE)

A variety of technological processes, including biomedicine and food technology, use ultrasound-assisted extraction (UAE) to obtain desired compounds. It is possible to improve the method by taking into account all the factors, including ultrasound power intensities, frequencies, wavelengths, and times [88]. In addition to reducing solvent consumption, temperature, and time, the technology also has low equipment costs and ease of use. This makes it primarily applicable to small businesses in the area [89]. One of the major disadvantages of UAE is the degradation of the sample's additives due to heating.

3.2.1.2 High Pressure Assisted Extraction (HPAE)

High pressure assisted extraction (HPAE) is a novel method for extracting active ingredients from plant materials. In order to complete this type of extraction, pressures ranging from 100 to 800 MPa must be applied. Since HPAE can alter the structural integrity of foods, it cannot be used to extract all compounds. These include protein denaturation and damage to the cell membrane. It is necessary to optimize the type and quantity of solvent, time, pressure, extraction temperature, and number of cycles in order to increase the extraction yield [90]. Among the benefits of HPAE are the improvement of solvent permeability in cells, the improvement of mass transfer rates, and the improvement of secondary metabolite diffusion. In addition, the extraction procedure can be carried out at room temperature, and yields from extraction are higher.

3.2.1.3 Microwave-Assisted Extraction (MAE)

Microwave-assisted extraction (MAE) combines microwave heating with traditional solvent extraction. In microwaves, electric and magnetic fields are present at frequencies between 300 and 300,000 MHz [91]. There are several benefits associated with this method, including high efficiency, rapid temperature rise, rapid extraction times, improved process monitoring, and reduced energy and cost consumption [92]. One of the disadvantages of MAE is that some compounds degrade due to the heat generated by irradiation. There are several factors that influence the performance of the MAE, including the strength of microwave radiation, the temperature, and the duration of extraction.

3.2.1.4 Enzyme Assisted Extraction (EAE)

Enzymes, which are protein molecules, are responsible for catalyzing chemical reactions. Enzymes have always been crucial to the food industry. They are often used to convert raw materials into better food products. This is because of their ability to accelerate reactions, such as the processing of starch in the dairy industry or the manufacture of predigested food [93]. Technological advancements, however, have led to the development of new applications and sources for enzymes. The advantages that microbial enzymes provide make them the preferred source; among these are their ease of production, profitability, and consistency [94]. The advantages of this method over more conventional ones include its high selectivity and efficiency, being an eco-friendly process with limited chemical and energy product consumption, good performance, and the potential for recycling the process. However, there are some disadvantages as well, including the price of the enzymes, the requirement for storage tanks that may necessitate prolonged incubation, the lack of information regarding the most effective or suitable enzyme compositions for cell disruption, and the inability to completely hydrolyze the bonds in the cellular membrane.

3.2.1.5 Supercritical Fluid Extraction (SFE)

In comparison with conventional extraction methods, supercritical fluid extraction (SFE) provides an environmentally friendly alternative to conventional extraction methods for the separation of natural chemical components from different natural sources [95].

4 Supercritical Fluids (SCFs)

4.1 History of Supercritical Fluids

Observing that the activity of the solvent altered at a specific temperature as well as pressure, Baron Charles Cagniard de la Tour discovered supercritical fluid (SCF) in 1822 [96]. As a result of his study of the influence of temperature and pressure on a sealed glass tube containing partially liquefied carbonic acid, Thomas Andrews coined the term “critical point” in 1869. In his description, the critical point is the characteristic temperature (Tc) and pressure (Pc) on the phase equilibrium curve where there are no two distinct phases [97]. Fundamental techniques for supercritical carbon dioxide extraction of natural components were developed by Zosel et al. [78]. In the 1980s, SFE developed on a large-scale industrial basis in Australia and the USA. The *Journal of SCFs* was additionally published in 1988, which was the first scientific journal in the field. Nowadays, this technique is used to manufacture numerous products available worldwide [87]. SCF refers to any substance that is

Fig. 2 A typical phase diagram showing the various states of carbon dioxide [87]

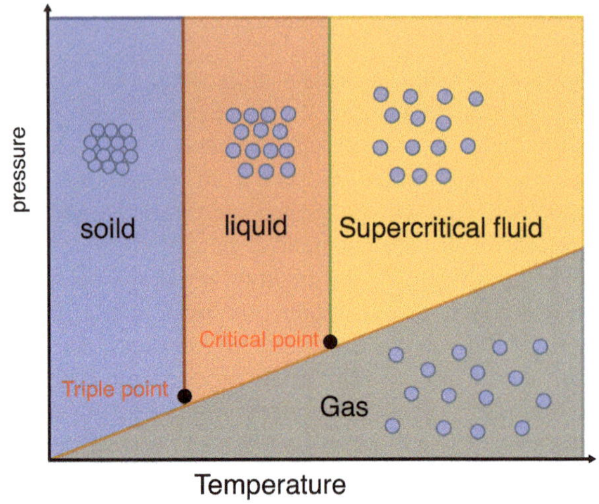

above its critical temperature and pressure. As a vapor and a liquid, a substance can coexist in equilibrium at the critical temperature and pressure. Currently, the fluid is expressed by its gaseous and liquid phases. As with gases, it diffuses through solids and dissolves materials (as with liquids) [99]. Figure 2 [87] shows a common phase diagram and indicates the various states of CO_2 at a particular pressure and temperature.

4.2 Properties of Supercritical Fluids

As chemical solvents that can be pressurized above their critical point, supercritical fluids are typically regarded as environmentally friendly. As a result of their distinct characteristics, supercritical fluids are frequently utilized in the extraction process, which produces excellent results. As a result of its behavior as a gas, an SCF assumes the shape of a container. When compared to gas molecules, molecules' mobility is noticeably higher. Due to its density, an SCF also has liquid properties, which affect its ability to dissolve. SCF parameters include density, viscosity, and diffusivity [86, 87]. An SCF's density and solvating power are largely determined by the relationship between pressure and temperature: As temperature remains constant, rising pressure increases the density and solvating power of the SCF. The density and solvent strength of an SCF decrease as temperature increases, while pressure remains the same. An SCF presents a viscosity that is similar to a gas but less than a liquid. The lack of viscosity, therefore, increases an SCF's ability to penetrate and permits elements to move within it. Additionally, the degree of viscosity of SCFs is significantly influenced by temperature but has little impact on the viscosity of liquids [99]. An SCF has a higher diffusivity than a liquid and a lower diffusivity than a gas. Considering these characteristics, SFE is considered a superb method for obtaining natural compounds [87]. Researchers are paying special attention to $scCO_2$ because of its eco-friendly and sustainable characteristics. SCF carbon dioxide CO_2 is widely used in different fields for extraction due to its critical characteristics (Tc = 31.1 °C; Pc = 73.8 bar), which makes it a suitable solvent for extracting bioactive compounds that are heat-sensitive [79]. The special solvent characteristics of $scCO_2$ have made it a preferred substance for extracting components from natural resources. The characteristics of CO_2 are shown in Fig. 3 [87].

4.3 Supercritical Fluid Extraction Principles

The SFE is based on the solvating characteristics of a SCF that is produced through increasing temperature as well as pressure above a substance's critical point [98]. There is no restriction on the type of sample that can be used during extraction. However, solid samples are more commonly used than liquid samples based on system settings. The yields of the SFE procedure can be influenced by the kind of SCF used and the process variables (temperature, the pressure, extraction duration, etc.), which in turn determine the characteristics and chemical components of the extracted material [100].

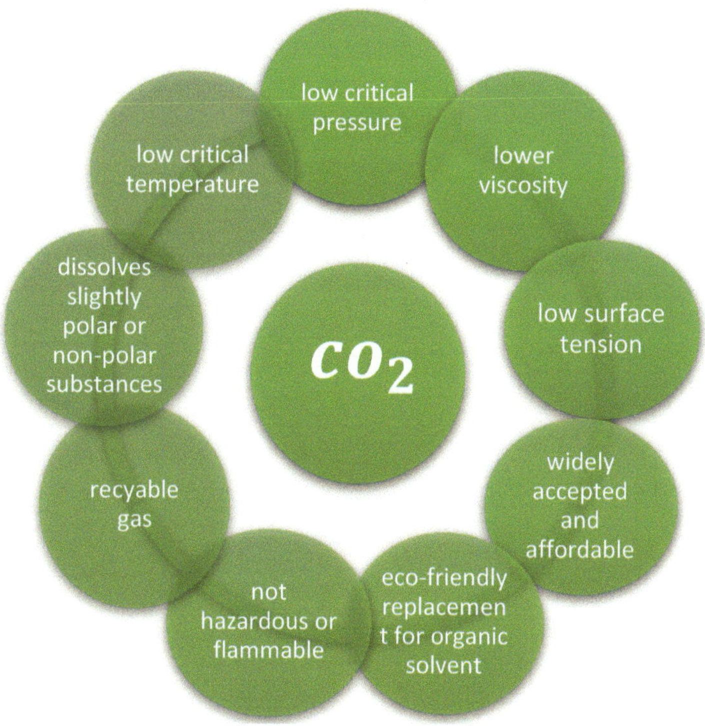

Fig. 3 The characteristics of CO_2 [87]

4.4 Supercritical Fluid Extraction Equipment

SFE equipment is divided into four categories according to vessel size: analytical (1–24 mL), bench (200–500 mL), pilot (1–50 L), and production (350+ L). All extraction systems have the same basic components, regardless of their complexity or cost. Such system consists of a CO_2 resource, a chiller unit, a CO_2 pump, valves, a heater controller, a back-pressure regulator, a vessel, a heating jacket, a modifier pump, and a collection vessel, as shown in Fig. 4 [101, 102].

4.5 Supercritical Fluid Extraction Mechanism

On numerous solid substrates, SFE has been used. It is possible to perform and execute the SFE method in two various modes: dynamic extraction and static extraction [101]. Dynamic extraction involves the continuous infusion of CO_2 into the plant matrix. Static extraction involves subjecting the plant model to a constant concentration of $scCO_2$ for the duration of a specific time. The static mode allows fluids, such as $scCO_2$, to enter the plant matrix and dissolve analytes. In the majority of SFE experiments, static and dynamic procedures are combined. In a static mode, $scCO_2$ can penetrate plant material and dissolve analytes, and the analytes are transferred from the vessel to the collecting system through the restrictor [101]. After the fluid and solubilized substances have been transported to the separators, they are collected using a tap located at the bottom of the separators. Upon decompression, the fluid is cycled or emptied into the surrounding media, after which the extract can be collected [102, 103].

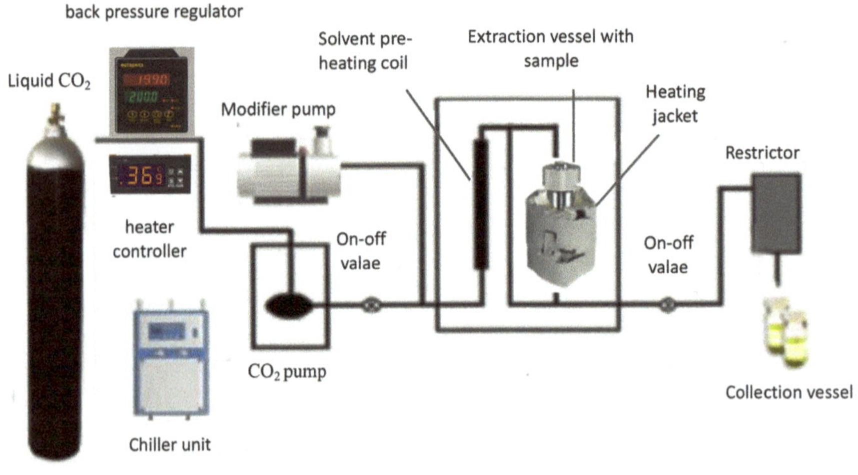

Fig. 4 Diagram of the supercritical fluid extractor used in laboratory [101, 102]

5 Variables Considered in the SFE and Their Effects

SFE extraction involves the selection of working factors based on the types of compounds to be extracted. However, the parameters of temperature, pressure, cosolvent concentration, extraction time, and particle size are the most notable and should be taken into consideration for effective extraction [104].

5.1 Temperature

There are two distinct results related to the extraction temperature at a stable pressure. As soon as the temperature rises, the solvent's density and solvating power decrease. Nevertheless, the vapor pressure of the required compounds rises with increasing temperature, and this enhances the degree of solubility and extraction outcome of the compound. Therefore, it could lead to the isotherms crossing, a process referred to as retrogradation where insufficient yields result from higher temperatures and a decrease in temperature makes prolific extract [87]. Considering the crossover aspect. Mezzomo et al. suggested that at pressures lower than the crossover pressure, the density impact predominates, whereas at higher pressures, the main mechanism affecting the extraction process is the solute vapor pressure [86]. Lower nonvolatile component extraction recovery is caused by a higher temperature. The instability of their volatile substances and their ability to dissolve in an SCF, however, have opposing properties. As temperature rises, solubility decreases while volatility rises. But as the temperature rises, the efficiency of the extraction increases. At higher extraction temperatures, several heat-sensitive materials could deteriorate and lose their purpose in biology [100]. Therefore, degradation must be avoided by maintaining the SFE degree of thermolabile substances between 35 and 60 °C [104].

5.2 Pressure

Due to the way in which pressure affects a substance's solubility [105], it is among significant factors affecting the SFE methodology. A back-pressure regulator (BPR) can be used to maintain the desired level of SCF pressure when using the SFE technique [101]. By increasing the pressure, the power of the solvent increases, and the extraction efficiency increases at a constant temperature, which enhances both the extraction yield and solubility of bioactive compounds. In proportion to the increase in pressure, the density of the SCF also increases. Thus, the solubility of the target compounds increases, and they are more readily recoverable [106]. Volatile substances within the selected compound are impacted by pressure. As a result, the recovery of volatile fractions increases with increasing pressure, while nonvolatile fraction recovery decreases. In case of increased pressure, the solvent diffusivity may decrease, and there may be less interaction with the raw materials pores, resulting in a slower dissolution of the solute. For any substances and elements, the use of high-pressure extraction is not suggested, due to the potential for compressing the raw material and lowering the extraction yield [107, 108].

5.3 Cosolvent

It has a significant solvent power that can be retained for the targeted substances when added to CO_2 in a variety of proportions as a cosolvent [109]. CO_2 is the most used SCF in a variety of industries, but due to its low polarity, it cannot frequently be used to extract polar and lipophilic compounds. One potential solution to overcome the issue is by enhancing the solvation capability of supercritical carbon dioxide ($scCO_2$) by using polar cosolvents such as ethanol, water, formic acid, and many other polar or nonpolar cosolvents and enhancing the extraction yield, solubility, and affinity of compounds that are difficult to dissolve (alkaloids, glycosidic and phenolic compounds) [86]. Methanol and ethanol are two of the most utilized cosolvents used at concentrations less than 10% of the CO_2 employed for the extraction [45]. To determine the best cosolvent for a particular extraction process, preliminary experiments should be conducted. It is possible to incorporate cosolvents or modifiers into the SFE procedure in two different ways: by simply combining the raw material and the modifier in the vessel or by blending the modifier with the CO_2 flow.

5.4 Extraction Time

The extraction time is critical for SFE and has a significant effect on the extract's composition because it increases the interaction between the supercritical solvent and the material, which facilitates a higher recovery yield. It is important to consider the extraction time at both the procedural and analytical levels when using the SFE method [110]. A rapid extraction process might yield an insufficient extraction process. Even so, both time and solvent will be wasteful if the extraction procedure takes too long [35]. The extraction period is affected by the fluid flow rate, which is shorter if there is an increase in flow rate [87]. The best time and fluid flow rate must be determined in advance to obtain the highest extract yield. To calculate the overall extraction time, both static and dynamic extraction stages are used. Using the SFE technique, extraction usually takes less than 2 h. An analysis of the extraction procedure was conducted based on the overall extraction detour (yield against extraction time), which explains information regarding the period required to create appropriate and successful extraction technique [86].

5.5 Flow Rate

The extraction's effectiveness is impacted by the SCF's flow rate [106]. Mass transfer is restricted by the state of equilibrium between a solid and a fluid. Increasing the flow rate at the start of the extraction process speeds up the recovery of the extracted material. In the final stages of extraction at minimal flows, recovery is constant. The thin film surface surrounding the solid substances becomes thinner as the flow rate rises. This results in a reduction in the solid particles' surrounding mass transfer resistance, increasing the overall extraction yield [86]. In most studies, CO_2 flow rates range from 1 to 10 L/min based on the substances' rates of solubility within $scCO_2$. To regulate the SCF rate of flow, a gas flow meter and a back-pressure regulator (BPR) are utilized [110].

5.6 Raw Matrix

When using $scCO_2$ extraction, a number of factors, including the origins of the raw material, moisture levels, particle shape, size, surface area, and porosity, play a significant role in affecting solubility and the mass transfer process [111]. Correctly choosing these components can result in a rapid and more effective extraction of the desired compounds [77]. Most often, the sample material that needs to be extracted needs to be dried to lower the moisture content, because the sample's water content could cause the extractable solute to interact with the solvent less efficiently, lowering the extraction yield. However, in some situations, the presence of water is necessary to allow for a good interaction between the solvent and the solute. The suggested moisture content should range from 4% to 14%. The extraction efficiency is increased by reducing the particle size because decreasing particle size shortens the solvent's diffusion path and rises the contact surface area, speeding up the extraction process [35, 45]. However, it's important to avoid using too few tiny particles, because they may raise the internal mass transfer resistance, which could lead to channeling within the column and a decline in extraction efficiency and yield [104, 112, 113].

6 Principal Benefits and Drawbacks of SFE

Due to its many distinctive characteristics, SFE is a viable alternative to conventional solvent extraction methods. Table 1 summarizes some of the benefits and some drawbacks of using SCFs for bioactive compound extraction [114, 115].

Table 1 Benefits and drawbacks of SFE [114, 115]

Benefits of SFE	Drawbacks of SFE
SCF has a low viscosity and a high spreading rate. Since they penetrate porous solid materials more readily than liquid solvents, they facilitate faster extraction SFE significantly reduces extraction time compared to conventional methods, from hours to minutes It is possible to continuously recirculate SCF into a sample to obtain a complete or quantitative extraction Compared to liquid solvents, SCFs exhibit superior selectivity because of their variable solvation ability, which can be adjusted by varying temperature and/or pressure Depressurization can separate the solute from the solvent, which reduces processing time Due to the fact that SFE is generally carried out at low temperatures, it is an excellent method for analyzing chemicals that are thermally labile. It is possible that brand-new natural components will be discovered because of this process SFE is environmentally friendly as it uses fewer or no toxic organic solvents It is possible to reuse and recycle SCFs to reduce the production of waste	While the polarity of the fluid is changed by cosolvents, additional purification is required As the process requires a lot of pressure, it is hard to continue putting solids into the raw material As compared to solvent extraction, this method extracts less material There is a high cost associated with these procedures There is a lack of equipment availability

7 Extraction of Natural Pigment Using Supercritical Fluids

In recent years, natural colorants have become more widely used because of their unique coloring properties and health benefits [116]. The quantity of pigment available from natural materials is scarce. As a result, a particular extraction method is needed to extract dye-containing ingredients from their original sources [117]. It is essential to determine the extraction technique to obtain high substantial quantity and quality extracts, as well as maintaining functional qualities like color. The extraction process should be inexpensive and energy-saving and utilize nontoxic solvents. Natural pigments can be extracted using the SFE technique very successfully. The primary natural pigments that are extracted by SFE are chlorophylls, carotenoids, etc. Figure 5 [118] presents the classification of natural pigment.

7.1 Extraction of Chlorophylls

Plant leaves contain chlorophyll as their primary pigment. Plant chlorophylls are pigments that are green, amphiphilic, and oil-soluble [34]. Cyclic pyrroles, which consist of four subunits, frequently interact with metal ions in their molecules [118]. Figure 6 [34, 119] illustrates two types of chlorophyll: chlorophyll a and chlorophyll b. Due to their high sensitivity to light, heat, acids, and oxygen, chlorophylls

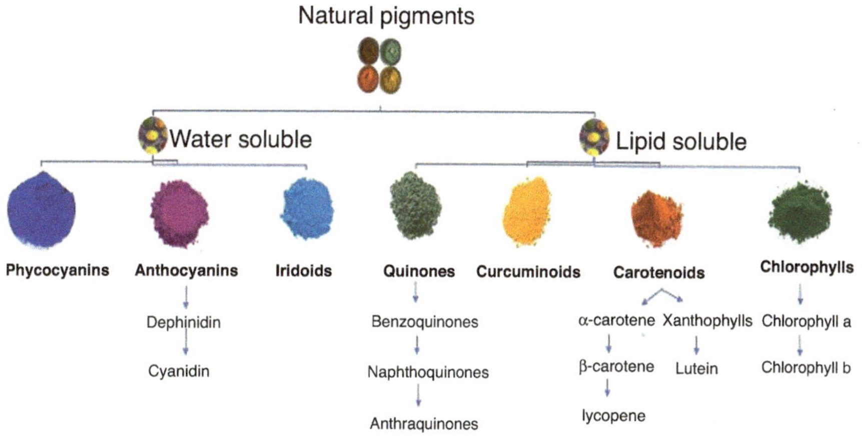

Fig. 5 Classification of natural pigment [118]

are easily degraded and change color [120]. The ideal conditions for the SFE technique for different raw plant materials have attracted the attention of researchers. Carotenoids and chlorophylls were extracted from olive pomace using SFE. It was found that freeze-dried samples produced the highest yields of carotenoids and chlorophyll, which effectively preserve phytocompounds [121].

According to SFE study by Carradori et al. [122], *Elaeagnus angustifolia* leaf extracts have higher concentrations of antioxidant pigments like chlorophyll. Chlorophyll a concentration in *Elaeagnus angustifolia* leaves was 45% higher than chlorophyll b. According to Wasilewski et al. [123], SFE factors affect the extraction of carotenoid from hop cones. The pigments are pyrrole derivatives that can be divided into two main groups: chlorophyll a (dark green) and chlorophyll b (light green). Depending on the conditions of extraction, the pigment amount may vary. As the extraction temperature increases, chlorophyll content improves. Consequently, the final extracted color is darker. When the temperature used for extraction exceeds 70 °C, the color of the extract may turn dark green or even brown. According to Lefebvre et al. [124], rosemary pigments were extracted using SFE. Sequential extraction and consecutive conditions were used to obtain different percentages that were rich in pigments. During the chlorophyll extraction process, 30% ethanol was used as a cosolvent. In addition, Derrien et al. [125] used SFE to extract spinach's lutein and chlorophyll. Compared to lutein, chlorophyll content was more significantly influenced by the cosolvent. Temperature had no effect on lutein and chlorophyll levels. As a result, the extraction of lutein and chlorophyll increased with higher pressure; however, the solvent's selectivity decreased. Optimal extraction parameters were 56 °C, 3.6 h of extraction time, 39 MPa of pressure, and 10% ethanol concentration. These factors resulted in a 50% chlorophyll yield and a 72% lutein yield. Morcelli et al. also used scCO_2 to extract chlorophylls and carotenoids from chlorella sorokiniana. According to [126], ethanol is necessary to extract chlorophylls effectively, as well as to alter the polarity of supercritical solvents in order

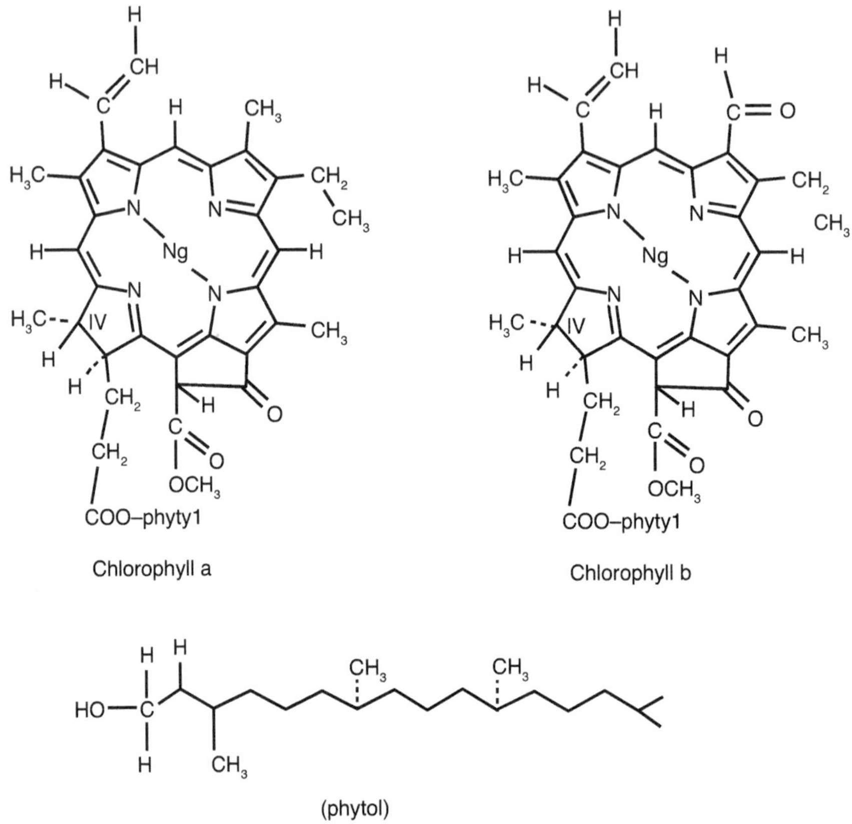

Fig. 6 Chemical structure of chlorophyll [34, 119]

to extract carotenoids. Additionally, ethanol's temperature, concentration, and pressure had an impact on the extraction process.

7.2 *Extraction of Anthocyanins*

In flowers and fruits, anthocyanins are pigments that are blue, purple, or red. In acidic conditions, anthocyanin appears as a red pigment, while in alkaline conditions, anthocyanin appears as a blue pigment. Anthocyanins have been shown to have beneficial effects against diabetes, cancer, inflammation, bacteria, and obesity [118]. Their structures consist of two aromatic rings connected by a three-carbon heterocyclic ring that contains oxygen. Anthocyanidin's conjugated double bonds are thought to be the chromophore. Figure 7 [127, 128] shows the general molecular

Fig. 7 Chemical structure of anthocyanin pigment [127, 128]

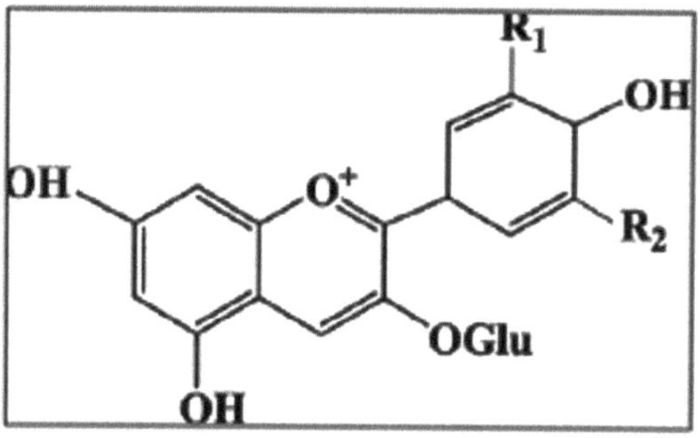

structure of anthocyanin. Additionally, anthocyanin's stability is influenced by structural factors, light, and pH. SFC could be used to extract anthocyanin pigment from various natural sources under specific conditions, such as temperature and pressure. Roselle and ethanol were extracted using $scCO_2$ and ethanol as a cosolvent. In Roselle, the deep color is primarily due to the presence of four anthocyanins, of which delphinidin 3-glucoside and cyanidin 3-glucoside are minor pigments and hybiscin and dephinidin 3-sambubioside are major pigments. At 8.90 MPa, 70 °C, and 9.49%, the extraction was effective, resulting in a maximum yield of 26.73% [129]. In their study [130], Idham et al. examined techniques to improve anthocyanin extraction and stability. The yield of anthocyanin increased with a rise in the cosolvent proportion; however, this increase was followed by a gradual decrease at cosolvent percentage of 10%. Anthocyanin content rises with increasing pressure but decreased at higher pressures. In the given temperature range (40–70 °C), anthocyanin content was not significantly affected by temperature. The investigated conditions with $scCO_2$ treatment were more successful in maintaining color stability and preserving anthocyanins at room temperature. Anthocyanins were extracted from juçara residues using $scCO_2$ and an acidified cosolvent using the SFE method, which increased the anthocyanin yield by making $scCO_2$ more polar. SFE has been demonstrated to be more effective for the extraction of anthocyanins [131]. According to Wo'zniak et al. [132], the cosolvent significantly affects the polarity of $scCO_2$ in the extraction of anthocyanins from chokeberry using SFE. Considerably less organic solvent was required for extraction when $scCO_2$ was used. Optimum parameters were determined for the extraction method. The ideal extraction factors were determined to be 10 MPa, 35 °C, and 80% m/m ethanol percentage. To extract anthocyanin from haka berry pulp paste, water and $scCO_2$ were used as cosolvents. For SFE extraction, pressure, temperature, and water content were all adjusted. Anthocyanin yield was highest at 65 °C, 45 MPa pressure, 15 min of static duration, and 20 min of dynamic period. A freeze-drying pretreatment was not necessary for SFE processing. The yield of anthocyanin extraction was raised to 52.7% when $scCO_2$ and water were used, and antioxidant activity was increased to 89.8% [133].

7.3 Extraction of Quinones

Quinones are cyclic diones with full conjugation ($C_6H_4O_2$) and enough conjugation to produce color [46]. There are natural quinonoids found in sea urchins, lichens, bacteria, fungi, flowering plants, insects, and fungi [134]. As shown in Fig. 8 [134–137], there are three common types of quinonoids: benzoquinones, naphthoquinones, and anthraquinones. Ubiquinones, plantoquinones, and menaquinones are additional quinonoids. There are a variety of hues associated with quinolones, ranging from yellow to red. Textiles with antimicrobial properties are finished with quinolones [134]. SFE was used to extract juglone from green walnut husks. According to the findings, pressure was the most important factor, while temperature and particle size had a bad impact on the amount of extracted juglone. Increasing the pressure increased the extraction efficiency. The three conditions of 35 °C, 15 MPa, and 375 m produced the highest juglone yields. It can be utilized as a dye in clothing made of natural or synthetic materials, skin-coloring products, hair dyes, and pharmaceuticals [138]. A further study revealed that the $scCO_2$ extract had significant antifungal and antioxidant properties and was rich in bioactive substances like phenolic acids and juglone. SFE with ethanol demonstrated an extractive yield of 27.18 g/100 g, which was comparable to the results of solvent (ethanol and methanol) extractions, which were 36.13 and 39.27 g/100 g, respectively. This is caused by the utilization of high pressures and the unique characteristics of CO_2 represented in the supercritical state [139].

7.4 Extraction of Carotenoids

Carotenoids are widely used in food, textiles, and cosmetics. They are a class of bioactive substances responsible for yellow-red pigmentation [140, 141]. Among the primary categories of carotenoids found in nature are α-carotene, β-carotene, lycopene, lutein, zeaxanthin, and β-cryptoxanthin. Food sources of lycopene, lutein, zeaxanthin, and cryptoxanthin include cereals and red peppers. Carotene and α-carotene are obtained from carrots, sweet potatoes, pumpkins, etc. [142]. A variety of raw materials were used to extract carotenoids using the SFE method.

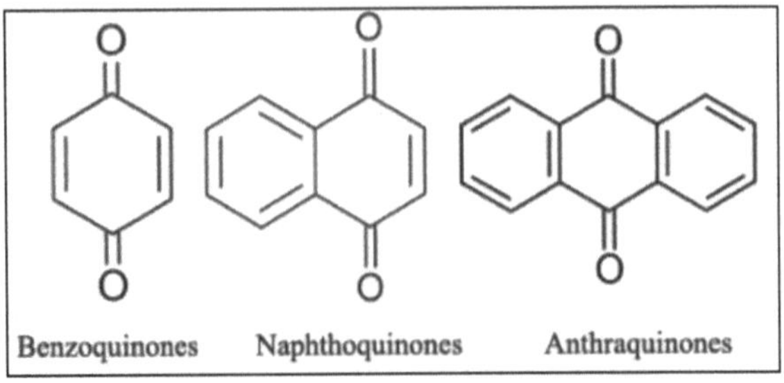

Fig. 8 The three basic types of quinonoids [134–137]

Fig. 9 The chemical structure of capsaicin [144]

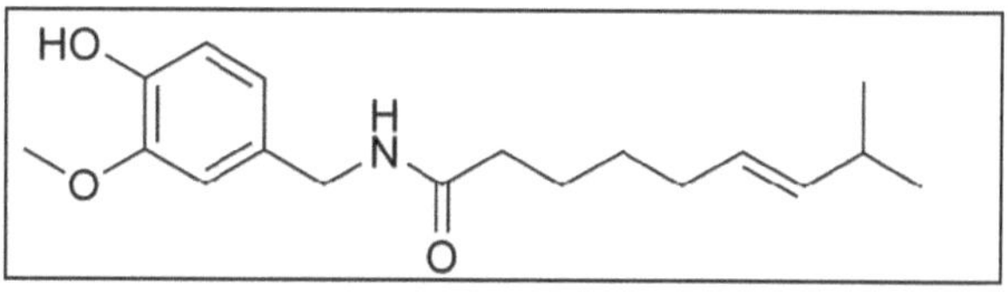

According to Shah et al. [143], extraction pressure, time, and particle size had a significant impact on the SFE of paprika oil. Additionally, process variables like temperature, pressure, and particle size had a big impact on the extraction of capsaicin from paprika, as shown in Fig. 9 [144]. However, only the extraction pressure affected the pigment yield from paprika. The ideal conditions for paprika oil's extraction, pigments, and capsaicin were 65, 65, and 57.04 °C; pressures of 357.8, 200.62, and 400 bar; extraction times of 90, 90, and 100.89 min; and particle sizes of 0.78, 0.50, and 0.89 mm.

Esquivel-Hernández et al. [145] examined the effects of different factors of SFE from *A. platensis* on the following (total extraction yield, pigment yield), carotenoids (lutein, beta-carotene) shown in Fig. 10 [146]. The highest content of bioactive metabolites including β-carotene, lutein, tocopherols, and γ-linoleic acid was obtained at P of 450 bar, CX 11/g/min, SX of 15 min, DX of 25 min, *T* of 60 °C, and Di of 35 g. In addition, this treatment showed the highest antioxidant content and most effective antibacterial action against *S. aureus, P. aeruginosa, E. coli* and *C. albicans*.

The SFE technique was utilized by Urbonaviciene et al. [147] for the extraction of lycopene pigment from tomatoes. The ideal extraction parameters for cis-lycopene isomers are 52 °C temperature, 55 MPa pressure, and 180 min of extraction time. In the scCO_2 extract, 62% of the total lycopene was in the form of the cis-lycopene isomer. The scCO_2 extraction factors could be changed in order to increase the amount of cis-lycopene isomers in the final product. Additionally, the SFE method was used to compare the red hues (oleoresins rich in lycopene) obtained from tomato, watermelon, and gac fruit. As shown in Fig. 11 [148], lycopene is abundant in all oleoresins [149].

De Andrade Lima et al. [150] reviewed the impact of different SFE parameters on the extraction of carotenoid compounds from carrot peels. The principal carotenoid in the samples was found to be β-carotene, which made up about 60% of the total carotenoid content, and α-carotene comes after that making up for about 30% of the total carotenoid content. The main factor affecting carotenoid recovery was pressure. Even though carotenoid degradation was caused by extremely high temperatures, the maximum carotenoid recovery yield was obtained at 59.0 °C and 34.9 MPa. Bobinaite et al. [151] explained the effects of sequential extraction with scCO_2 and pressurized solvents on the extraction of carotenoids from rowanberry pomace. β-Carotene, which made up an average of 52% of the total carotenoid content in the samples, was the main carotenoid. Additionally, it was found that the most significant aspect that affects carotenoid recovery is pressure. Wang et al. [152] utilized the SFE technique for pigment extraction from pumpkins. The results

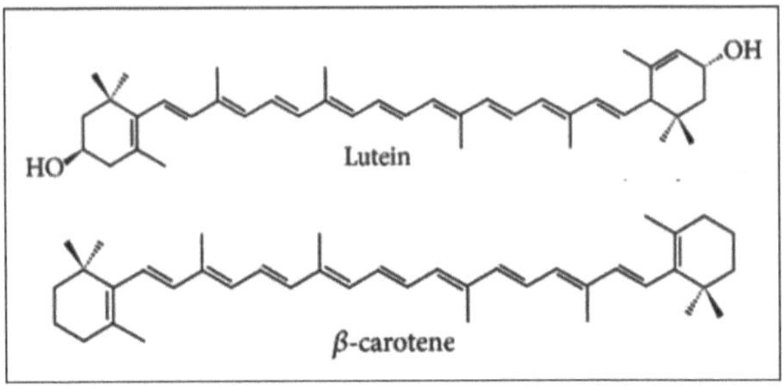

Fig. 10 The chemical structure of lutein, β-carotene [146]

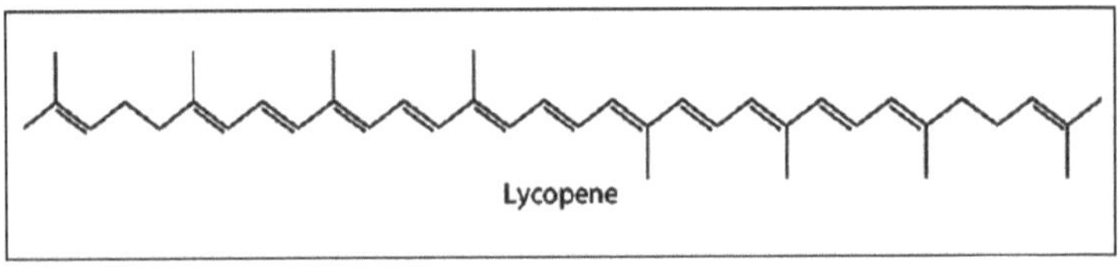

Fig. 11 The chemical structure of lycopene [148]

assumed that the mass ratio of pumpkin flesh to seeds had a significant impact on the yield of oil, β-carotene, and α-tocopherol. The ideal factors were achieved at 47.75 °C, 30 MPa, and a seed mass of 67%.

7.5 *Extraction of Curcuminoids*

Turmeric contains water-insoluble substances known as curcuminoids. Figure 12 [153] shows the chemical structure of curcuminoids. Curcumins are composed of two benzenemethoxy rings connected by an unsaturated chain. Curcuminoids are yellow-orange in color. Anti-inflammatory, antimutagenic, and anticancer are some of the bioactive qualities of this phenolic class [118]. There has been previous research conducted on the SFE of curcumin derived from *Curcuma longa* rhizomes using water in the two-step procedure. The maximum overall yield was obtained with $scCO_2$ extraction, subsequent to ethanol, or the use water for extraction. In the two-step procedure, large quantities of curcumin were obtained when ethanol was used as the solvent. Although the SFE contained very little curcumin, it demonstrated strong antimalarial activity [154].

7.6 *Extraction of Phycocyanins*

The blue fluorescent pigments known as phycocyanins are developed via eukaryotic algae as well as cyanobacteria. They have fewer applications in the pharmaceutical as well as food industries due to their sensitive nature to light and heat. They possess

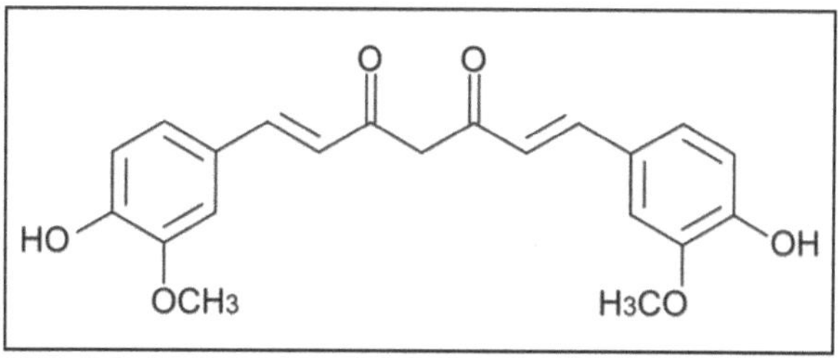

Fig. 12 The chemical structure of curcuminoids [153]

cholesterol-lowering, anticancer, antiviral, anti-inflammatory, and antioxidant properties [118]. The ideal $scCO_2$ conditions were determined to be 60 °C and 24.13 MPa. Because of C-phycocyanin, the SFE extract was supposed to have a variety of health benefits, including anti-inflammatory, antiviral, and antioxidant properties [155].

7.7 Extraction of Black Sesame Pigment

Sesame dregs were used to produce black sesame pigment. SFE is used in this process. The highest significant yield was achieved with the following parameters: 30 °C, 4 MPa, and 10.80 L/min [156]. There are several industries that use black sesame pigment, including pharmacology, cosmetics, and others.

7.8 Extraction of Iridoids

Iridoids are blue pigments that are water-soluble. Anti-inflammatory, antioxidant, and antimicrobial properties are associated with this class of bioactive substances [118]. Saengsa et al. extracted plummericin, an iridoid color, from the *Momordica charantia* vine using SFE. The factors of extraction were 50 °C, 25 MPa, and 5 L/min in the presence of cosolvents like ethanol and ethyl acetate.

8 Application of Extracted Pigment on Different Fabrics

Natural dyes are widely used for dyeing fabrics to improve their environmental properties [157]. When using natural dye, there are two steps to take into consideration: mordanting and dye application. The mordanting gets the fabric ready for the dyes and makes it possible for the dye molecules to bond to the fiber. However, some natural dyes do not require mordants to produce bright colors. Without the need for auxiliaries, some plant dyes already have properties that will bond the color to the fiber. Although not always necessary, the utilization of mordants can still have benefits, especially when attempting to achieve a range of shades, increase dye absorption, and enhance colorfastness [158]. Depending on the required color, suitable dyes or resources that contain dye are chosen. In general, shades of yellow are produced using flowers and leaves that contain flavonoids. For the color red, anthraquinone dyes derived from both plant and animal sources can be used. Typically, indigo is used to develop blue color shades. With the right dye and mordant combination or by combining two compatible dyes, secondary colors like orange can be produced. On the other hand, in order to create a secondary color, green or blue must be obtained. Indigo is used as a base dye, followed by the additional dye [3]. There are a limited number of researches which are listed below using the extracted pigment from natural sources on dyeing textiles.

This study examines the potential of various extraction methods to recover natural dyestuffs. The resulting extracts showed various UV spectra that indicated different chemical compounds and were considered for further investigation. Additionally, traditional dyeing techniques have been used to investigate these extracts' dyeability. The results exhibited bright shades produced using plants and by-products, which are regarded as a rich source of natural, affordable dyes. The natural extracts produced with supercritical CO_2 in particular enhance the dyeability of fibers. So, supercritical extraction might be a suitable environmentally friendly method to possibly recover natural dyestuffs [158].

8.1 Application of Extracted Pigment on Silk Fabrics

The dyeing abilities and durability of natural dyes derived from *Chlorella vulgaris* sp. were studied by Rahim et al. [159]. A natural dye, *Chlorella vulgaris* sp., was found to produce bright shades on silk fabrics. $scCO_2$ extraction is also suggested as a good extraction technique due to its use of environmentally friendly technology and high extract yield. Apart from the color change during washing, which is rated as poor, dyed silk fabrics mordanted with alum received a good to an excellent rating in comparison with dyed silk fabrics mordanted with iron. An additional study used $scCO_2$ to extract natural dyes from *Sargassum* sp. used to dye silk. As a function of extract yield, pressure and temperature were the most critical parameters for $scCO_2$ extraction. To achieve the best yield of 0.27%, 4500 psi and 65 °C were the ideal extraction conditions. UV-vis analysis of the extracts revealed the presence of carotenoids at 673 nm and chlorophyll at its maximum absorption spectrum, 410 nm. These pigments are responsible for the color of seaweed. Several mordants were used to dye silk fabrics using the pre-mordanting and simultaneous mordanting techniques. Despite relatively low ratings for light fastness, the analysis indicated reasonable to excellent fastness properties. Due to the same dye source being used for all dyed samples, the color strength (K/S) results were similar. Those silk fabrics that were simultaneously mordanted with cinnamon and ferrous sulfate had the highest K/S values [160].

8.2 *Application of Extracted Pigment on Wool Fabrics*

According to Abou Elmaaty et al. [161], SFE technology was used to remove alizarin pigment from *R. tinctorum*. Optimal process parameters for alizarin were determined by investigating the effects of heating rate, pressure, and flow rate. In addition to time, the most significant variable affecting alizarin extraction yield was the cosolvent ratio. Considering the recovery of alizarin and reducing environmental impact, the following SFE conditions were preferred: 90% CO_2, 10% Me, 65 °C, 250 bar, 45 min, and 9 mL/min. *R. tinctorum* extract (RE), obtained under favorable conditions at various concentrations (2–6%), was used to dye the wool fabrics. The antibacterial qualities of dyed textiles and RE were examined. *P. aeruginosa* was more resistant to dyed fabric and RE than other bacterial strains. The environmental effects of organic solvents can be reduced by using SFE to extract natural colorants with supercritical technology.

9 Conclusions

The natural dyes have a high level of color fastness and can produce a wide variety of bright shades. The use of natural dyes in the textile industry and the regeneration of natural dyes as an alternative to synthetic dyes are currently being studied on a global scale, resulting in sustainable development, which satisfies present needs without impacting the ability of upcoming generations to fulfill their own needs. By utilizing the advantages of SCFs, which have characteristics similar to those of liquids and gases, the SFE method enhances extraction. It is highly effective, does not use any toxic solvents, maintains a constant fluid flow, and can extract heat-sensitive pigments. A number of natural dyes and pigments have been effectively extracted using SFE. The purpose of extracting dyes from natural sources is to prevent environmental pollution and allergic reactions. Additionally, their antibacterial and antioxidant properties contribute to their beneficial effects. Each class of natural pigment (quinones, curcuminoids, carotenoids, and others) has unique SFE requirements, such as temperature, pressure, flow rate, time, etc. By using the optimum conditions for the SFE, the highest yield of pigment was obtained. Research and reports on dyeing textiles with natural dyes are still insufficient. Therefore, there is a need for a lot of investigations to produce a database and knowledge base along with the making of suitable colors for various textiles.

Abbreviations

BPR	Back-pressure regulator
EAE	Enzyme assisted extraction
HPAE	High pressure assisted extraction

MAE	Microwave-assisted extraction
Pc	Critical pressure
$scCO_2$	Supercritical carbon dioxide
SCF	Supercritical fluid
SCFs	Supercritical fluids
SFE	Supercritical fluid extraction
Tc	Critical temperature
UAE	Ultrasound-assisted extraction

References

1. Iqbal, S., & Ansari, T. N. (2021). Extraction and application of natural dyes. In *Sustainable practices in the textile industry* (pp. 1–40). Scrivener Publishing LLC.
2. Marsili, R., & Callahan, D. (1993). Comparison of a liquid solvent extraction technique and supercritical fluid extraction for the determination of alpha- and beta-carotene in vegetables. *Journal of Chromatographic Science, 31*, 422–428.
3. Shi, J., Mittal, G., Kim, E., et al. (2007). Solubility of carotenoids in supercritical CO_2. *Food Review International, 23*, 341–371.
4. Reverchon, E., Donsi, G., & Osseo, L. (1993). Modeling of supercritical fluid extraction from herbaceous matrices. *Industrial and Engineering Chemistry Research, 32*, 2721–2726.
5. Mendes, R., Fernandes, H., Coelho, J., et al. (1995). Supercritical CO_2 extraction of carotenoids and other lipids from Chlorella vulgaris. *Food Chemistry, 53*(1), 99–103.
6. Montero, G. A., Smith, C. B., Hendrix, W. A., et al. (2000). Supercritical fluid technology in textile processing: An overview. *Industrial and Engineering Chemistry Research, 39*(12), 4806–4812.
7. Katewaraphorn, J., & Aldred, A. (2016). A study of microcapsules containing psidium guajava leaf extract for antibacterial agent on cotton fabric. *International Journal of Chemical Engineering and Applications, 7*(1), 27–31.
8. Verma, M., Singh, S., & Singh, R. (2016). Ecofriendly dyeing of cotton fabric after biopolymer treatment by using leaves of Psidium guajava. *International Journal of Innovative Research in Science, Engineering and Technology, 5*(9), 16518–16523.
9. Sani, A., & Mohseni, S. (2014). Effect of type and concentration of solvent on chemical properties of saffron edible extract. *Nutrition & Food Science, 44*(1), 2–5.
10. Chia, P. W., Lim, B. S., Tan, K. C., et al. (2019). Water extract of onion peel for the synthesis of bisindolylmethanes. *Journal of King Saud University, Science, 31*(4), 642–647.
11. Narayanaswamy, V., Ninge Gowda, K. N., & Sudhakar, R. (2013). Dyeing and color fastness of natural dye from Psidium guajuva on silk. *Journal of Natural Fibers, 10*(3), 257–270.
12. Kamal, M. M., Helmy, H. M., & Hawary, N. S. (2014). Some studies on dyeing properties of cotton fabrics with Crocus sativus (saffron flowers) using an ultrasonic method. *Journal of Natural Fibers, 6*, 151–170.
13. Ahmed, M. A., Eltahir, Y. A., & Saeed, H. A. (2018). Eco-friendly utilization of banana plant extract for dyeing of textile materials and related purposes. *International Journal of Recent Scientific Research, 9*(5), 26465–26473.
14. Saleh, S. M., Abd-elhady, Y. A., & Farag, M. M. (2009). Cotton fabric dyeing with natural colorants extracted from banana peel. *Egyptian Journal of Agricultural Research, 87*(1), 159–170.
15. Ponnusami, V., Vikram, S., & Srivastava, S. (2008). Guava (Psidium guajava) leaf powder: Novel adsorbent for removal of methylene blue from aqueous solutions. *Journal of Hazardous Materials, 152*(1), 276–286.

16. Ali, N. F., El-Khatib, E. M., El-Mohamedy, R. S., et al. (2014). Antimicrobial activity of silk fabrics dyed with saffron dye using microwave heating. *International Journal of Current Microbiology and Applied Sciences, 3*(10), 140–146.
17. Helmy, H. M. (2020). Extraction approaches of natural dyes for textile coloration. *Journal of Textiles, Coloration and Polymer Science, 17*(2), 65–76.
18. Chilukoti, G., Venkatesh, B., & Chandana, V. (2019). Optimisation of dyeing and mordanting parameters on cotton fabrics treated with allium cepa as a natural dye source. *Journal of the Textile Association, 79*(4), 333–339.
19. Broadbent, A. D. (2001). An introduction to textiles, dyes and dyeing. In *Basic principles of textile coloration* (Vol. 132, pp. 1–19). Society of Dyers and Colorists. ISBN 0-901956-76-7.
20. Patel, B. H. (2011). Natural dyes. In *Handbook of textile and industrial dyeing* (Vol. 1, pp. 395–424). Woodhead Publishing. ISBN 978-1-84569-695-5.
21. Bhandari, N., Bist, K., Ghimire, J., et al. (2020). Feasibility study of the Euphorbia pulcherrima plant extract as natural dye with different mordants for fabric dyeing. *Journal of Science and Technology, 25*(1), 30–36.
22. Affat, S. S. (2021). Classifications, advantages, disadvantages, toxicity effects of natural and synthetic dyes. *Journal of Science, 8*(1), 130–135.
23. Rana, S., Pichandi, S., Parveen, S., et al. (2014). Natural plant fibers: Production, processing, properties and their sustainability parameters. In *Roadmap to sustainable textiles and clothing* (pp. 1–35). Springer.
24. Ebrahim, S., Mosaad, M., Othman, H., et al. (2022). A valuable observation of ecofriendly natural dyes for valuable utilisation in the textile industry. *Journal of Textiles, Coloration and Polymer Science, 19*(1), 25–37.
25. Zayed, M., Othman, H., Ghazal, H., et al. (2022). A valuable observation on natural plants extracts for valuable functionalization of cotton fabric (an overview). *Egyptian Journal of Chemistry, 65*(4), 499–524.
26. El-Sayed, G., Othman, H., & Hassabo, A. (2021). An overview on the eco-friendly printing of jute fabrics using natural dyes. *Journal of Textiles, Coloration and Polymer Science, 18*(2), 239–245.
27. AlAshkar, A., & Hassabo, A. (2021). Recent use of natural animal dyes in various field. *Journal of Textiles, Coloration and Polymer Science, 18*(2), 191–210.
28. Hamdy, D. M., Othman, H. A., & Hassabo, A. G. (2021). Various natural dyes from different sources. *Journal of Textiles, Coloration and Polymer Science, 18*(2), 171–190.
29. Hamdy, D. M., Othman, H. A., & Hassabo, A. G. (2021). Various natural dyes using plant palette in coloration of natural fabrics. *Journal of Textiles, Coloration and Polymer Science, 18*(2), 121–141.
30. Ragab, M. M., & Hassabo, A. G. (2021). Various uses of natural plants extracts for functionalization textile-based materials. *Journal of Textiles, Coloration and Polymer Science, 18*(2), 143–158.
31. Reda, E., Ghazal, H., Othman, H., et al. (2022). An observation on the wet processes of natural fabrics. *Journal of Textiles, Coloration and Polymer Science, 19*(1), 71–97.
32. Habib, N., Akram, W., Adeel, S., et al. (2022). Environmental-friendly extraction of Peepal (Ficus Religiosa) bark-based reddish-brown tannin natural dye for silk coloration. *Environmental Science and Pollution Research, 29*(23), 35048–35060.
33. Butola, B., & Roy, A. (2018). Chitosan polysaccharide as a renewable functional agent to develop antibacterial, antioxidant activity and colourful shades on wool dyed with tea extract polyphenols. *International Journal of Biological Macromolecules, 120*, 1999–2006.
34. Sharma, M., Usmani, Z., Gupta, V. K., et al. (2021). Valorization of fruits and vegetable wastes and by-products to produce natural pigments. *Critical Reviews in Biotechnology, 41*, 535–563.
35. Lee, Y. H., Hwang, E. K., & Kim, H. D. (2009). Colorimetric assay and antibacterial activity of cotton, silk, and wool fabrics dyed with peony, pomegranate, clove, Coptis chinenis and gallnut extracts. *Materials, 2*, 10–21.

36. Numan, M., Bashir, S., Mumtaz, R., et al. (2018). Therapeutic applications of bacterial pigments: A review of current status and future opportunities. *3 Biotech, 8*, 1–15.
37. Rehman, F., Sanbhal, N., Naveed, T., et al. (2018). Antibacterial performance of Tencel fabric dyed with pomegranate peel extracted via ultrasonic method. *Cellulose, 25*, 4251–4260.
38. Gupta, D., Jain, A., & Panwar, S. (2005). Anti-UV and anti-microbial properties of some natural dyes on cotton. *Indian Journal of Fibre & Textile Research, 30*, 190.
39. Yusuf, M., Shahid, M., Khan, M., et al. (2015). Dyeing studies with henna and madder: A research on effect of tin (II) chloride mordant. *Journal of Saudi Chemical Society, 19*, 64–72.
40. Inprasit, T., Motina, K., Pisitsak, P., et al. (2018). Dyeability and antibacterial finishing of hemp fabric using natural bioactive neem extract. *Fibers and Polymers, 19*, 2121–2126.
41. Fernandes, F., Pereira, E., Círic, A., et al. (2019). Ocimum basilicum var. purpurascens leaves (red rubin basil): A source of bioactive compounds and natural pigments for the food industry. *Food & Function, 10*, 3161–3171.
42. Moreira, M. D., Melo, M. M., Coimbra, J. M., et al. (2018). Solid coffee waste as alternative to produce carotenoids with antioxidant and antimicrobial activities. *Waste Management, 82*, 93–99.
43. Abdel-Shafi, S., Al-Mohammadi, A., Sitohy, M., et al. (2019). Antimicrobial activity and chemical constitution of the crude, phenolic-rich extracts of Hibiscus sabdariffa, Brassica oleracea and Beta vulgaris. *Molecules, 24*, 4280.
44. Pargai, D., Jahan, S., & Gahlot, M. (2020). Functional properties of natural dyed textiles. In *Chemistry and technology of natural and synthetic dyes and pigments* (pp. 1–19). IntechOpen. ISBN 978-1-83968-758-7.
45. Silva, A., Silva, S., Carpena, M., et al. (2020). Macroalgae as a source of valuable antimicrobial compounds: Extraction and applications. *Antibiotics, 9*, 642.
46. Phan, K., Raes, K., Van Speybroeck, V., et al. (2021). Non-food applications of natural dyes extracted from agro-food residues: A critical review. *Journal of Cleaner Production, 301*, 126920.
47. Sanku, L., & Padma, A. (2021). Identifying the viability of natural dye sources from India: A review. *International Research Journal of Modernization in Engineering Technology and Science, 2*, 736–744.
48. Saxena, S., & Raja, A. (2014). Natural dyes: Sources, chemistry, application and sustainability issues. In *Roadmap to sustainable textiles and clothing* (pp. 37–80). Springer. ISBN 978-981-287-064-3.
49. Duke, J. A. (1981). Legume species. In *Handbook of legumes of world economic importance* (pp. 5–310). Springer. ISBN 978-1-4684-8153-2.
50. Jothi, D. (2008). Extraction of natural dyes from African marigold flower (Tagetus erecta) for textile colouration. *Autex Research Journal, 8*(2), 49–53.
51. Kale, S., Naik, S., & Gaikwad, R. (2007). Cosmos flowers – An effective source of natural dye for wool. *The Journal of Textile Association, 68*, 164–167.
52. Sanku, L., Padma, A., Lakshm, V., et al. (2021). Exploring dye extraction methods and preservatives for natural dyes. *International Journal of Pharmacognosy and Phytochemical Research, 10*(1), 137–142.
53. Thu Dao, T., Webb, H., & Malherbe, F. (2021). Optimization of pectin extraction from fruit peels by response surface method: Conventional versus microwave-assisted heating. *Food Hydrocolloids, 113*, 106475.
54. Garcia-Vaquero, M., Rajauria, G., & Tiwari, B. (2020). Conventional extraction techniques: Solvent extraction. In *Sustainable seaweed technologies* (pp. 171–189). Elsevier. ISBN 9780128179437.
55. Samanta, A., Agarwal, P., Singhee, D., et al. (2009). Application of single and mixtures of red sandalwood and other natural dyes for dyeing of jute fabric: Studies on colour parameters/colour fastness and compatibility. *Journal of the Textile Institute, 100*(7), 565–587.

56. Merdan, N., Eyupoglu, S., & Duman, M. N. (2017). Ecological and sustainable natural dyes. In *Textiles and clothing sustainability* (Textile Science and Clothing Technology) (pp. 1–41). Springer. ISBN 978-981-10-2184-8.
57. Adeel, S., Salman, M., Zahoor, A., et al. (2020). An insight into herbal-based natural dyes: Isolation and applications. In *Recycling from waste in fashion and textiles: A sustainable and circular economic approach* (pp. 423–456). Wiley-Scrivener.
58. Adeel, S., Rehman, F.-U., Rafi, S., et al. (2019). Environmentally friendly plant-based natural dyes: Extraction methodology and applications. In *Plant and human health* (Vol. 2, pp. 383–415). Springer. ISBN 978-3-030-03343-9.
59. Rastogi, D., Gulrajani, M., & Gupta, P. (2000). Application of lac dye on cationised cotton. *Colourage, 47*(4), 36–40.
60. Saxena, S., Iyer, V., Shaikh, A., et al. (1997). Dyeing of cotton with lac dye. *Colourage, 44*(11), 23–28.
61. Mongkholrattanasit, R., Saiwan, C., Rungruangkitkrai, N., et al. (2015). Ecological dyeing of silk fabric with lac dye by using padding techniques. *Journal of the Textile Institute, 106*(10), 1106–1114.
62. Mahmud-Ali, A., Fitz-Binder, C., & Bechtold, T. (2012). Aluminium based dye lakes from plant extracts for textile coloration. *Dyes and Pigments, 94*(3), 533–540.
63. Wangatia, L., Tadesse, K., Moyo, S., et al. (2015). Mango bark mordant for dyeing cotton with natural dye: Fully eco-friendly natural dyeing. *International Journal of Textile Science, 4*(2), 36–41.
64. Akilandeswari, P., & Pradeep, B. (2016). Exploration of industrially important pigments from soil fungi. *Applied Microbiology and Biotechnology, 100*(4), 1631–1643.
65. Venil, C. K., Zakaria, Z. A., & Ahmad, W. A. (2013). Bacterial pigments and their applications. *Process Biochemistry, 48*(7), 1065–1079.
66. Chattopadhyay, S., Pan, N., Roy, A., et al. (2013). Development of natural dyed jute fabric with improved colour yield and UV protection characteristics. *Journal of the Textile Institute, 104*(8), 808–818.
67. Katarzyna, S., & Kowalinski, J. (Eds.). (2008). Light fastness properties and UV protection factor of naturally dyed linen, hemp and silk. In *Proceedings, flaxbast 2008: International conference on flax and other bast PLANTS*, Saskatoon, Canada.
68. Sarkar, A. K. (2004). An evaluation of UV protection imparted by cotton fabrics dyed with natural colorants. *BMC Dermatology, 4*(1), 1–8.
69. Ibrahim, N., El-Gamal, A., Gouda, M., et al. (2010). A new approach for natural dyeing and functional finishing of cotton cellulose. *Carbohydrate Polymers, 82*(4), 1205–1211.
70. Sun, S.-S., & Tang, R.-C. (2011). Adsorption and UV protection properties of the extract from honeysuckle onto wool. *Industrial and Engineering Chemistry Research, 50*(8), 4217–4224.
71. Ibrahim, N. A., El-Zairy, W. M., El-Zairy, M. R., et al. (2013). Enhancing the UV-protection and antibacterialproperties of polyamide-6 fabric by natural dyeing. *Interactions, 8*, 9.
72. Anttha, K., & Prasad, S. (2007). Developing multiple natural dyes from flower parts of Gulmohur. *Current Science, 92*(12), 1681–1682.
73. Bechtold, T., Turcanu, A., Ganglberger, E., et al. (2003). Natural dyes in modern textile dyehouses—How to combine experiences of two centuries to meet the demands of the future? *Journal of Cleaner Production, 11*(5), 499–509.
74. Oztav, F. (2009). *The investigation of usage of the plant of alkanet (Alkanna tinctoria) as cellulosic and protein fiberic dyestuff* [Doctor of Philosophy Thesis]. Gaziosmanpaşa Universty.
75. Borges, M., Tejera, R., Díaz, L., et al. (2012). Natural dyes extraction from cochineal (Dactylopius coccus). New extraction methods. *Food Chemistry, 132*(4), 1855–1860.
76. Al-Alwani, M. A., Mohamad, A. B., Kadhum, A. A., et al. (2015). Effect of solvents on the extraction of natural pigments and adsorption onto TiO_2 for dye-sensitized solar cell applications. *Spectrochimica Acta Part A: Molecular and Biomolecular Spectroscopy, 138*, 130–137.
77. Budisa, N., & Schulze-Makuch, D. (2014). Supercritical carbon dioxide and its potential as a life-sustaining solvent in a planetary environment. *Life, 4*, 331–340.

78. Zosel, K. (1981). *Process for the decaffeination of coffee* (U.S. Patent No. 4,260,639). U.S. Patent and Trademark Office, Washington, DC.
79. Chemat, F., Vian, M., & Cravotto, G. (2012). Green extraction of natural products: Concept and principles. *International Journal of Molecular Sciences, 13*, 8615–8627.
80. Zhang, Q. W., Lin, L. G., & Ye, W. C. (2018). Techniques for extraction and isolation of natural products: A comprehensive review. *Chinese Medicine, 13*(1), 1–26.
81. Herrero, M., Mendiola, J. A., Cifuentes, A., et al. (2010). Supercritical fluid extraction: Recent advances and applications. *Journal of Chromatography A, 1217*(16), 2495–2511.
82. Zia, K., Adeel, S., Aslam, H., et al. (2019). Influence of ultrasonic radiation on extraction and green dyeing of mordanted cotton using neem bark extract. *Journal of Industrial and Engineering Chemistry, 77*, 317–322.
83. Adeel, S., Azeem, M., Rehman, F. U., et al. (2019). Sustainable utilization of ultrasonic radiation in extraction and dyeing of wool fabric using logwood (Haematoxylum campechianum) extracts. *Journal of Textile and Apparel, Technology and Management, 29*(2), 181–185.
84. Adeel, S., Zia, K., Abdullah, M., et al. (2019). Ultrasonic assisted improved extraction and dyeing of mordanted silk fabric using neem bark as source of natural colourant. *Natural Product Research, 33*(14), 2060–2072.
85. Ticha, M., Meksi, N., Attia, H., et al. (2017). Ultrasonic extraction of Parthenocissus quinquefolia colorants: Extract identification by HPLC-MS analysis and cleaner application on the phytodyeing of natural fibres. *Dyes and Pigments, 141*, 103–111.
86. Bell, A. T. (1997). Chemistry and chemical engineering. In *Practical applications of fuzzy technologies* (Vol. 52, pp. 301–318). Springer. ISBN 9781493990597.
87. Uwineza, P., & Waśkiewicz, A. (2020). Recent advances in supercritical fluid extraction of natural bioactive compounds from natural plant materials. *Molecules, 25*, 3847.
88. Chemat, F., Rombaut, N., Sicaire, A., et al. (2017). Ultrasound assisted extraction of food and natural products. Mechanisms, techniques, combinations, protocols and applications. A review. *Ultrasonics Sonochemistry, 34*, 540–560.
89. Boonkird, S., Phisalaphong, C., & Phisalaphong, M. (2008). Ultrasound-assisted extraction of capsaicinoids from Capsicum frutescens on a lab- and pilot-plant scale. *Ultrasonics Sonochemistry, 15*, 1075–1079.
90. Khan, S., Aslam, R., & Makroo, H. (2019). High pressure extraction and its application in the extraction of bio-active compounds: A review. *Journal of Food Process Engineering, 42*, e12896.
91. Wang, J., Zhang, J., Zhao, B., et al. (2010). A comparison study on microwave-assisted extraction of Potentilla anserina L. polysaccharides with conventional method: Molecule weight and antioxidant activities evaluation. *Carbohydrate Polymers, 80*, 84–93.
92. Sarfarazi, M., Jafari, S., Rajabzadeh, G., et al. (2020). Evaluation of microwave-assisted extraction technology for separation of bioactive components of saffron (Crocus sativus L.). *Industrial Crops and Products, 145*, 111978.
93. Chaudhary, S., Sagar, S., Kumar, M., et al. (2015). The use of enzymes in food processing: A review. *South Asian Journal of Food Technology and Environment, 1*(4), 2394–5168.
94. Raveendran, S., Parameswaran, B., Ummalyma, S. B., et al. (2018). Applications of microbial enzymes in food industry. *Food Technology and Biotechnology, 56*, 16–30.
95. Echave, J., Pereira, A., Carpena, M., et al. (2020). Capsicum seeds as a source of bioactive compounds: Biological properties, extraction systems, and industrial application. In *Capsicum* (pp. 123–158). IntechOpen. ISBN 978-1-83880-941-6.
96. Branch, J. A., Bartlett, P. N., et al. (2015). Electrochemistry in supercritical fluids. *Philosophical Transactions of the Royal Society A – Mathematical Physical and Engineering Sciences, 373*, 20150007.
97. Kate, A., Singh, A., Shahi, N., et al. (2016). Novel eco-friendly techniques for extraction of food based lipophilic compounds from biological materials. *Natural Products Chemistry & Research, 4*(5), 1000231.

98. Kulazynski, M., Stolarski, M., Faltynowicz, H., et al. (2016). Supercritical fluid extraction of vegetable materials. *Chemistry & Chemical Technology, 10*, 637–644. [CrossRef].
99. Gopaliya, P., Kamble, P., Kamble, R., et al. (2014). A review article on supercritical fluid chromatography. *International Journal of Pharma Research & Review, 3*, 59–66.
100. Sharif, K., Rahman, M., Azmir, J., et al. (2014). Experimental design of supercritical fluid extraction—A review. *Journal of Food Engineering, 124*, 105–116.
101. Bhusnure, O. G., Gholve, S. B., Giram, P. S., et al. (2015). Importance of supercritical fluid extraction techniques in pharmaceutical industry: A review. *Indo American Journal of Pharmaceutical Sciences, 5*, 6876.
102. Lam, C., Sin, L., Bee, S., et al. (2020). A review of natural products from plants using carbon dioxide supercritical fluid extraction. *Journal of the Institution of Engineers, Malaysia, 81*(1). https://doi.org/10.54552/v81i1.70
103. Chandrakant, K. K., Pravin, D. J., Bharat, H. S., et al. (2011). A overview on supercritical fluid extraction for herbal drugs. *Pharmacology, 2*, 575–596.
104. Reverchon, E., & de Marco, I. (2006). Supercritical fluid extraction and fractionation of natural matter. *Journal of Supercritical Fluids, 38*, 146–166.
105. Khaw, K. Y., Parat, M. O., Shaw, P. N., et al. (2017). Solvent supercritical fluid technologies to extract bioactive compounds from natural sources: A review. *Molecules, 22*, 1186.
106. Molino, A., Mehariya, S., di Sanzo, G., et al. (2020). Recent developments in supercritical fluid extraction of bioactive compounds from microalgae: Role of key parameters, technological achievements and challenges. *Journal of CO2 Utilization, 36*, 196–209.
107. Pourmortazavi, S. M., & Hajimirsadeghi, S. S. (2007). Supercritical fluid extraction in plant essential and volatile oil analysis. *Journal of Chromatography A, 1163*, 2–24.
108. Kargılı, U., & Aytaç, E. (2022). Supercritical fluid extraction of cannabinoids (THC and CBD) from four different strains of cannabis grown in different regions. *Journal of Supercritical Fluids, 179*, 105410.
109. Zaghloul, D., Abou Elmaaty, T., Nakamura, K., et al. (2021). Influence of additive organic base on dyeing of cotton fabric under supercritical carbon dioxide using fluorotriazine reactive disperse dye and investigation of optimal dyeing conditions. *Journal of Supercritical Fluids, 174*, 105243.
110. Ahangari, H., King, J., Ehsani, A., et al. (2021). Supercritical fluid extraction of seed oils—A short review of current trends. *Trends in Food Science and Technology, 111*, 249–260.
111. Wrona, O., Rafińska, K., Możeński, C., et al. (2017). Supercritical fluid extraction of bioactive compounds from plant materials. *Journal of AOAC International, 100*(6), 1624–1635.
112. Sánchez-Camargo, A., Mendiola, A., Báñez, E., et al. (2014). Supercritical fluid extraction. In *Reference module in chemistry, molecular sciences and chemical engineering*. Elsevier.
113. Pereira, C. G., & Meireles, A. A. (2020). Supercritical fluid extraction of bioactive compounds: Fundamentals, applications and economic perspectives. *Food and Bioprocess Technology, 3*, 340–372.
114. Sahena, F., Zaidul, I., Jinap, S., et al. (2009). Application of supercritical CO_2 in lipid extraction—A review. *Journal of Food Engineering, 95*, 240–253.
115. Rodríguez-mena, A., Ochoa-martínez, L., Marina, S., et al. (2022). Natural pigments of plant origin: Classification, extraction and application in foods. *Food Chemistry, 398*, 133908.
116. Ghosh, S., Sarkar, T., Das, A., et al. (2022). Natural colorants from plant pigments and their encapsulation: An emerging window for the food industry. *LWT, 153*, 112527.
117. Uddin, M. A., Rahman, M. M., Haque, A. N., et al. (2022). Textile colouration with natural colourants: A review. *Journal of Cleaner Production, 349*, 131489.
118. Zielinski, A. A., Sanchez-Camargo, A., et al. (2021). High-pressure fluid technologies: Recent approaches to the production of natural pigments for food and pharmaceutical applications. *Trends in Food Science and Technology, 118*, 850–869.
119. Singh, A. K., Rana, H. K., & Pandey, A. K. (2020). Analysis of chlorophylls. In *Recent advances in natural products analysis* (pp. 635–650). Elsevier. ISBN 9780128164556.

120. Ngamwonglumlert, L., Devahastin, S., & Chiewchan, N. (2017). Natural colorants: Pigment stability and extraction yield enhancement via utilization of appropriate pretreatment and extraction methods. *Critical Reviews in Food Science and Nutrition, 57*, 3243–3259.
121. Difonzo, G., Aresta, A., Cotugno, P., et al. (2021). Supercritical CO_2 extraction of phytocompounds from olive pomace subjected to different drying methods. *Molecules, 26*, 598.
122. Carradori, S., Cairone, F., Garzoli, S., et al. (2020). Phytocomplex characterization and biological evaluation of powdered fruits and leaves from Elaeagnus angustifolia. *Molecules, 25*, 2021.
123. Wasilewski, T., Czerwonka, D., Piotrowska, U., et al. (2018). Use of hop cone extract obtained under supercritical CO_2 conditions for producing antibacterial all-purpose cleaners. *Green Chemistry Letters and Reviews, 11*, 419–428.
124. Lefebvre, T., Destandau, E., & Lesellier, E. (2021). Sequential extraction of carnosic acid, rosmarinic acid and pigments (carotenoids and chlorophylls) from rosemary by online supercritical fluid extraction-supercritical fluid chromatography. *Journal of Chromatography. A, 1639*, 461709.
125. Derrien, M., Aghabararnejad, M., & Gosselin, A. (2018). Optimization of supercritical carbon dioxide extraction of lutein and chlorophyll from spinach by-products using response surface methodology. *LWT, 93*, 79–87.
126. Morcelli, A., Cassel, E., Vargas, R., et al. (2021). Supercritical fluid (CO_2+ethanol) extraction of chlorophylls and carotenoids from Chlorella sorokiniana: COSMO-SAC assisted prediction of properties and experimental approach. *Journal of CO2 Utilization, 51*, 101649.
127. Laleh, G., Frydoonfar, H., Heidary, R., et al. (2006). The effect of light, temperature, pH and species on stability of anthocyanin pigments in four *Berberis* species. *Pakistan Journal of Nutrition, 5*(1), 90–21.
128. Tadesse, S., Abebe, A., & Chebude, Y. (2012). Natural dye-sensitized solar cells using pigments extracted from Syzygium guineense. *Journal of Photonics for Energy, 2*(1), 027001.
129. Idham, Z., Nasir, H., Yunus, M., et al. (2017). Optimisation of supercritical CO_2 extraction of red colour from roselle (Hibiscus sabdariffa Linn.) calyces. *Chemical Engineering Transactions, 56*, 871–876.
130. Idham, Z., Putra, N., Aziz, A., et al. (2021). Improvement of extraction and stability of anthocyanins, the natural red pigment from roselle calyces using supercritical carbon dioxide extraction. *Journal of CO2 Utilization, 56*, 101839.
131. Garcia-Mendoza, M. P., Espinosa-Pardo, F., et al. (2017). Extraction of phenolic compounds and anthocyanins from Juçara (Euterpe edulis Mart.) residues using pressurized liquids and supercritical fluids. *Journal of Supercritical Fluids, 119*, 9–16.
132. Wozniak, L., Marszalek, K., Skapska, S., et al. (2017). The application of supercritical carbon dioxide and ethanol for the extraction of phenolic compounds from chokeberry pomace. *Applied Sciences, 7*, 322.
133. Jiao, G., & Kermanshahi pour, A. (2018). Extraction of anthocyanins from haskap berry pulp using supercritical carbon dioxide: Influence of co-solvent composition and pretreatment. *LWT, 98*, 237–244.
134. Nambela, L., Haule, L., & Mgani, Q. (2020). A review on source, chemistry, green synthesis and application of textile colorants. *Journal of Cleaner Production, 246*, 119036.
135. Smith, P. W., & Tatchell, A. R. (2016). *Aromatic chemistry: Organic chemistry for general degree students* (pp. 1–20). Elsevier. ISBN 9780080129488.
136. Ben Ghezala, H., Chaouali, N., Gana, I., et al. (2015). Toxic effects of Rhamnus alaternus: a rare case report. *Case Reports in Emergency Medicine, 2015*, 182951.
137. Lopez, L., et al. (2014). Naphthoquinones: Biological properties and synthesis of lawsone and derivatives-a structured review/Naftoquinonas: Propiedades biológicas y síntesis de lawsona y derivados-una revisión estructurada. *Vitae, 21*(3), 248.
138. Ramezani, N., Raji, F., Rezakazemi, M., et al. (2020). Juglone extraction from walnut (Juglans regia L.) green husk by supercritical CO_2: Process optimization using Taguchi method. *Journal of Environmental Chemical Engineering, 8*, 103776.

139. Romano, R., Aiello, A., Meca, G., et al. (2021). Recovery of bioactive compounds from walnut (Juglans regia L.) green husk by supercritical carbon dioxide extraction. *International Journal of Food Science and Technology, 56*, 4658–4668.
140. Bendich, A. (1989). Carotenoids and the immune response. *The Journal of Nutrition, 119*(1), 112–115.
141. Chew, B. P., & Park, J. S. (2004). Carotenoid action on the immune response. *The Journal of Nutrition, 134*, 257S–261S.
142. Maiani, G., Castón, M., Catasta, G., et al. (2009). Carotenoids: Actual knowledge on food sources, intakes, stability and bioavailability and their protective role in humans. *Molecular Nutrition & Food Research, 53*, 194–218.
143. Shah, N. A., Prasad, R. V., & Patel, B. B. (2020). Optimization of supercritical fluid extraction of paprika (Cv. Reshampatti) oil, capsaicin and pigments. *Flavour and Fragrance Journal, 35*, 469–477.
144. Shang, K., Amna, T., Amina, M., et al. (2017). Influence of capsaicin on inflammatory cytokines induced by lipopolysaccharide in myoblast cells under in vitro environment. *Pharmacognosy Magazine, 13*(1), S26.
145. Esquivel-Hernández, D., Rodríguez-Rodríguez, J., Cuéllar-Bermúdez, S., et al. (2017). Effect of supercritical carbon dioxide extraction parameters on the biological activities and metabolites present in extracts from Arthrospira platensis. *Marine Drugs, 15*, 174.
146. Achir, N., Servent, A., & Soto, M. (2022). Feasibility of individual carotenoid quantification in mixtures using UV-Vis spectrophotometry with multivariate curve resolution alternating least squares (MCR-ALS). *Journal of Spectroscopy, 2022*, 4509523.
147. Urbonaviciene, D., & Viskelis, P. (2007). The cis-lycopene isomers composition in supercritical CO_2 extracted tomato by-products. *LWT – Food Science and Technology, 85*, 517–523.
148. De Mejia, E., Zhang, Q., Penta, K., et al. (2020). The colors of health: Chemistry, bioactivity, and market demand for colorful foods and natural food sources of colorants. *Annual Review of Food Science and Technology, 11*(1), 145–182.
149. Bruno, A., Durante, M., Marrese, P., et al. (2018). Shades of red: Comparative study on supercritical CO_2 extraction of lycopene-rich oleoresins from gac, tomato and watermelon fruits and effect of the α-cyclodextrin clathrated extracts on cultured lung adenocarcinoma cells' viability. *Journal of Food Composition and Analysis, 65*, 23–32.
150. De Andrade Lima, M., Charalampopoulos, D., & Chatzifragkou, A. (2018). Optimisation and modelling of supercritical CO_2 extraction process of carotenoids from carrot peels. *Journal of Supercritical Fluids, 133*, 94–102.
151. Bobinaitė, R., Kraujalis, P., Tamkutė, L., et al. (2020). Recovery of bioactive substances from rowanberry pomace by consecutive extraction with supercritical carbon dioxide and pressurized solvents. *Journal of Industrial and Engineering Chemistry, 85*, 152–160.
152. Wang, X., Wang, C., Zha, X., et al. (2017). Supercritical carbon dioxide extraction of β-carotene and α-tocopherol from pumpkin: A Box-Behnken design for extraction variables. *Analytical Methods, 9*, 294–303.
153. Farazuddin, M., Dua, B., Zia, Q., et al. (2014). Chemotherapeutic potential of curcumin-bearing microcells against hepatocellular carcinoma in model animals. *International Journal of Nanomedicine, 9*, 1139.
154. Martinez-Correa, H. A., Paula, J. T., Kayano, A. C., et al. (2017). Composition and antimalarial activity of extracts of Curcuma longa L. obtained by a combination of extraction processes using supercritical CO_2, ethanol and water as solvents. *Journal of Supercritical Fluids, 119*, 122–129.
155. Dejsungkranont, M., Chen, H., & Sirisansaneeyakul, S. (2017). Enhancement of antioxidant activity of C-phycocyanin of spirulina powder treated with supercritical fluid carbon dioxide. *Agriculture and Natural Resources, 51*, 347–354.
156. Bai, L., Cheng, X., Xu, J., et al. (2019). Black sesame pigment extract from sesame dregs by subcritical CO_2: Extraction optimization, composition analysis, binding copper and antioxidant protection. *LWT, 100*, 28–34.

157. Abou Elmaaty, T., Sayed-Ahmed, K., Elsisi, H., et al. (2022). Optimization of extraction of natural antimicrobial pigments using supercritical fluids: A review. *Processes, 0*(10), 111.
158. Dhouibi, N., Baaka, N., Charradi, R., et al. (2021). Multi-fiber dyeing improvement using natural supercritical CO_2 extracts. *Fibers and Polymers, 22*(7), 1874–1882.
159. Che Rahim, A., Ahmad, M., Ismail, K., et al. (2018). Natural dyes from chlorella vulgaris sp extracted via supercritical fluid extraction dyed on silk fabric. *International Journal of Engineering Research & Technology, 7*(4.14), 326–330.
160. Talib, N. B. (2021). *Natural dyes from Sargassum sp. extracted using supercritical carbon dioxide and its application on silk coloration* [Doctor of Philosophy (Science) AS950]. Universiti Teknologi Mara.
161. Abou Elmaaty, T., Sayed-Ahmed, K., Magdi, M., et al. (2023). An eco-friendly method of extracting alizarin from Rubia tinctorum roots under supercritical carbon dioxide and its application to wool dyeing. *Scientific Reports, 13*(1), 1–10.

Natural Dyes as Corrosion Inhibitors

Priya Vashishth, Bindu Mangla, Himanshi Bairagi, Rajni Narang, and Harish Moudgil

Abstract The development of sustainable corrosion inhibitors is mostly introduced by the organic inhibitors since these inhibitors are environmentally acceptable, biodegradable, and free of heavy metals and other potentially hazardous materials. Moreover, natural dyes are biodegradable and less harmful to the environment than artificial dyes. These are classified into four categories: those derived from plants, animals, minerals, and microbes and these are utilized in a variety of applications, including food, medicine, handicrafts, and leather processing, as well as in the field of corrosion. The adoption of corrosion inhibitors to protect metals and their alloys when they come into contact with a challenging environment is a common method. Numerous corrosion mitigation agents, encompassing organic, inorganic, nanomaterial-based, natural, and polymeric variants, serve as proficient inhibitors. These inhibitors are distinguished by the presence of heteroatoms and conjugation within their molecular structures. Regrettably, a considerable portion of synthetic inhibitors are characterized by their elevated cost and deleterious toxicity, engendering notable apprehensions regarding their environmental impact and sustainability. The majority of studies and reviews of the literature explored natural dyes as corrosion inhibitors, whose classification, inhibitory mechanism, production techniques, uses, and contributions are discussed here. This chapter analyzes the existing trends and history of natural dyes as corrosion inhibitors to limit the rate of corrosion and discusses the prevention techniques that helped to solve this devastating issue.

Keywords Corrosion inhibitors · Adsorption · Natural dyes · Prevention techniques

P. Vashishth · B. Mangla (✉) · H. Bairagi · R. Narang
Corrosion Testing Research Lab, Department of Chemistry, J.C. Bose University of Science and Technology, YMCA Faridabad, Haryana, India

H. Moudgil
Department of Chemistry, School of Basic Sciences, Central University of Haryana, Mahendergarh, Haryana, India

S. S. Muthu (ed.), *Natural Dyes and Sustainability*, Sustainable Textiles: Production, Processing, Manufacturing & Chemistry,
https://doi.org/10.1007/978-3-031-47471-2_12

1 Introduction

Natural dyes were extensively utilized to color textiles from antiquity until the eighteenth century. Natural dyes, as the name implies, are made from natural resources. To color various textile fabrics, coloring compounds manufactured from substances that are naturally found in flora, mammal, mineral, and microbiological sources have been used. The European market for natural dyes and pigments was worth USD 1255 million in 2020, and it is expected to grow at a CAGR of 10.1% between 2021 and 2028. Natural dyes are considered eco-friendly as they are renewable and biodegradable; they are also skin friendly and may also provide health benefits to the wearer. Natural dyes can be used for dyeing almost all types of natural fibers. Recent research shows that they can also be used to dye some synthetic fibers. Demand for dyes and pigments in textiles, foods and beverages, paints and coatings, and cosmetic applications is likely to drive market growth. Furthermore, the negative effects of producing synthetic dyes and pigments, such as pollution of aquatic bodies due to chemical dumping, have increased demand for natural dyes and pigments. Natural dyes and pigments depend less on raw materials that are derived from crude oil resources because they are generated from naturally occurring substances. Natural dyes can be made from fruits, plant leaves, red wines, vegetables, green tea, and fruits along with several things. These raw ingredients are then transformed into plant-based, animal-based, and mineral-based dyes, as well as pigments including carotenes, quinolines, flavonoids, chlorophyll, and phycoerythrin, which are widely utilized in a variety of products including colorants, fabrics, meals and beverages, skincare products, and other applications. Natural dyes are regaining popularity globally, and given the present emphasis on environmental preservation, they offer tremendous economic appeal. There is also an increasing understanding of their nonhazardous nature. Natural dyes have a highly complex chemical structure. Apart from synthetic dyes, they are frequently a combination of highly associated chemical compounds rather than a specific group. They are classified according to the fundamental chemical additives present (Fig. 1):

A survey conducted by the German Ministry of Food, Agriculture, and Forestry found that 90,000 metric tonnes of dyes could be produced per year. For the creation of natural colors, agricultural lands must be used. Projections indicate that a land expanse spanning from 250 to 500 million acres within the agricultural domain will be a requisite to facilitate the production of 100 million metric tonnes of botanical material essential for the synthesis of one million tonnes of natural pigments. This land allocation corresponds to 10–20% of the global aggregate grain cultivation area. Notably, the United States relies significantly on imports of natural colorants. Specifically, the annual import volume of colorants derived from botanical and zoological sources for these chromatic agents approximates 3500 metric tonnes, constituting approximately 0.4% of the consumption attributed to synthetic dyes [1]. A prominent industry in the field, the Allergo Company, has articulated an overarching objective aimed at substituting 1% of synthetic dyes with their natural counterparts. Their strategic blueprint estimates a requisition of 0.2% of the available arable land

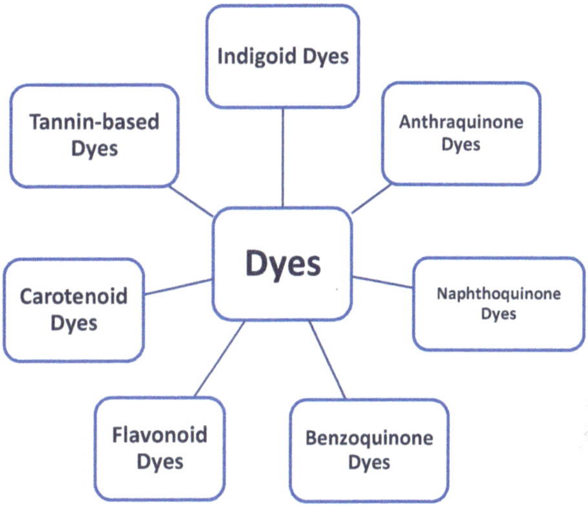

Fig. 1 Classification of dyes

for this endeavor. This transition entails multifaceted undertakings such as the cultivation of precursor materials, the refinement of the dyeing apparatus to align with industrial requisites, and the dissemination of technological knowledge to prominent textile conglomerates. A focal point of investigation has been the establishment of economically viable pathways to supplant synthetic dyes with natural alternatives. In pursuit of this, the Allergo Company has pioneered extraction methodologies yielding unadulterated liquid chromophores, subsequently converted into powder form [2]. The waste by-product originating from the extraction process is cyclically repurposed as a carbon adjunct in composting operations. Additionally, the company has engineered an ecologically benign mordant, further underscoring its commitment to sustainable dyeing practices.

Natural dye extraction is a challenging approach since natural coloring substances do not exist as a single molecule entity and the plant composite also comprises a wide range of other phytoconstituents. Prior to performing an extraction method, the coloring components' nature and solubility should be assessed. There are various types of extraction for natural dyes as shown in Fig. 2:

Also, the employment of an indicator within the food industry stands to yield advantages in the realms of food safety enhancement, real-time surveillance, and the dissemination of pertinent details pertaining to packaged food commodities. Phytochemical components with antioxidant action include phenolic acids (flavonoids), nitrogen compounds (photosynthetic pigment compounds), tocotrienols, carotene, and tocopherol. These chromatic agents exhibit both benign safety profiles and ecologically harmonious attributes, concurrently possessing medicinal and therapeutic efficacy. The cultivation of botanical sources for dye extraction serves as an impetus for afforestation endeavors, thereby engendering notable ecological equilibrium. These pigments find multifarious utility across domains encompassing

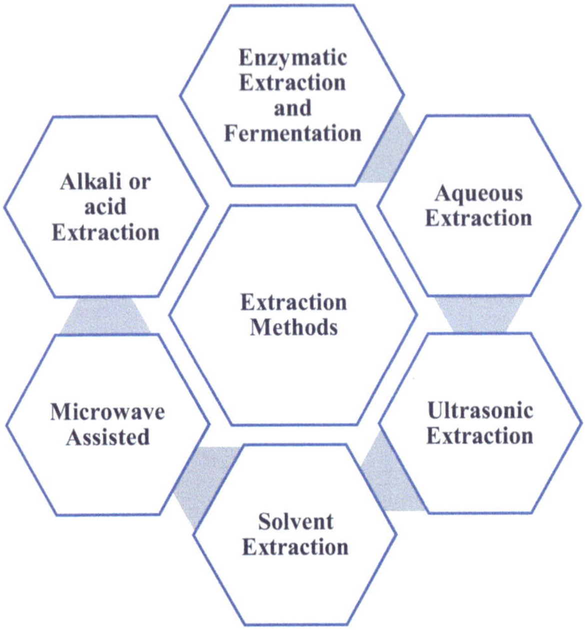

Fig. 2 Different extraction methods for natural dyes

alimentation, pharmaceuticals, artisanal crafts, hide treatment, and corrosion mitigation.

Corrosion is generally perceived in terms of the degradation of a metal due to the consequences of a corrosive medium [3] and is attracting considerable interest in the destructive result of a chemical reaction between a metal or metal alloy and its environment. The Indian government has spent approximately 3.5% of the nation's gross domestic product to minimize corrosion impacts [4]. In accordance with antecedent surveys, India, alongside other global entities, is amassing financial resources aimed at satiating the escalating demand for corrosion inhibitors. Corrosive processes have engendered deterioration of the exteriors of myriad infrastructural entities, encompassing thoroughfares, spans, edifices, industrial processing facilities, and sanitation systems, thereby affecting virtually all commonplace metallic commodities. Diverse manifestations of corrosion phenomena manifest within aqueous surroundings [5]. The phenomenon of corrosion is profoundly influenced by an amalgamation of factors, including metal reactivity, accessibility of contaminants, ambient oxygen levels, humidity fluctuations, and the presence of gases such as sulfur dioxide and carbon dioxide. Additionally, the presence of electrolytes, the inherent grain structure of the metal, compositional attributes during alloying, and the thermal conditions observed during fabrication collectively exert substantial influence over this phenomenon [6]. In the year 1824, Sir Humphry Davy delivered a discourse before the Royal Society in London, presenting an assemblage of scholarly manuscripts. Through these writings, he substantiated the capacity of zinc and iron anodes to impede the corrosion-induced degradation of copper cladding adorning the timber hulls of British naval warships [7].

Because of the number of corrosion mechanisms and material failures, corrosion types are classified according to their main characteristics. In the case of water-cooled systems with electrochemical deterioration, there are some common types of corrosion (Fig. 3):

Carbon steel and its alloys are the most valuable metals in various fields that suffer localized and uniform corrosion in different aggressive media, such as aqueous, acidic, basic, alkali, and alkaline solutions, containing chlorides with the passive oxide layer's destruction, forming pits. Mild steel (low carbon steel) is employed in conventional fields such as automobiles, infrastructure, laboratories, pipelines, the petroleum industry, and other related sectors in which corrosion resistance is required. On a daily basis, our existence is confronted with an array of deleterious consequences engendered by this phenomenon. These encompass the squandering of precious resources, the cessation of industrial operations, product adulteration, exorbitant maintenance expenditures, compromised efficiency, and the implementation of costly and excessive precautions. Acids are well known for their widespread application in industries, laboratories, and other areas such as pickling, cleaning, descaling, etc. Because of the aggressive nature of these acidic solutions, the corrosion process takes place on the metal surfaces. Mineral acids, with a particular emphasis on hydrochloric acid, find prevalent usage in industrial processes encompassing activities such as acid cleansing, acid pickling, acid descaling, and the acidization of oil wells [8]. Due to the potently corrosive nature of these acids, the exploration of corrosion mitigation consistently remains a subject of profound theoretical and pragmatic significance [9].

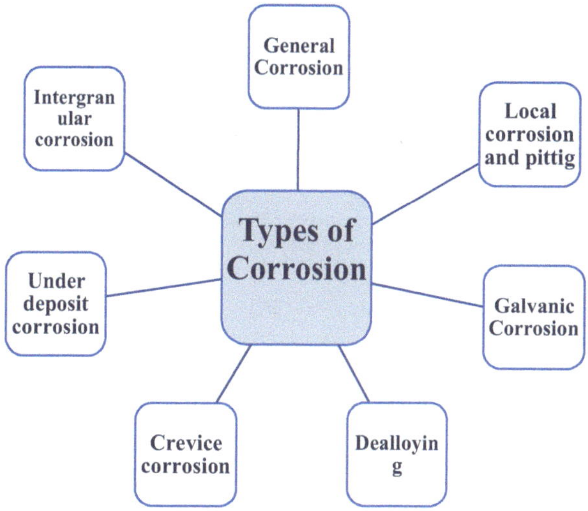

Fig. 3 Types of corrosion

To mitigate corrosion, the implementation of corrosion inhibitory methodologies offers a measure of efficacy. A diverse array of strategies exists for safeguarding metals against corrosive processes. Over the past few decades, numerous corrosion investigations and prophylactic technologies have been devised and applied across both laboratory and industrial contexts. Principally, five fundamental avenues of corrosion mitigation have emerged, encompassing the judicious selection of materials and the formulation of structural designs [10], cathodic and anodic protection, notably in the forms of impressed current and sacrificial anodes, which constitute pivotal modalities within the domain of cathodic prevention. Simultaneously, the employment of safeguarding coatings, ranging from sacrificial layers and organic/polymeric coatings to ceramic coatings, rust- and corrosion-resistant oils, and gas-phase deposition methodologies, stands as another indispensable avenue for corrosion mitigation [11]. Compared to some recent reviews, articles, and prevention techniques, the present study offers up-to-date knowledge of the different types of protective coatings on metals (Fig. 4).

Corrosion inhibition serves as the most practicable, cost-effective, and efficient approach to managing deterioration on the surface of the metal in different media by preventing metal dissolution and acid consumption [12]. These inhibitors operate through the mechanism of molecular adsorption, wherein the compound exerts corrosion inhibition by regulating both anodic and cathodic processes. The protonated entities adhere to cathodic sites on the surface, attenuating the liberation of hydrogen gas. Consequently, the pace of corrosion diminishes correspondingly [13]. Inhibitory agents can be administered either in the form of solutions or as protective coatings, employing diverse methodologies. Corrosion inhibitors encompass compounds that engage with the metal surface or ambient gases in order to curtail the electrochemical reactions inherent to the corrosion process. These agents are harnessed to mitigate the impact of corrosive surroundings upon metallic substrates. In

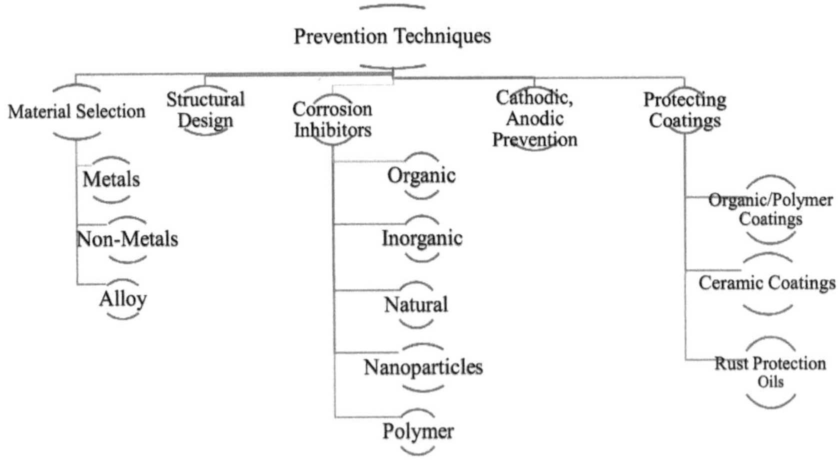

Fig. 4 Different types of protective techniques

different industries, the various compounds including organic and inorganic substances are used as inhibitors. The compounds generally containing atom having a lone pair of electrons such as nitrogen (N), oxygen (O), sulfur (S), and phosphorus (P) and a mediately with an aromatic ring, unsaturation that are observed to be a significant component of extremely efficient inhibitors to minimize the corrosion rate of metals in acidic environment [14–21]. In general, the efficacy of inhibition hinges upon the proclivity for establishing strong coordinate bonds with the metal substrates. Consequently, both phenomena adhere to an analogous sequence, namely, $O < N < S < P$ [22]. The risk inherent with these substances is apparent during the fabrication of such products or in their utilization. The stability and environmental concerns related to corrosion inhibitors have long been a global concern. As a consequence, it is a great juncture to gain knowledge about protection against corrosion using low-cost, readily accessible, and environment benign inhibitors.

Natural compounds are commonly accessible and renewable in nature, making them environmentally friendly. Plant-derived compounds are currently a prominent alternative for protection against corrosion since they are beneficial to the environment, commonly available, and regenerative in nature [23]. Moreover, these inhibitors are biodegradable in nature and bereft of any heavy metals or toxic compounds. Therefore, finding naturally occurring substances as corrosion inhibitors is a subject of great practical significance. Natural inhibitors are a rich source of ingredients that have very high inhibition efficiency. They are capable in forming of a protective layer to retard the effects of reactive species like carbon dioxide, oxygen, chlorine, moisture, etc. For the reduction of the corrosion process, weight loss studies, electrochemical studies, characterization, surface studies, etc. are the most popular techniques, by which the analysis of corrosion rate, inhibition efficiency, and such other parameters can be determined (Fig. 5).

2 Electrochemical Aspects of Corrosion

Most metals that we commonly use are unstable in the environment as they were obtained from corresponding ores by synthetic reduction via a chemical change. As a consequence, such materials undergo a rapid reaction to convert into a stable state. The chemical processes that do not necessitate any external environment or chemical reagent, including energy or a catalyst, are known as spontaneous reaction processes. In spontaneous reactions, a change in Gibbs free energy (ΔG) is negative, and a change in enthalpy (ΔH) is also negative, while a change in entropy (ΔS) is positive [24]. In addition to oxidation and reduction, corrosion process also includes a redox reaction in which one type of metal or component of a metal functions as an anode and the other as a cathode. While electrons are lost at the anode and mass is lost, electrons are gained at the cathode and corrosion products are deposited.

The first corrosion occurs when pure iron undergoes anodic oxidation processes when it is exposed to a potent acid like hydrochloric acid. Bubbles forcefully form as a result of the reaction. The particulars are as follows:

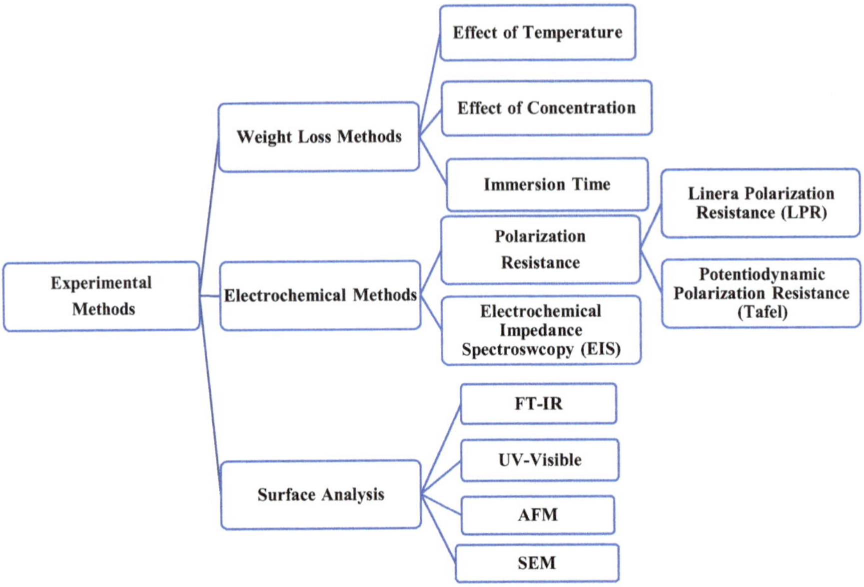

Fig. 5 Experimental methods to determine the different parameters

$$Fe + 2H^{+} \rightarrow Fe^{2+} + H_2$$

$$Fe^{2+} + 2HCl \rightarrow FeCl_2 + 2H^{+}$$

3 Techniques for Corrosion Measurement

A quantitative technique called the weight loss method is used to quantify and monitor the internal or external corrosion in metal components. Since, it is the most appropriate and efficient method, it is used to determine the efficacy of inhibition (IE %) and rate of corrosion (C_r). In order to quantify them and find out the weight loss of metal and alloys after exposure, they were thoroughly cleaned in deionized water and acetone accompanied by a rigorous drying phase. The inhibition efficiency (IE %), surface coverage (θ), and corrosion rate (C_r) were used to quantify the inhibitory performance using the following formulas [25]:

$$\mathrm{IE\%} = \frac{W_o - W_i}{W_o} \times 100$$

$$\theta = \frac{W_o - W_i}{W_o}$$

$$C_r = \frac{534 \times W}{A \times t \times d}$$

where IE is the inhibition efficiency, C_r is the corrosion rate, θ is the surface coverage, W_0 is the weight loss of metal specimen without inhibitor, W_i is the weight loss of metal specimen with inhibitor, W indicates the weight loss in mg, A is the area of the metal specimen in cm^2, t is the time of exposure in hours, and d is the density of the metal specimen in g/cm^3.

The linear polarization resistance (LPR) method has emerged as the most widely used electrochemical technique, and for good reason. Its nondestructive nature allows to accurately measure without causing any harm or damage. This makes LPR an invaluable tool in various applications where preserving the integrity of the materials being tested is crucial [26] and the application is quick and can be tested in the field using portable instrumentation [27]. The main concept behind LPR is to disturb the corrosion balance on the surface of a metal sample by using a small DC electrical pulse [27]. A standard half cell is used to measure how the equilibrium reacts to this disruption [28]. Electrochemical impedance spectroscopy (EIS) is a powerful technique extensively used in corrosion research due to its wide range of applications. This method provides valuable insights into the effectiveness of corrosion inhibitors, offering crucial information about the level of protection they provide. In order to gather valuable information about the system under study, a technique is used where either an alternating current voltage (in the case of potentiostatic EIS) or an alternating current (in the case of galvanostatic EIS) is applied. This approach enables us to obtain an alternating current or voltage response, providing us with crucial insights into its behavior. With the choice between a two- or three-electrode setup, along with an essential equipment such as a potentiostat–galvanostat and a frequency response analyzer, we have the flexibility to perform this procedure in a manner that suits our needs. This versatility guarantees precise measurements and effective analysis, ultimately leading to optimal results [29]. Based on the pattern of the Nyquist plot obtained by the investigation, an equivalent circuit depicts the electrochemical cell consisting the specimen, deposited inhibitor, and the electrolyte media. An increase in the value of Rct value and a decrease in the value of Cdl with the inclusion of inhibitor concentrations revealed efficient corrosion prevention [30].

Electrochemical-based corrosion monitoring tests are incredibly valuable for gathering information about corrosion rates and the mechanisms of corrosion protection. Given that corrosion is a complex electrochemical phenomenon, these tests provide us with crucial insights into its behavior and how to effectively combat it [31]. The fundamental of polarization techniques is altering the current or potential on a sample under study and logging the resulting potential or current change. When it comes to powering your devices, there is an option to select either a direct current (DC) or an alternating current (AC) source, depending on the specific needs and preferences [32]. Several significant and popular methods, including Tafel extrapolation, electrochemical impedance, and linear polarization resistance, have been discussed. The anodic and cathodic reactions in the electrochemical cell are depicted by a Tafel curve, which is a current–potential diagram. In this procedure, the working electrode's potential was adjusted over a range at a predetermined pace, and the response in current that resulted was recorded. The total current resulting from both anodic and cathodic reactions can be visually represented by a smooth curving line. This line intersects with the linear segments of the logarithmic Tafel plot, creating a point that estimates the corrosion current (i_{corr}) and corrosion potential (E_{corr}). This intuitive representation helps in understanding and analyzing the electrochemical processes involved in corrosion.

The inhibitory performance (IE %) of the inhibitor was calculated using current densities (I_{corr}) by the following formula [33, 34]:

$$\mathrm{IE}\% = \frac{\left(I^{\circ}_{\mathrm{corr}} - I^{'}_{\mathrm{corr}}\right)}{I^{\circ}_{\mathrm{corr}}} \times 100$$

where I°_{corr} is the corrosion current density in blank solution and $I^{'}_{corr}$ is the corrosion current density at various concentrations of inhibitor.

Atomic force microscopy (AFM), scanning electron microscopy (SEM), and transmission electron microscopy are used to study surface morphology of metal outer surface before and after inhibition. FTIR and FTIR-ATR detect the type of functional group and bonding between a metal surface and an inhibitor. Kinetic and thermodynamic parameters evaluate the feasibility of corrosion. Quantum chemical calculation method and density functional theory (DFT) with high accuracy are used for the calculation of the molecular structure parameters of inhibitors including electronegativity, frontier orbitals (HOMO and LUMO), chemical hardness, chemical softness. X-ray fluorescence (XRF) is helpful in finding composition analysis of steel. EDAX (energy dispersive X-ray analysis) is employed to identify elemental composition of material. UV–Vis spectroscopy helps to determine the type of transition in inhibitor molecule and to ensure the formation of protective layer.

4 Corrosion Inhibitor

Corrosion inhibition is the most practical, economical, and convenient technique to control corrosion on metal surfaces in different media by preventing metal dissolution and acid consumption [35]. These inhibitors work on the mechanism of molecular adsorption, i.e., the compound inhibits corrosion by controlling both anodic and cathodic reactions. The protonated species are adsorbed on the cathodic sites of the surface and decrease the evolution of hydrogen gas. Hence, the rate of corrosion also decreases. Inhibitors can be applied as a solution or as a protective coating using different techniques [36]. Corrosion inhibitors are very widely used to reduce corrosion processes in the environment and are the substances that, when added in small amounts to aggressive environments, reduce the rate of corrosion [37]. The successful adsorption of an inhibitor on a metal surface relies on several crucial factors. These include the nature of the metal itself, the surface charge it carries, the specific adsorption mode, and its corresponding chemical structure, as well as the composition of the electrolyte solution being used. By understanding and considering these factors, one can effectively enhance and optimize inhibitor adsorption on metal surfaces for various applications.

Most organic molecules tend to adsorb to metal surfaces, but in general, the most desirable properties when seeking corrosion inhibition relate to the interaction of a double/triple bond or lone pair of electrons (N, S, P, or O) [38]. Most organic molecules tend to adsorb to metal surfaces, but in general, the most desirable properties when seeking corrosion inhibition relate to the interaction of a double/triple bond or lone pair of electrons (N, S, P or O) with an empty metal d orbital or a combination of some of the previous mechanisms [14, 15, 17–19, 21, 39]. The positive impact of electron-donating groups, such as -OH, -NH_2, -CH_3, -OCH3, etc., on the inhibition performance of organic molecules has been widely observed. These groups enhance the overall effectiveness and efficiency of the compounds in inhibiting various processes, whereas the inhibition efficiency of organic molecules tends to decrease when electron-withdrawing substituents, such as -NO_2, -CN, COOH, and -$COOC_2H_5$, are present. Electron-donating substituents, such as -OH, -NH_2, -CH_3, -OCH_3, and others, boost electron density at the inhibitor's adsorption site, increasing interactions between the inhibitor molecule and the metallic surface. Electron-withdrawing substituents, on the other hand, such as -NO_2, -CN, COOH, -$COOC_2H_5$, and so on, reduce the electron density at the inhibitor's adsorption site, hence promoting the adsorption propensity on metallic surfaces.

Corrosion inhibitors are classified based on how they prevent corrosion: cathodic, anodic, or mixed. Cathodic corrosion inhibitors lessen corrosion potential by blocking cathodic processes, including oxygen reduction and hydrogen evolution [40]. Anodic corrosion inhibitors enhance corrosion potential by interacting with reactive areas on the metal surface and passivating them. Inhibitors that are neither cathodic nor anodic are classified as mixed-type inhibitors [41]. By physisorption, chemisorption, or film formation, these inhibitors can preserve the metal surface. Physisorption is driven by the electrostatic interaction of inhibitor molecules with the metal surface. Chemisorption, on the other hand, is caused by donor–acceptor interactions between vacant metal surface orbitals and free electron pairs in the inhibitor. Different authors classify inhibitors in different ways [42]. Some authors prefer top-group inhibitors according to their chemical function as follows:

4.1 Inorganic Inhibitor

Inorganic inhibitors are compounds in which the active agent is an inorganic chemical. One of the easiest and basic ways to enhance a metal's passivity is to introduce electropositive metal salts to a corrosive media. However, the preventive metal ion must possess a redox potential greater than the one to be shielded and potentially greater than that essential for releasing protons in order for the preventive metal ion to be discharged on the surface of the metal in need of preservation. Cathodic depolarization occurs through the accumulation of the insulating metal on the surface of the metal resistant to corrosion, followed by the formation of an adhering deposit. Palladium (Pd), platinum (Pt), iridium (Ir), rhodium (Rh), mercury (Hg), and rhenium are some of the metals that can be used for this purpose (Re). Various inorganic anions include chromates (CrO_4^{2-}), molybdate (MoO_3), silicates (SiO_4^{4-}), phosphate (H_2PO_3), and nitrate (NO_2^-), which offer passivation protection to the metal surfaces by incorporating into the oxide layer [43].

4.2 Organic Inhibitor

Organic inhibitors work by producing a layer on the metal's surface and can serve as anodic, cathodic, or mixed inhibitors. This protective film forms as a result of strong interactions such as orbital adsorption, chemisorption, and electrostatic adsorption, which prevent corrosive species from attacking the metal surface [44]. This adsorption is generally one molecule layer thick and is unable to access the bulk of metals [45]. The adsorption process is governed by physicochemical variables such as functional groups, steric factors, aromaticity, π-orbital character of donating electrons, electron density at donor atoms, and the electronic structure of the molecules [46]. An essential organic inhibitor will typically comprise polar functional groups in the molecule with S, O, or N atom, as well as a hydrophobic moiety that will repel aqueous corrosive species away from the metal surface [47].

4.3 Polymer Inhibitor

Polymers are considered as potential corrosion inhibitors since they have the desired impact on the environment and economy with superior corrosion inhibition properties than simple organic molecules because they attain inherent stability [48]. The polymers anticipate two advantages: (i) displace many water molecules by a single polymeric chain from the metal interface, which makes the process entropically favorable, and (ii) desorption of polymers occurs by the presence of various multiple binding sites and giving a process having a low rate. Polymers, their composites, and blends with functional groups like –OH, $-NH_2$, –COOH -S, etc. are water soluble and reported as corrosion inhibitors in various aggressive media through complexation of polymers with metal ions [49], giving multiple applications. Complexes of polymers and their composites provide a larger surface area on the metal surface, covering the interface by forming a layer and protecting the surface against corrosion.

Gum arabic is widely used as a stabilizer, emulsifier, and viscosity modifier in the food and paint industries, and its potential as a corrosion inhibitor has been investigated in acidic and alkaline environments. For example, Chen Shen et al. studied the inhibitory effect of gum arabic in oil field industry on carbon steel 1018 under brine conditions through electrochemical studies and found that it is a mixed inhibitor that follows chemisorption and Langmuir adsorption isotherms, achieving the highest inhibition efficiency at 94% at room temperature [50], and this form of gum also works in alkaline environment on aluminum surface, i.e., on AA1060 type, which gives the highest inhibition efficiency at 75.86% at 40 °C by thermometric method, and their inhibition effect can be brought about by the presence of oligosaccharides, polysaccharides, glucoproteins and arabinogalactan constituents since they have oxygen and nitrogen in their structure which promote the adsorption on the aluminum surface [51].

4.4 Nanoparticle as Inhibitor

For many years, nanostructured materials have been establishing themselves as the modern generation of high-performance materials, going from automotive engineering to bioengineering, owing to a vast array of unique properties due to their ideal grain size, larger surface-to-volume ratio, and high grain boundary volume fraction [52]. They are used from self-cleaning glass to high-performance components for aerospace applications [53]. Nanocoating provides materials with improved strength, durability, optical qualities, and thermal stability. The use of nanomaterials as corrosion inhibitors is the most effective and economical method to prevent the metal from corrosion and increase its durability [54]. Nanomaterials have a high advantage to show higher corrosion inhibition properties because they have large surface-to-volume ratio. These materials stop the reaction of the surface and mitigate the rate of corrosion by blocking active sites of the metal surface, providing hardness, strength, durability, optical qualities, thermal stability, etc. [55].

Ag nanoparticles with PVP and PEGSH as thin film on carbon steel alloy acted as good corrosion inhibitors in 1 M HCl studied by A. M. Atta et al. They investigated by using UV–Vis spectroscopy, Fourier transform infrared (FTIR) spectroscopy, scanning electron microscopy (SEM)/EDAX, TEM, potentiodynamic polarization (PDP), electrochemical impedance spectroscopy (EIS), and DLS measurement. The results that were obtained from electrochemical measurement indicate that the Ag nanoparticles with PVP exhibits 90.95% inhibition efficiency [53]. Gamal A. El-Mahdy et al. studied the corrosion inhibition of steel in 1 M HCl by using magnetite (PAMPS-Na-co-St/Fe_3O_4) nanocomposite. The techniques employed for the study were FTIR, TEM, thermogravimetric analysis (TGA), electrochemical impedance spectroscopy (EIS), and potentiodynamic polarization (PDP). The obtained results revealed that this nanocomposite was a good corrosion inhibitor for steel in HCl, and the maximum inhibition efficiency was found to be 99.7% [56]. Thin-film coatings of nanomaterials through mixing or blending with organic materials, such as polyurethane [57], polyamide [58], polyester [59], PVC [60], acrylics [61], alkyds, and epoxies [62], play a crucial role as protective layers by delaying the transition of corrosive species and different environments such as chlorine and hydroxyl ions, water, oxygen, pollutants, and pigments, which have an affinity to react with the material surface [63].

Ekemini Itueri et al. investigated onion mesocarp extract–Ni nanocomposites as corrosion inhibitors on X60 steel. By using EDAX, FTIR spectroscopy, UV–Vis spectroscopy, DLS, XRD, SEM, TEM, weight loss measurements, EIS, and H_2 evolution technique, results were obtained, and these results revealed that onion mesocarp extract–Ni nanocomposites were good corrosion inhibitors for X60 steel in 1 M HCl, which achieved the higher inhibition efficiency at 88.1% [64]. S. John et al. studied the inhibition action of chitosan/TiO_2 nanocomposites dip coating on mild steel in 0.1 M HCl, and they did the experiment by using FTIR, XRD, EIS, TGA, and contact angle measurement techniques. The obtained results revealed that nanocomposites acted as good corrosion inhibitors as these nanocomposites showed the inhibition efficiency of 84.69% in 0.1 M HCl [65]. K. L. Palanisamy et al. studied the corrosion inhibition of mild steel with carrier oil, i.e., olive oil, stabilized of iron oxide nanoparticles incorporated into a paint. By using weight loss measurements and scanning electron microscopy (SEM), results were obtained, and these results revealed that the nanocomposites of iron oxide with olive oil showed the higher inhibition efficiency of 80.88%, which proved that these are good corrosion inhibitors [66].

4.5 Green Inhibitor

Corrosion inhibitors are extensively used for the protection of metals and equipment, and they are required to be acceptable, nontoxic, and eco-friendly due to environmental concerns. The products obtained from the plants have become a preferable choice at large for corrosion prevention as these products are environment friendly, easily available, and renewable in nature [20]. Moreover, these inhibitors are biodegradable in nature and bereft of any heavy metal or toxic compounds. Therefore, finding naturally occurring substances as corrosion inhibitors are a subject of great practical significance. Natural inhibitors are a rich source of ingredients that have very high inhibition efficiency. They are capable of forming protective layer to retard the effect of reactive species like carbon dioxide, oxygen, chlorine, moisture, etc. [67]. Victoria Vorobyova et al. studied the corrosion inhibition of steel by tomato pomace extract in NaCl medium. Gravimetric measurements, potentiodynamic polarization, and electrochemical impedance spectroscopy (EIS) were used in this study. The results showed that tomato pomace extract was an effective corrosion inhibitor for steel in sodium chloride media exhibiting a maximum inhibition efficiency of 98% at 500 ppm in 48 hours of immersion time [68]. Rajesh Haldhar et al. studied the corrosion inhibition effects of the aerial parts of *Solanum surattense* which showed 93% efficiency at an inhibitor concentration of 500 mg/L [69]. Harish Kumar et al. studied the corrosion inhibition of mild steel in 0.5 M HCl by *Morus nigra* (mulberry leaves). The techniques employed for the study were

FTIR, UV–Vis spectroscopy, NMR, weight loss measurement, electrochemical impedance spectroscopy (EIS), potentiodynamic polarization (PDP), and DFT. The obtained results revealed that the *Morus nigra* proved to be a good corrosion inhibitor for steel in HCl and showed a maximum inhibition efficiency of 91.67% and follows both physical and chemical modes of adsorption [70]. *Allium sativum* L. acted as a corrosion inhibitor on mild steel studied by Harish Kumar et al. They did the gravimetric, electrochemical, and DFT analyses to assess the inhibition performance in 5 M HCl. The inhibition efficiency was found to be 94.76% following the physio-chemisorption mechanism [71]. Among the various categories of such inhibitors, here we focused one, the natural dye obtained from various parts of plants.

4.5.1 Natural Dye as a Corrosion Inhibitor

A dye is a colored substance with a particular affinity for the medium on which it is treated. The dye is generally applied in an aqueous solution, and a mordant may be applied to maximize the durability of fiber. There are two types of dyes: natural and synthetic. Natural dyes are mostly made from natural materials. These are manufactured from renewable resources. In contrast, synthetic dyes are created by using a range of chemicals, synthetic materials, and other substances. The importance of color in textile products cannot be emphasized. Colors have attracted mankind from the dawn of time. Natural vegetable dyes have been used by humans since the beginning of humanity. These colors can be seen in Egyptian mummies as well as in Mughal period texts. Since the invention of synthetic dyes in 1856, the use of natural colors has been dropping [72, 73].

Synthetic dyes have several advantages over natural dyes, including shade constancy and color brilliance. Natural colorants are regaining popularity around the world, and in the current climate of environmental preservation, they offer huge commercial potential. There is also a growing awareness of their non-hazardous nature. Natural resource depletion, ecological imbalance, pollution, and our polluted environment as a result of the widespread use of dangerous chemicals, particularly synthetic colors, have prompted us to investigate safer alternatives on a national and international scale. These characteristics have expanded the variety of natural dye applications. Synthetic colors are toxic and harm both human skin and the environment. As a result, many European governments have prohibited their use. As a result, natural dyes for textile dyeing must be reintroduced. Natural dyes are safe for the environment and biodegradable. Growing plants to obtain natural dyes promotes natural regeneration, which leads to greater ecological balance [74].

One of the areas of green chemistry, where an eco-friendly chemical frequently applied to natural products, offering innovations in terms of sinking environmental collision, is related to metal surface protection. With the advent of green chemistry, the concept of green corrosion prevention is getting more and more attention. As plants are an excellent reservoir of endogenous products in the form of extracts or pure sustainably sourced compound (plant molecule) that are environmentally acceptable, easily accessible, and eco-friendly in nature. The primary component that enables plants and plant wastes to function as corrosion inhibitors are the classes of phytochemicals that block the active site on metal surfaces by creating a protective film on it, causing inhibition effects due to inhibitory molecule adsorption above the metal surface. Plant extracts are made up of a very complex combination of plant components from many chemical families such as anthocyanins [75], flavonoids [76], tannins [77], terpenoids [21], alkaloids, carotenoids [23], saponins [78], etc. These plant molecules are rich in π electrons (benzene ring, double/triple bond) and contain electrically negative functional groups in their chemical structure and heteroatom such as oxygen, sulfur, and nitrogen [79].

Berberine was investigated as a natural source inhibitor for mild steel in 1 M H_2SO_4 using weight loss measurement, electrochemical methods, and scanning electronic microscopy [80]. The inhibitory impact of berberine on hot-dip coated steels in diluted HCl was evaluated utilizing quantum chemistry analysis, mass-loss tests, electrochemical measurements, and scanning electron microscopy (SEM) observation [81]. The stem alkaloid extract of many types of holy basil was found to be efficient for aluminum corrosion inhibition in HCl solution [82]. Ethanol solution of *Ocimum basilicum* L. oil was investigated as a corrosion inhibitor for aluminum in 0.5 M HCl utilizing weight loss measurements, potentiodynamic polarization, and EIS methods [83]. By using Tafel extra polarization and electrochemical impedance spectroscopy, the effects of date palm (*Phoenix dactylifera* L.) fruit juice on the 7075-type aluminum (AA7075) alloy in 3.5% NaCl solution were examined. A minor cathodic inhibitory effect of PDL fruit juice was discovered, and as the concentration of PDL fruit juice increased, the effectiveness of the inhibitory effect also increased [84]. Conversely, there is little evidence available on berberine's effectiveness as a corrosion inhibitor for alloys of aluminum 7075. This inspired us to research berberine as a corrosion inhibitor of 7075 aluminum alloy in 3.5% NaCl solution. The adhesion of the inhibitors on the metal surface impacts the corrosion inhibition performance of natural products/plant extracts. Plant extracts/natural substances include a variety of phytochemical elements such as alkaloids, polyphenols, steroids, and others, as well as a high concentration of heteroatoms. As a result, these heteroatoms can direct their electron pairs to the metal's unoccupied d-orbital and establish a coordinate bond with the metal surface. In this manner, the environmentally beneficial inhibitors adsorb on the metal and protect it. Several methodologies can be employed to investigate the adsorption of environmentally friendly inhibitors on the metal surface. SEM and AFM are the most often utilized methods. The surface morphologies of metals can be examined in different environments using SEM analysis, which clearly shows the creation of a protective coating on the metal surface in the presence of inhibitors. In the instance of the AFM study, we may compare metal surface roughness, which simply indicates whether or not the inhibitors adsorb on the metal surface.

To elucidate the mechanism of adsorption of inhibitors, the adsorption mechanisms of the preventing species must be pre-determined whether it is ionic or molecular in nature. The dominant mode of adsorption would be determined by some factors including the composition of the extract, type of acid anion, etc. The chemical compounds present in the inhibitor contain a conjugated double bond along with an oxygen hetero atom, which satisfies the general properties of a corrosion inhibitor. Parameters obtained from thermodynamic and kinetic studies justify the physical adsorption of tomato stem extract on the surface of mild steel in acid solutions. The chemical compounds is possibly protonated in acid solutions, and protonated species are expected to be improperly adsorbed since corroding mild steel samples have a positive surface charge in both acid solutions. The first step would be to adsorb the acid anions of sulfate and chloride, and these anions create an excess negative charge to the solution which stimulates the cations to absorb more entirely. As a consequence of van der Waals interaction, the protonated inhibitor may then accumulate on the negatively charged surface of the metal and form coordinate bonds by partial transfer of electrons from the O-hetero atom and multiple bonds to the unfilled d-orbitals of Fe. The compound present in the inhibitor might be combined with metal ions formed on a metal surface due to unshared pair of electrons of O-atom to create the metal inhibitor complexes. Through the action of van der Waals forces, these complexes adhere to the mild steel surface and create a protective layer that prevents corrosion on mild steel.

5 Conclusion

Metal corrosion in an acidic media is inescapable, and it has a significant impact on the global economy. Metal corrosion control is becoming increasingly significant from a technical, aesthetic, environmental, and economic perspective. This review study is primarily concerned with the corrosion inhibition of mild steel, as many important items are constructed of mild steel, which disintegrates when exposed to powerful acids. It can be deduced from the preceding observations that plant extracts are effective corrosion inhibitors in acidic media for preventing metal dissolution. Numerous synthetic, organic, and inorganic corrosion inhibitors are available, but green corrosion inhibitors should be used due to climate concerns. Because of the progress of the concept of green chemistry and economic development, the adoption of environmentally friendly alternatives is gaining importance in the fields of engineering and research. Because of their eco-friendliness, natural plant extracts efficiently replace widely used corrosion inhibitors. The extract is made from several plant parts, including roots, bark, leaves, and stems. The extract of the plant can be made in either water or any organic solvent: an aqueous extract is recommended. The adsorption of a substantial fraction of the plant components on metallic surfaces followed the Langmuir adsorption isotherm; nevertheless, the Temkin and Freundlich adsorption was also reported in a number of observations. Weight loss measurements, as well as several electrochemical techniques, such as potentiodynamic polarization, electrochemical impedance spectroscopy, quantum chemical analysis, gravimetric analysis, and the surface analysis approach, are the most useful strategies in investigating corrosion inhibition.

References

1. Gaikwad, K., & Eng, L. H.-A. C., and undefined. (2022). Natural dyes from plants for smart packaging and printing applications. *researchgate.net*. https://doi.org/10.35248/2090-4568.22.12.246.
2. Pimpodkar, N., et al. (2014). Rhoeo syathacea and Allamanda cathartic extract as a natural indicator in acidometry-alkalimetry. *indianjournals.com, 4*, 82–84. Accessed 27 Jan 2023. [Online]. Available: https://www.indianjournals.com/ijor.aspx?target=ijor:ajpa&volume=4&issue=2&article=005
3. Tasić, Z., Mihajlović, M., & Reports, A. S.-S., and undefined. (2019). Ibuprofen as a corrosion inhibitor for copper in synthetic acid rain solution. nature.com. Accessed 8 Dec 2022. [Online]. Available: https://www.nature.com/articles/s41598-019-51299-2
4. Adejoro, I. A., Ojo, F. K., & Obafemi, S. K. (2015). Corrosion inhibition potentials of ampicillin for mild steel in hydrochloric acid solution. *Journal of Taibah University for Science, 9*(2), 196–202. https://doi.org/10.1016/j.jtusci.2014.10.002
5. Price, S. J., & Figueira, R. B. (2017). Corrosion protection systems and fatigue corrosion in offshore wind structures: Current status and future perspectives. *Coatings, 7*(2), 1–51. https://doi.org/10.3390/coatings7020025
6. Alfakeer, M., Abdallah, M., & Fawzy, A. (2020). Corrosion inhibition effect of expired ampicillin and flucloxacillin drugs for mild steel in aqueous acidic medium. *International Journal of Electrochemical Science, 15*(4), 3283–3297. https://doi.org/10.20964/2020.04.09
7. Habashi, F. (2003). History of corrosion research. *CIM Bulletin, 96*(1067), 88–94.
8. Solomon, M. M., Umoren, S. A., Quraishi, M. A., & Salman, M. (2019). Myristic acid based imidazoline derivative as effective corrosion inhibitor for steel in 15% HCl medium. *Journal of Colloid and Interface Science, 551*, 47–60. https://doi.org/10.1016/J.JCIS.2019.05.004
9. Gupta, M., Mishra, J., & Pitre, K. S. (2013). Corrosion and inhibition effects of mild steel in hydrochloric acid solutions containing organophosphonic acid. *International Journal of Corrosion, 2013*, 1–6. https://doi.org/10.1155/2013/582982
10. Abdeen, D. H., El Hachach, M., Koc, M., & Atieh, M. A. (2019). A review on the corrosion behaviour of nanocoatings on metallic substrates. *Materials, 12*(2). https://doi.org/10.3390/ma12020210
11. Corrosion, V. S.-U. P., & undefined. (2014). *Corrosion processes and the use of corrosion inhibitors in managing corrosion in underground pipelines*. Elsevier.
12. Kumar Shukla, S., Singh, A. K., & Quraishi, M. A. (2011). *Corrosion inhibition and adsorption properties of N-phenylhydrazine-1,2-dicarbothioamide on mild steel in hydrochloric acid.*
13. Shukla, S. K., Quraishi, M. A., & Prakash, R. (2008). A self-doped conducting polymer 'polyanthranilic acid': An efficient corrosion inhibitor for mild steel in acidic solution. *Corrosion Science, 50*(10), 2867–2872. https://doi.org/10.1016/j.corsci.2008.07.025
14. Ghasemi, O., Danaee, I., Rashed, G. R., Avei, M. R., & Maddahy, M. H. (2013). Adsorption and inhibition effect of synthesized N,N′-bis(3-hydroxybenzaldehyde)-1,3-propandiimine on the corrosion of mild steel in 1 M HCl. *Journal of Dispersion Science and Technology, 34*(7), 985–995. https://doi.org/10.1080/01932691.2012.712006
15. Kumar, S., Quraishi, S. Æ. M. A., & Drug, Á. E. I. S. Á. (2009). Ceftriaxone: A novel corrosion inhibitor for mild steel in hydrochloric acid, 1517–1523. https://doi.org/10.1007/s10800-009-9834-1
16. Singh, A., Shukla, S., Singh, M., & M. Q.-M. C. and, undefined. (2011). *Inhibitive effect of ceftazidime on corrosion of mild steel in hydrochloric acid solution*. Elsevier. Accessed 3 Sept 2022. [Online]. Available: https://www.sciencedirect.com/science/article/pii/S0254058411002574
17. Shukla, S. K., Singh, A. K., & Quraishi, M. A. (2012). Triazines: Efficient corrosion inhibitors for mild steel in hydrochloric acid solution, *7*, 3371–3389.
18. Singh, A. K., Shukla, S. K., & Quraishi, M. A. (2011, November). *Corrosion behaviour of mild steel in sulphuric acid solution in presence of ceftazidime.*

19. Moretti, G., Guidi, F., & Grion, G. (2004). Tryptamine as a green iron corrosion inhibitor in 0.5 M deaerated sulphuric acid, *46*, 387–403. https://doi.org/10.1016/S0010-938X(03)00150-1
20. Xhanari, K., Finšgar, M., Knez Hrnčič, M., Maver, U., Knez, Ž., & Seiti, B. (2017). Green corrosion inhibitors for aluminium and its alloys: A review. *RSC Advances, 7*(44), 27299–27330. https://doi.org/10.1039/c7ra03944a
21. Bothi, P., Kaleem, A., Abdul, A., Osman, H., & Awang, K. (2013). Neolamarckia cadamba alkaloids as eco-friendly corrosion inhibitors for mild steel in 1 M HCl media. *Corrosion Science, 69*, 292–301. https://doi.org/10.1016/j.corsci.2012.11.042
22. Shalabi, K. (2004). Adsorption and corrosion inhibition of atropa belladonna extract on carbon steel in 1 M HCl solution. *International Journal of Electrochemical Science, 9*(3), 1468–1487.
23. Butnariu, M. (2016). Methods of analysis (extraction, separation, identification and quantification) of carotenoids from natural products. *Journal of Ecosystem & Ecography, 6*(2), 1–19. https://doi.org/10.4172/2157-7625.1000193
24. Bairagi, H., Vashishth, P., Narang, R., Shukla, S. K., & Mangla, B. (2023). Experimental and computational studies of a novel metal oxide nanoparticle/conducting polymer nanocomposite (TiO2/PVP) as a corrosion inhibitor on low-carbon steel in diprotic acidic medium. *Industrial and Engineering Chemistry Research, 62*(28), 10982–11000. https://doi.org/10.1021/ACS.IECR.3C00099
25. Shukla, S. K., Singh, A. K., Ahamad, I., & Quraishi, M. A. (2009). Streptomycin: A commercially available drug as corrosion inhibitor for mild steel in hydrochloric acid solution. *Materials Letters, 63*(9–10), 819–822. https://doi.org/10.1016/j.matlet.2009.01.020
26. Zou, Y., Wang, J., Science, Y. Z.-C., and undefined. (2011). *Electrochemical techniques for determining corrosion rate of rusted steel in seawater*. Elsevier. Accessed 27 Jan 2023. [Online]. Available: https://www.sciencedirect.com/science/article/pii/S0010938X10004427
27. Andrade, C., C. A.-C. and building materials, and undefined. (1996). *Corrosion rate monitoring in the laboratory and on-site*. Elsevier. Accessed 27 Jan 2023. [Online]. Available: https://www.sciencedirect.com/science/article/pii/0950061895000445
28. Sadowski, L. (2010). New non-destructive method for linear polarisation resistance corrosion rate measurement. *Archives of Civil and Mechanical Engineering, 10*(2), 109–116. https://doi.org/10.1016/S1644-9665(12)60053-3
29. Orazem, M., & Jersey, B. T.-N., and undefined. (2008). Electrochemical impedance spectroscopy. *iopscience.iop.org*. Accessed 27 Jan 2023. [Online]. Available: https://iopscience.iop.org/article/10.1149/2.20192if/pdf#page=50
30. Gray, L., B. A.-J. of protective coatings & linings, and undefined. (2003). EIS: Electrochemical impedance spectroscopy. *trid.trb.org*. Accessed 27 Jan 2023. [Online]. Available: https://trid.trb.org/view/734445
31. Tiu, B. D. B., & Advincula, R. C. (2015). Polymeric corrosion inhibitors for the oil and gas industry: Design principles and mechanism. *Reactive and Functional Polymers, 95*, 25–45. https://doi.org/10.1016/J.REACTFUNCTPOLYM.2015.08.006
32. Goni, L., M. M.-C. inhibitors, and undefined. (2019). Green corrosion inhibitors. books.google.com. Accessed 26 Jan 2023.
33. Shukla, S., E. E.-Int. J. Electrochem. Sci, and undefined. (2012). Inhibitive effect of N'-(7-chloroquinolin-4-yl)-N, N-diethyl-pentane-1, 4-diamine towards mild steel/sulphuric acid solution interface. *Citeseer, 7*, 12121–12133. Accessed 3 Sept 2022.
34. Ji, G., Shukla, S. K., Ebenso, E. E., & Prakash, R. (2013). Argemone mexicana leaf extract for inhibition of mild steel corrosion in sulfuric acid solutions. *International Journal of Electrochemical Science, 8*(8), 10878–10889.
35. Taghavikish, M., Dutta, N., N. R. C.- Coatings, and undefined. (2017). Emerging corrosion inhibitors for interfacial coating. *mdpi.com*. https://doi.org/10.3390/coatings7120217.
36. Narang, R., Vashishth, P., Bairagi, H., Shukla, S. K., & Mangla, B. (2023). Electrochemical and surface study of an antibiotic drug as sustainable corrosion inhibitor on mild steel in 0.5 M H2SO4. *Journal of Molecular Liquids, 384*. https://doi.org/10.1016/j.molliq.2023.122277

37. Vashishth, P., Bairagi, H., Narang, R., Shukla, S. K., & Mangla, B. (2022). Thermodynamic and electrochemical investigation of inhibition efficiency of green corrosion inhibitor and its comparison with synthetic dyes on MS in acidic medium. *Journal of Molecular Liquids, 365*, 120042. https://doi.org/10.1016/j.molliq.2022.120042
38. Papavinasam, S. Corrosion inhibitors. In: Revie RW,... - Google Scholar." https://scholar.google.com/scholar?hl=en&as_sdt=0%2C5&q=Papavinasam+S.+Corrosion++Inhibitors.+In%3A+Revie+RW%2C+editor.+Uhlig%27s++Corrosion+Handbook.+3rd+ed.+New+York%3A++Wiley%3B+2000.+pp.+1021-1033.&btnG=. accessed Jan. 27, 2023).
39. Shukla, S. K., & Quraishi, M. A. (2009). Cefotaxime sodium: A new and efficient corrosion inhibitor for mild steel in hydrochloric acid solution. *Corrosion Science, 51*(5), 1007–1011. https://doi.org/10.1016/J.CORSCI.2009.02.024
40. Vashishth, P., Bairagi, H., Narang, R., Kumar, H., & Mangla, B. (2023). Efficacy of biomass-derived nanocomposites as promising materials as corrosion inhibitors. In *Antiviral and antimicrobial coatings based on functionalized nanomaterials* (pp. 285–303). Elsevier. https://doi.org/10.1016/B978-0-323-91783-4.00007-3
41. Dariva, C., A. G.-D. in corrosion protection, and undefined 2014. (2014). Corrosion inhibitors–principles, mechanisms and applications. books.google.com. https://doi.org/10.5772/57255
42. Mohamed, R. (2014). Corrosion in oil and gas industry: A perspective on corrosion inhibitors. *Journal of Material Science & Engineering, 3*(3), 110. https://doi.org/10.4172/2169-0022.1000e110
43. Martinez, R., Olivares-Xomelt, O., & Likhanova, N. v. (2014). Environmentally friendly corrosion inhibitors. *books.google.com*. https://doi.org/10.5772/57252.
44. Kelland, M. A. (2010). Production chemicals for the oil and gas industry. *Chromatographia, 72*(1–2), 199–199. https://doi.org/10.1365/S10337-010-1557-2
45. Arthur, D. E., Jonathan, A., Ameh, P. O., & Anya, C. (2013). A review on the assessment of polymeric materials used as corrosion inhibitor of metals and alloys. *International Journal of Industrial Chemistry, 4*(1), 2. https://doi.org/10.1186/2228-5547-4-2
46. Granese, S., Rosales, B., Oviedo, C., Science, J. Z.-C., and undefined. (1992). The inhibition action of heterocyclic nitrogen organic compounds on Fe and steel in HCl media. Elsevier. Accessed 27 Jan 2023. [Online]. Available: https://www.sciencedirect.com/science/article/pii/0010938X92901823
47. Ali, A., Sabirneeza, F., Geethanjali, R., & Subhashini, S. (2015). Polymeric corrosion inhibitors for iron and its alloys: A review. *Taylor & Francis, 202*(2), 232–244. https://doi.org/10.1080/00986445.2014.934448
48. Arukalam, I. O. (2014). Durability and synergistic effects of KI on the acid corrosion inhibition of mild steel by hydroxypropyl methylcellulose. *Carbohydrate Polymers, 112*, 291–299. https://doi.org/10.1016/J.CARBPOL.2014.05.071
49. Mobin, M., & Rizvi, M. (2016). Inhibitory effect of xanthan gum and synergistic surfactant additives for mild steel corrosion in 1 M HCl. *Carbohydrate Polymers, 136*, 384–393. https://doi.org/10.1016/j.carbpol.2015.09.027
50. Shen, C., Alvarez, V., Koenig, J. D. B., & Luo, J. L. (2019). Gum Arabic as corrosion inhibitor in the oil industry: Experimental and theoretical studies. *Corrosion Engineering Science and Technology, 54*(5), 444–454. https://doi.org/10.1080/1478422X.2019.1613780
51. Umoren, S. A., Obot, I. B., Ebenso, E. E., Okafor, P. C., Ogbobe, O., & Oguzie, E. E. (2006). Gum arabic as a potential corrosion inhibitor for aluminium in alkaline medium and its adsorption characteristics. *Anti-Corrosion Methods and Materials, 53*(5), 277–282. https://doi.org/10.1108/00035590610692554/FULL/HTML
52. Saji, V. S., & Thomas, J. (2007). Nanomaterials for corrosion control. *Current Science, 92*(1), 51–55.
53. Atta, A. M., Allohedan, H. A., El-Mahdy, G. A., & Ezzat, A. R. O. (2013). Application of stabilized silver nanoparticles as thin films as corrosion inhibitors for carbon steel alloy in 1M hydrochloric acid. *Journal of Nanomaterials, 2013*. https://doi.org/10.1155/2013/580607

54. Alaskry, M., & Soliman, S. NANOMAtERiAlS AS CORROSiON INhibitORS. *academia.edu*. Accessed 25 July 2022. [Online]. Available: https://www.academia.edu/download/37195256/nanomaterials_as_corrosion_inhibitors.pdf
55. Kolahalam, L. A., Kasi Viswanath, I. V., Diwakar, B. S., Govindh, B., Reddy, V., & Murthy, Y. L. N. (2019). Review on nanomaterials: Synthesis and applications. *Materials Today Proceedings, 18*, 2182–2190. https://doi.org/10.1016/J.MATPR.2019.07.371
56. El-Mahdy, G. A., Atta, A. M., & Al-Lohedan, H. A. (2014). Synthesis and evaluation of poly(sodium 2-acrylamido-2-methylpropane sulfonate-co-styrene)/magnetite nanoparticle composites as corrosion inhibitors for steel. *Molecules, 19*(2), 1713–1731. https://doi.org/10.3390/MOLECULES19021713
57. Rahman, M. M., Zahir, M. H., Haq, M. B., al Shehri, D. A., & Kumar, A. M. (2018). Corrosion inhibition properties of waterborne polyurethane/cerium nitrate coatings on mild steel. *Coatings, 8*(1), 34. https://doi.org/10.3390/COATINGS8010034
58. Javidparvar, A. A., Naderi, R., & Ramezanzadeh, B. (2019). Epoxy-polyamide nanocomposite coating with graphene oxide as cerium nanocontainer generating effective dual active/barrier corrosion protection. *Composites Part B: Engineering, 172*, 363–375. https://doi.org/10.1016/J.COMPOSITESB.2019.05.055
59. Chen, C. J., et al. (2021). Low-mass liquid crystalline materials blended in recycled thermoplastic polyester elastomer for corrosion inhibitor application. *Polymers, 13*(18), 3188. https://doi.org/10.3390/POLYM13183188
60. Petelska, A. D., Kazimierska-Drobny, K., Janicka, K., Majewski, T., & Urbaniak, W. (2019). Multifunctional protective PVC-ZnO nanocomposite coatings deposited on aluminum alloys by electrospinning. *Coatings, 9*(4), 216. https://doi.org/10.3390/COATINGS9040216
61. da Silva, J. E. P., de Torresi, S. I. C., & Torresi, R. M. (2005). Polyaniline acrylic coatings for corrosion inhibition: The role played by counter-ions. *Corrosion Science, 47*(3 SPEC. ISS), 811–822. https://doi.org/10.1016/J.CORSCI.2004.07.014
62. Chen, B., et al. (2020). A novel and green method to synthesize a epoxidized biomass eucommia gum as the nanofiller in the epoxy composite coating with excellent anticorrosive performance. *Chemical Engineering Journal, 379*, 122323. https://doi.org/10.1016/J.CEJ.2019.122323
63. Umoren, S. A., & Solomon, M. M. (2019). Protective polymeric films for industrial substrates: A critical review on past and recent applications with conducting polymers and polymer composites/nanocomposites. *Progress in Materials Science, 104*, 380–450. https://doi.org/10.1016/J.PMATSCI.2019.04.002
64. Ituen, E., Singh, A., & Yuanhua, L. (2020). Inhibitive effect of onion mesocarp extract-nickel nanoparticles composite on simultaneous hydrogen production and pipework corrosion in 1 M HCl. *International Journal of Hydrogen Energy, 45*(18), 10814–10825. https://doi.org/10.1016/J.IJHYDENE.2020.01.225
65. John, S., Salam, A., Baby, A. M., & Joseph, A. (2019). Corrosion inhibition of mild steel using chitosan / TiO2 nanocomposite coatings. *Progress in Organic Coating, 129*, 254–259. https://doi.org/10.1016/J.PORGCOAT.2019.01.025
66. Vijayaraghvan, D., Palanisamy, K. L., Devabharathi, V., & Sundaram, N. M. (2015). Corrosion inhibition studies of mild steel with carrier oil stabilized of iron oxide nanoparticles incorporated into a paint bio-nanocomposites for bone regeneration view project Corrosion inhibition studies of mild steel with carrier oil stabilized of iron oxide nanoparticles incorporated into a paint. *Article in International Journal of ChemTech Research, 7*(4), 1661–1664.
67. Yusoff, Z. b. (2016). Piper nigrum (black pepper) as green inhibitor to prevent corrosion. *Petrochemical Engineering Department, 1*(1), 26–35.
68. Vorobyova, V., Skiba, M., & Andrey, K. (2022). Tomato pomace extract as a novel corrosion inhibitor for the steel in industrial media: The role of chemical transformation of the extract and proinhibition effect. *Journal of Molecular Structure, 1264*, 133155. https://doi.org/10.1016/J.MOLSTRUC.2022.133155

69. Haldhar, R., et al. (2021). Investigation of plant waste as a renewable biomass source to develop efficient, economical and eco-friendly corrosion inhibitor. *Journal of Molecular Liquids, 335*. https://doi.org/10.1016/J.MOLLIQ.2021.116184
70. Kumar, H., Sharma, S., & Kumari, R. (2022). Corrosion inhibition and adsorption mechanism of Morus nigra on mild steel in acidic medium: A sustainable and green approach. *Vietnam Journal of Chemistry, 2022*, 60. https://doi.org/10.1002/VJCH.202100166
71. Kumar, H., Yadav, V., Anu, S. K. S., & Kang, N. (2021). Adsorption and inhibition mechanism of efficient and environment friendly corrosion inhibitor for mild steel: Experimental and theoretical study. *Journal of Molecular Liquids, 338*, 116634. https://doi.org/10.1016/J.MOLLIQ.2021.116634
72. Zieba-Palus, J., & Trzcińska, B. M. (2011). Establishing of chemical composition of printing ink. *Journal of Forensic Sciences, 56*(3), 819–821. https://doi.org/10.1111/J.1556-4029.2011.01734.X
73. Khankaew, S., Boonsupthip, W., … C. P.-… I. S. on, and undefined 2013. (2013). Screening of naturally-derived pH dyes from plant extract powders as colorimetric bio-indicator and possible application in intelligent packaging. *researchgate.net*. Accessed 27 Jan 2023.
74. Gaikwad, K. K., & Lee, Y. S. (2016). Novel natural phenolic compound-based oxygen scavenging system for active packaging applications. *Journal of Food Measurement and Characterization, 10*(3), 533–538. https://doi.org/10.1007/S11694-016-9332-1
75. Teng, Y., et al. (2021). Anthocyanin as sustainable and non-toxic corrosion inhibitor for mild steel in HCl media: Electrochemical, surface morphology and theoretical investigations. *Journal of Molecular Liquids, 344*, 117721. https://doi.org/10.1016/J.MOLLIQ.2021.117721
76. Chihi, S., Gherraf, N., Alabed, B., & Hameurlain, S. (2015). Inhibition effect of flavonoid extract of *Euphorbia guyoniana* on the corrosion of mild steel in H_2SO_4 medium. *Journal of Fundamental and Applied Sciences, 1*(2), 31. https://doi.org/10.4314/jfas.v1i2.4
77. Dargahi, M., Olsson, A. L. J., Tufenkji, N., & Gaudreault, R. (2015). Green technology: Tannin-based corrosion inhibitor for protection of mild steel. *Corrosion, 71*(11), 1321–1329. https://doi.org/10.5006/1777
78. Ikeuba, A. I., & Okafor, P. C. (2019). Green corrosion protection for mild steel in acidic media: Saponins and crude extracts of Gongronema latifolium. *Pigment and Resin Technology, 48*(1), 57–64. https://doi.org/10.1108/PRT-03-2018-0020
79. Miralrio, A., & Vázquez, A. E. (2020). Plant extracts as green corrosion inhibitors for different metal surfaces and corrosive media: A review. *PRO, 8*(8). https://doi.org/10.3390/PR8080942
80. Li, Y., Zhao, P., Liang, Q., Hou, B., Appl. Surf. Sci.... - Google Scholar. https://scholar.google.com/scholar?hl=en&as_sdt=0%2C5&q=Y.+Li%2C+P.+Zhao%2C+Q.+Liang%2C+B.+Hou%2C+Appl.+Surf.+Sci.+252+%282005%29+1245.&btnG=. Accessed 30 Jan 2023.
81. Ju, H., & Ju, Y. . (2012). Berberine as an environmental-friendly inhibitor for hot-dip coated steels in diluted hydrochloric acid. *Elsevier, 28*(9), 809–816. https://doi.org/10.1016/S1005-0302(12)60135-2.Y. L.-J. of M. S. & Technology, and undefined 2012
82. Kumpawat, N., Chaturvedi, A., & Upadhyay, R. K. (2012). Study on corrosion inhibition efficiency of stem alkaloid extract of different varieties of holy basil on aluminium in HCl solution. *koreascience.or.kr, 56*(4). https://doi.org/10.5012/jkcs.2012.56.4.401
83. Halambek, J., Žutinić, A., K. B. E. Science, and undefined 2013. (2013). Ocimum basilicum L. oil as corrosion inhibitor for aluminium in hydrochloric acid solution. *electrochemsci.org, 8*, 11201–11214. Accessed 30 Jan 2023. [Online]. Available: http://electrochemsci.org/papers/vol8/80911201.pdf
84. Gerengi, H. (2012). Anticorrosive properties of date palm (Phoenix dactylifera L.) fruit juice on 7075 type aluminum alloy in 3.5% NaCl solution. *Industrial and Engineering Chemistry Research, 51*(39), 12835–12843. https://doi.org/10.1021/IE301771U

Evaluation of Colorimetric and Color Fastness Properties of Marble-Printed Silk Fabrics with the Use of Various Mordants and After-Treatments

Habibe Kahvecioğlu Sarı, Fatma Filiz Yıldırım, Ozan Avinc, and Arzu Yavas

Abstract Marble printing, also known as Ebru, is a traditional art of decorating paper and textile fabrics. In this study, unmordanted and premordanted silk fabrics were marble-printed with organic natural red ocher and synthetic red ink for comparison. Here, silk fabrics were firstly premordanted with one natural mordant (gallnut) and two chemical mordants [potassium bichromate and iron (II) sulfate] for comparison. Moreover, unmordanted silk fabrics were used as control fabrics. After the premordanting process, the premordanted samples were marble-printed with organic natural red ocher and synthetic red ink (Pebeo Marbling Vermillion 02 Red Ink). Then, the marble-printed (Ebru) silk fabric samples were posttreated with various steam fixation (at 102 °C for 15 minutes and at 102 °C for 30 minutes) and thermofixation (at 130 °C for 4 minutes and at 150 °C for 4 minutes) processes. Moreover, different control fabrics were included and evaluated in this study. Here, various posttreatments were carried out for fixing the natural dye onto the silk fabrics. Fixation parameters of different fixation processes were chosen to understand the effect of the fixation temperature and time and the effects of steam on the marble-printed fabrics. After the posttreatment processes, the washing, light, rub (dry and wet), perspiration (acidic and alkaline), water fastness, and colorimetric properties of premordanted, marble-printed, and fixed silk fabric samples were eval-

H. K. Sarı
Pamukkale University Denizli Vocational School of Technical Sciences, Department of Handicrafts, Denizli, Türkiye

F. F. Yıldırım
Pamukkale University Acıpayam Vocational School of Higher Education, Department of Textile, Clothing, Footwear and Leather, Shoe Design & Production Program, Denizli, Türkiye

Pamukkale University, Faculty of Engineering, Textile Engineering Department, Denizli, Türkiye

O. Avinc · A. Yavas (✉)
Pamukkale University, Faculty of Engineering, Textile Engineering Department, Denizli, Türkiye
e-mail: aozerdem@pau.edu.tr

S. S. Muthu (ed.), *Natural Dyes and Sustainability*, Sustainable Textiles: Production, Processing, Manufacturing & Chemistry,
https://doi.org/10.1007/978-3-031-47471-2_13

uated. Iron (II) sulfate premordanted and then marble-printed silk fabrics exhibited the highest color strength (*K*/*S*) values. The highest *K*/*S* value was 19.7, and this value belongs to iron (II) sulfate premordanted then marble printed silk fabrics thermofixed at 130 °C for 4 minutes. Overall, the utilization of iron mordant generally resulted in the highest color strength values leading to darker color shades on silk fabrics. The thermofixation processes after marble printing generally led to the highest color strength values. Marble-printed silk fabric samples generally displayed good to excellent washing, light, perspiration (acidic and alkaline), and water fastness results. This study showed that the premordanting process, type of mordant used, fixation method applied after marble printing, and application conditions of these fixation methods greatly affect the final colorimetric and color fastness characteristics of marble-printed silk fiber fabrics.

Keywords Marble printing · Ebru technique · Silk · Organic natural red ocher · Synthetic red ink · Gallnut · Mordant · Color · Fastness

1 Introduction

Sustainability is one of the most important issues today. The more we can reduce the possible damage to the environment from production processes, the better it will be for our world. Hence, it is very valuable to integrate sustainable and renewable natural products into production processes and increase their use. Coloration is one of the processes that can create the most important environmental problem in the textile industry. Successful application of natural and natural-based products in the coloration of textile products will contribute to the sustainability efforts that the whole world is trying to realize. As is known, many different techniques and colorants can be utilized in textile coloring. In this current research, evaluation of colorimetric and color fastness characteristics of marble-printed (Ebru) silk fiber fabrics using various mordants and different after-treatments was carried out. Ebru (Turkish art of marbling) is a method by which very different designs can be transferred to paper and fabrics. In this current study, unmordanted and premordanted silk fiber fabrics were marble-printed (Ebru) with organic natural red ocher and synthetic red ink for comparison. Here, silk fabrics were firstly premordanted with one natural mordant (gallnut) and two chemical mordants [potassium bichromate and iron (II) sulfate] for comparison. So, in this study, there are several aspects that contribute to sustainability efforts. The fabric used in this study was made of silk fiber, which is a natural and sustainable fiber. A natural, renewable, and sustainable gallnut was utilized as a mordant using the premordanting technique. In addition, organic natural red ocher was used in the coloring process of the silk fabric made with the marbling style (Ebru). Of course, these natural products were also compared with chemical products. In this book chapter, firstly, brief information about "Ebru" (Turkish Art of Marbling) and its history is given, and then information about the materials used in the Ebru (Turkish art of marbling) technique is presented. Afterwards, after giving information about the materials, methods, processes, and tests used in this study, the outcomes of this study were compared and related information about the results was given.

In general, "Ebru" (Turkish Art of Marbling) is a decorative art made by transferring it onto paper. The pattern is created on the surface of the water in a marbling tray. Nonacidic and nonfading natural pigments are spread to the surface. Because the viscosity of the water is increased by adding gum tragacanth, carrageenan algae, or similar additives, the dyes/pigments do not sink into the bottom of the marbling tray [1]. Nowadays, "Ebru" as a name is used for both "Ebru art" and "Ebru paper." In this respect, ebruism and marbling, which derive in accordance with the Turkish rules, are also included in the dictionary dictated by the Tanzimat period [2]. There is no definite information about when and where the art of Ebru starts. However, there are studies of some historians about the beginning of the "Ebru" in Central Asia. It is called "Ebri" or "Ebru" in Anatolia of Turkey. The art of Ebru was widely used during the Seljuk and Ottoman periods. It is known in Europe as the "Turkish paper" at the beginning of the seventeenth century [3]. During the Ottoman period, many Ebru masters were trained. The works of these masters decorated the rich persons' walls or were sent as gifts to important people. Eventually, these beautiful pastel- and light color-patterned papers have begun to be used as grounds in state documents and official correspondence. The principal intention of this application is to provide aesthetic value and to prevent tampering attempt, which is based on the understanding of the logic of the complex patterns on checks, notes, and paper money [3].

Our traditional artwork "Ebru" has begun to attract the attention of aristocrats and adventurers in Europe since the end of the sixteenth century and always remembered with the word "Turk." Many people interested in Ebru's history accept that this art came to Europe through East and the Ottomans. A French aristocrat Pierre de L'Estoile presented books covered with Ebru to his friends. A British traveler George Sandys reported an interesting art from Istanbul when he came in 1610, saying that "Turks generally color their papers with mysterious numbers, mostly chamolet-like designs, by sinking their papers into the water" [2]. The art of marbling has preserved its existence in Anatolia for centuries [4–8].

Nowadays, the art of Ebru is continued with its traditional structure, thanks to the numbered Ebru masters dedicated to this work. In the past, Ebru paper-painted art is used; however, today Ebru can be applied to paper, fabric, wood, porcelain, and glass. It can be applied on nearly all kinds of materials.

Materials Used in Ebru (Turkish Art of Marbling)

In making Ebru (Turkish art of marbling), marbling tray, brush, colorant, paper, gall, water, thickener (tragacanth, seaweed, etc.), comb, and biz with various thicknesses (*used for pattern-making, the handle is made of wood and the tip is usually metal*) are used. In the following section, the materials used in the art of Ebru are introduced [9].

Marbling Tray: This marbling tray is the size of the paper or product to be marbling or slightly larger. This marbling tray is a container in which prepared water with gum tragacanth can be placed. The depth of the tray can be between 3 and 5 cm. It can be made of materials such as metal, glass, wood, galvanized, or plastic.

Water: Generally, drinking water (drinking water with low hardness, low lime content and low chlorine) or distilled pure water can be used in Ebru (marbling) production.

Brush: The brushes used in Ebru are traditionally made of a horse's tail and 8- to 12-mm thick hair. Round brushes are also used. A separate brush is required for each color [1].

Paper: In Ebru, any paper with absorbent properties can be used. The most ideal is the acid-free paper, and it is recommended to be used during the mastery as it is expensive.

Colorants: Traditionally natural dyes are used in marbling. Obtaining natural dyes is very time-consuming and laborious. Instead, ready-made dyestuffs made of metal oxide or soil can be used. Additionally, nowadays pigment dyes, oil paints, synthetic dyes, gouache, acrylic, and ink-type dyestuffs can also be used in Ebru construction.

Gall: This is the secret of marbling. The gall is a surface-active substance that reduces the surface tension. If there is no gall existence in the marbling tray, colorants may sink into the bottom of the tray. Its adhesive properties help the colorant adhere to the paper and also ensure that the colors do not mix on the surface.

Thickeners: It is difficult to give a pattern in the marbling process. Movements are sometimes uncontrollable. The viscosity of the liquid used should be higher than that of the water. In order to achieve this, a wide variety of substances have been used, such as thickeners [7].

Gum Tragacanth: This is an ossified secretion in contact with the air of the liquid coming out of the body of the Astragalus that grows wildly in the mountains. It is the gum occurring in the body of some Astragalus (Leguminosae) species [9].

In this study, the "Ebru" marbling method is used for printing silk fabrics. Here, unmordanted and premordanted silk fabrics were marble-printed with organic natural red ocher and synthetic red ink for comparison. Here, silk fabrics were firstly premordanted with one natural mordant (gallnut) and two chemical mordants [potassium bichromate and iron (II) sulfate] for comparison. Moreover, unmordanted silk fabrics were used as control fabrics. After the premordanting process, the premordanted samples were marble-printed with organic natural red ocher and synthetic red ink (Pebeo Marbling Vermillion 02 Red Ink). Then, the marble-printed (Ebru) silk fabric samples were posttreated using various steam fixation (at 102 °C for 15 minutes and at 102 °C for 30 minutes) and thermofixation (at 130 °C for 4 minutes and at 150 °C for 4 minutes) processes. Moreover, different control fabrics were also included and evaluated in this study. Here, various posttreatments were carried out for fixing the natural dye onto the silk fabrics. Fixation parameters of different fixation processes were chosen to understand the effect of the fixation temperature and time and the effects of steam on marble-printed fabrics. After the posttreatment processes, the washing, light, rub (dry and wet), perspiration (acidic and alkaline), water fastness, and colorimetric properties of premordanted, marble-printed, and fixed silk fabric samples were evaluated.

2 Experimental

2.1 *Materials*

Marble printing processes were carried out using 100% silk fiber plain woven fabrics made from silk fibers that are natural, renewable, and sustainable. Potassium bichromate ($K_2Cr_2O_7$ – Tekkim, chemical mordant), iron (II) sulfate (ferrous sulfate) ($FeSO_4{\cdot}7H_2O$ – Tekkim, chemical mordant), and gallnut (natural mordant) were used in the premordanting of 100% silk fabrics. Here, gallnut, used as a natural mordant, is natural, renewable, and sustainable. All chemical mordants were of technical grade, and gallnut was obtained from nature. Organic natural red ocher was used as a natural red colorant, and Pebeo Marbling Vermillion 02 Red Ink was used as a synthetic red ink in marble printing (Ebru). Organic natural red ocher was obtained from nature. So, in this study, there are several aspects that contribute to sustainability efforts.

2.2 *Methods*

2.2.1 Pretreatment With Mordants (Premordanting)

Fabric samples were premordanted with 1/40 liquor ratio at 100 °C in 60 minutes in ATAC LAB DYE HT Laboratory Type dyeing machine before marble printing (Ebru). In the premordanting process, 4% potassium bichromate, 4% iron (II) sulfate (ferrous sulfate), and 20% gallnut were used. Afterwards, premordanted silk fabric samples were washed and air-dried.

2.2.2 Marble Printing (Ebru) Process

Marble printing (Ebru) of silk fabrics with organic natural red ocher and synthetic red ink was performed. Before the marble printing process, primarily, tragacanth was prepared. Powder tragacanth was used as a thickener in the marbling application. About 43 g of powdered tragacanth was gradually added using a spoon a little to the 5 lt of pH 7.2 and was mixed using a mixer for 10 minutes. The tragacanth was drained using a cloth to the 35 × 50 tray, and the mass was covered with newspaper and left to stand for 1 hour. After 1 hour, tragacanth was ready for use.

In Ebru making, natural red ocher pigment and synthetic red ink have been used. "Kızılboya (natural red ocher)" is obtained from red soil containing iron oxide [7]. The use of red pigment has also become a tradition. In our study, natural red ocher (soil) was used. About 40 cc of organic natural red ocher pigment was added to the sufficient amount of water, and the desired consistency of the dye was achieved by mixing it using a brush. A few drops (0.5 ml) of gall were added to the prepared dye solution, and then the solution was stirred well. Moreover, 45 ml of Pebeo Marbling Vermillion Synthetic Red Ink was used for marble printing. Dye solutions were tested by dripping them on the surface of the water. If the dye solution is not spread on the water well enough, more gall drops were added to the dye solution and thus the surface tension of the dye solution can be provided.

The "gel-git" Ebru was applied on the fabrics. For this purpose, the "battal" Ebru was first performed, followed by the "gel-git." Colorant solution was sprinkled using a brush on the surface of the tragacanth. Thus, the surface of the tragacanth was coated with synthetic ink. This technique is called "battal" marbling. Then, firstly the "gel-git" design was made with a length of 1 cm parallel to the long edge of the tray using a pricker. Then the same process was applied parallel to the short edge of the marbling tray. We waited for a few seconds for the fabric to absorb the colorant. The fabric was then lifted up from the marbling tray. In our study, marbled fabrics were washed under tap water and hung to dry in a shaded place away from sunlight.

2.2.3 After-Treatment Process

After the marble-printing (Ebru) process, various posttreatments were performed. Although the posttreatments were performed to fix the colorants onto the fabrics, some control samples have been selected. The initial control samples were only rinsed with tap water for 30 minutes. The control samples of the second type were printed, stored 24 hours under ambient conditions, and then rinsed with running tap water for 30 seconds. The control samples of the third type were stored 1 month under ambient conditions and then rinsed with running tap water for 30 seconds. Thermofixation of samples was performed at 130 °C for 4 minutes and at 150 °C for 4 minutes. Then all thermofixed samples were rinsed. Steam fixation was performed at 102 °C for 15 minutes and at 102 °C for 30 minutes, then all samples were rinsed. Fixation parameters of the process were chosen to understand the effect of the fixation temperature, time, and effects of steam on the marble-printed fabrics.

2.2.4 Colorimetric Measurements

The color strength (K/S) values of marble-printed (Ebru) silk fabrics were measured using a DataColor SpectraFlash 600 (Datacolor International. Lawrenceville. NJ. USA) spectrophotometer under illuminant D65, using 10° Standard observer. Lightness (L^*), a^*, b^*, C^*, and $h°$ coordinates were obtained in relation to the CIE system, and the color yield value of *K/S* was calculated by using the Kubelka–Munk equation. Equation (1) given at below:

$$K/S = (1-R)^2/2R \quad (1)$$

2.2.5 Color Fastness Properties

The wash, rub, and light fastness of the marble-printed samples with natural dye were investigated. The wash fastness test was implemented in a M228 Rotawash machine (SDL ATLAS, UK) at 40 °C with sodium perborate in accordance with the ISO 105:C06 A2S test. Dry and wet rub fastness properties were obtained according to the ISO 105: X12 protocol using James Heal Model 670 hand-driven crock master. The light fastness test was conducted in accordance with ISO 105: B02: color fastness to artificial light (Xenon arc lamp). The overall fastness characteristics were assessed using ISO gray scales in the light box.

3 Results and Discussion

3.1 Colorimetric Properties of Marble-Printed (Ebru) Samples

The posttreatment types of marble-printed samples with or without mordant are given in Table 1.

3.1.1 Colorimetric Properties of Organic Natural Red Ocher and Synthetic Ink Marble-Printed (Ebru) Silk Fabric Samples Without Mordant

3.1.1.1 Colorimetric Properties of Organic Natural Red Ocher Marble-Printed Silk Fabric Samples Without Mordant

The colorimetric data (*K/S values*) obtained from organic natural red ocher marble-printed silk fabrics are shown in Figs. 1, 2, 3, and 4 and Table 2.

When the marble-printed fabrics without mordant were examined, the highest *K/S* outcome of unmordanted substrates was 1.8 after thermofixation at 150 °C for 4 minutes (Table 2).

When the reflectance curves of the silk fiber fabrics marble-printed with organic natural red ocher without mordant were examined, the reflectance curves of the silk fabrics were close to each other with one exception (C1) (Fig. 1). All the treated fabrics exhibited red color shades. So, all $h°$ (hue angle) values of all marble-printed silk fabrics were smaller than 90°, and this indicates that the color shades of these

Table 1 Posttreatment conditions of the marble-printed silk fabrics

Samples[a]	
C1	Marble printing with organic natural red ocher/synthetic red ink, immediate rinsing
C2	Marble printing with organic natural red ocher/synthetic red ink 24 hours storage under ambient conditions, rinsing
C3	Marble printing with organic natural red ocher/synthetic red ink, 1 month storage under ambient conditions, rinsing
TF1	Marble printing with organic natural red ocher/synthetic red ink, thermofixation at 130 °C for 4 minutes, rinsing
TF2	Marble printing with organic natural red ocher/synthetic red ink, thermofixation at 150 °C for 4 minutes, rinsing
SF1	Marble printing with organic natural red ocher/synthetic red ink, steam fixation at 102 °C for 15 minutes, rinsing
SF2	Marble printing with organic natural red ocher/synthetic red ink, steam fixation at 102 °C for 30 minutes, rinsing

[a]Rinsing: rinsing with running tap water for 30 seconds
All mordanted and unmordanted fabrics

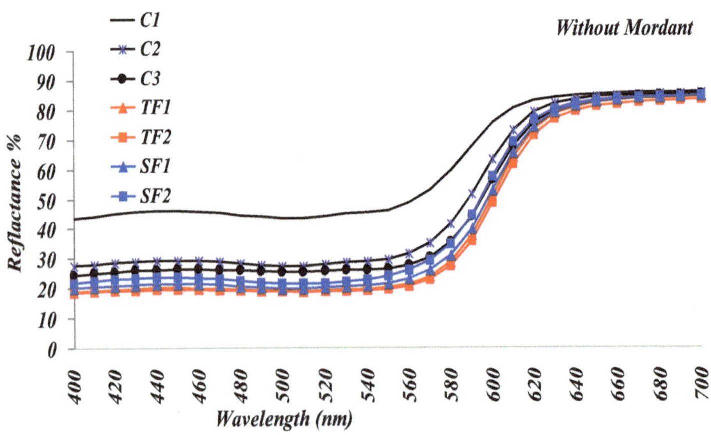

Fig. 1 Reflectance curves of organic natural red ocher marble-printed samples without mordant

Fig. 2 Yellowness (b^*) and Redness (a^*) values of organic natural red ocher marble-printed samples without mordant

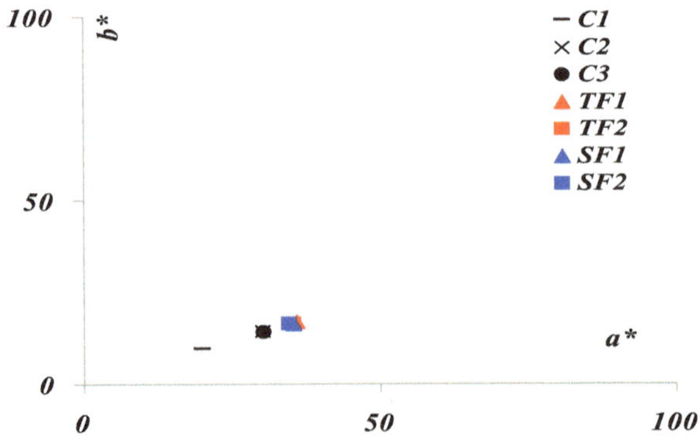

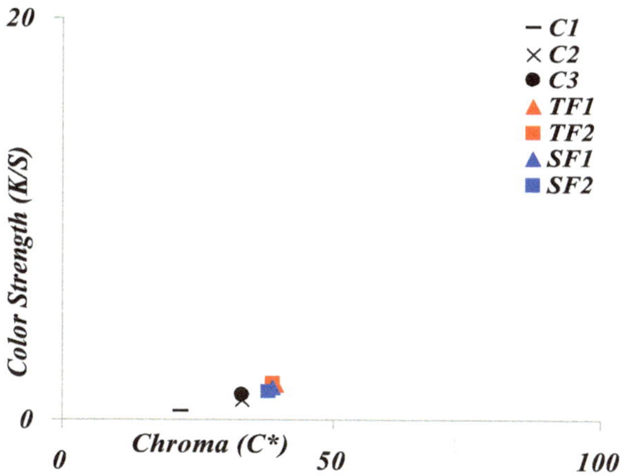

Fig. 3 Color strength (*K/S*) and chroma (*C**) outcomes of organic natural red ocher marble-printed samples without mordant inclusion

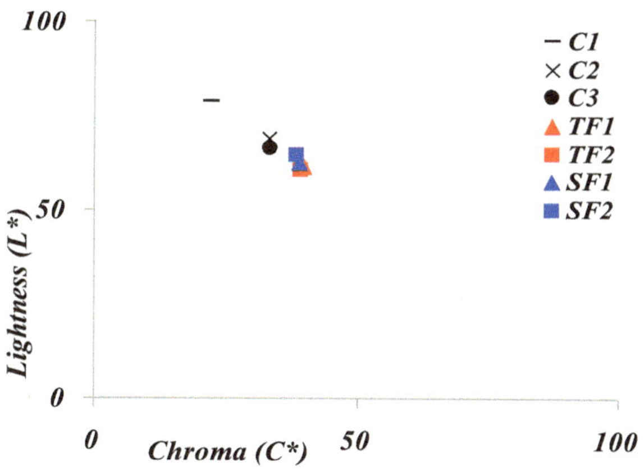

Fig. 4 Lightness (*L**) and chroma (*C**) values of organic natural red ocher marble-printed samples without mordant

Table 2 Colorimetric characteristics of organic natural red ocher marble-printed samples without mordant

Without mordant						
Fabrics	*K/S*	*L**	*a**	*b**	*h*°	*C**
C1	0.4	78.59	19.84	9.38	25.3	21.94
C2	1	68.86	29.94	14.2	25.4	33.15
C3	1.2	66.13	30.15	14.1	25.1	33.28
TF1	1.7	61.26	35.76	16.8	25.2	39.51
TF2	1.8	60.48	35.28	16.4	25	38.92
SF1	1.6	62.53	35.15	16.4	25.1	38.8
SF2	1.4	64.58	34.43	16.3	25.3	38.07

fabrics are in reddish tones (Table 2). This finding is in parallel with the reflectance curves (Fig. 1 and Table 2). When the a^*(*redness*) – b^*(*yellowness*) values (Fig. 2 and Table 2) of the organic natural red ocher marble-printed silk fabrics were examined, it was observed that nonfixed marble-printed silk fabrics (C1, C2, C3) exhibited the lowest $a^* - b^*$ values. Both a^* and b^* values increased after thermofixation and steamfixation in comparison with nonfixed counterparts, leading to yellower and redder appearances (Fig. 2 and Table 2). When the *K/S-Chroma* graph (Fig. 3) of organic natural red ocher marble-printed fabrics without mordant were examined, it was observed that the C^* values of marble-printed and then thermofixed (TF1 and TF2) and marble-printed and then steamfixed (SF1 and SF2) silk fabrics were higher than those of nonfixed marble-printed silk fabrics (C1, C2, C3) (Table 2 and Figs. 3 and 4). When the $L^* - C^*$ graph (Fig. 4) of silk fabrics marble-printed with organic natural red ocher without using mordant were examined, it was observed that the L^* (lightness) value changed according to the processing conditions and displayed a decreasing trend, leading to darker color shades. The colorimetric results of thermofixed (TF1 and TF2) and steamfixed (SF1 and SF2) silk fabrics marble-printed with organic natural red ocher without mordant were comparable and more or less similar.

Overall, as can be seen from Figs. 1, 2, 3, and 4 and Table 2, marble-printed silk fabrics with organic natural red ocher subjected to thermofixation (TF1 and TF2) and steamfixation (SF1 and SF2) after-treatments exhibited higher color strength (*K/S*), higher chroma (C^*), lower lightness (L^*), and higher a^* (redder appearance) and b^* (yellower appearance) values when compared with nonfixed marble-printed silk fabrics (C1, C2, C3). In other words, the thermofixation or steamfixation process after marble printing with organic natural red ocher without using mordant caused higher color strength, darker colors, higher chroma, and redder and yellower color shades when compared with nonfixed marble-printed silk fabrics with organic natural red ocher without using mordant.

3.1.1.2 Colorimetric Properties of Synthetic Red Ink Marble-Printed Silk Fabric Samples Without Mordant

The colorimetric properties of silk fabrics marble-printed with synthetic red ink are shown in Figs. 5, 6, 7, and 8 and Table 3.

From Tables 2 and 3, synthetic red ink marble-printed silk fabrics without any mordant displayed significantly higher color strength, higher chroma (C^*), and higher a^* and b^* values than organic natural red ocher marble-printed silk fabrics without any mordant leading to darker color shades. As in the case of marble printing with organic natural red ocher without using mordant, marble-printed silk fabrics with synthetic red ink (without mordant inclusion) subjected to thermofixation (TF1 and TF2) and steamfixation (SF1 and SF2) after-treatments exhibited higher color strength (*K/S*), higher chroma (C^*), lower lightness (L^*), and higher a^* (redder appearance) and b^* (yellower appearance) values when compared with nonfixed marble-printed silk fabrics with synthetic red ink (without mordant) (C1, C2, C3)

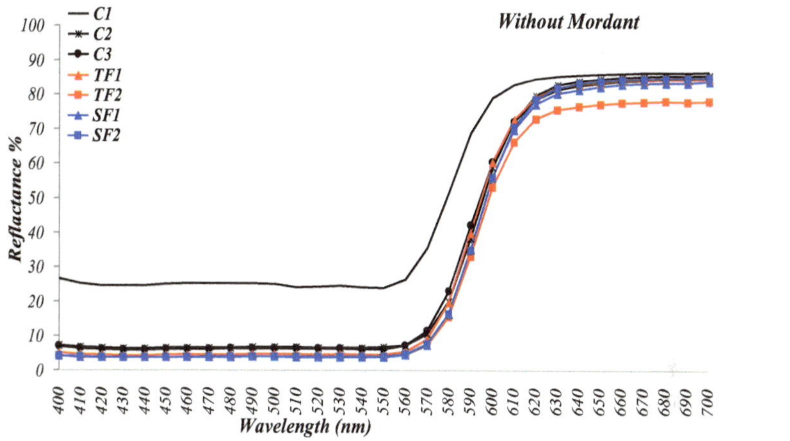

Fig. 5 Reflectance curves of synthetic red ink marble-printed silk fabric samples without mordant

Fig. 6 Yellowness (b^*) and redness (a^*) values of synthetic red ink marble-printed silk fabric samples without mordant

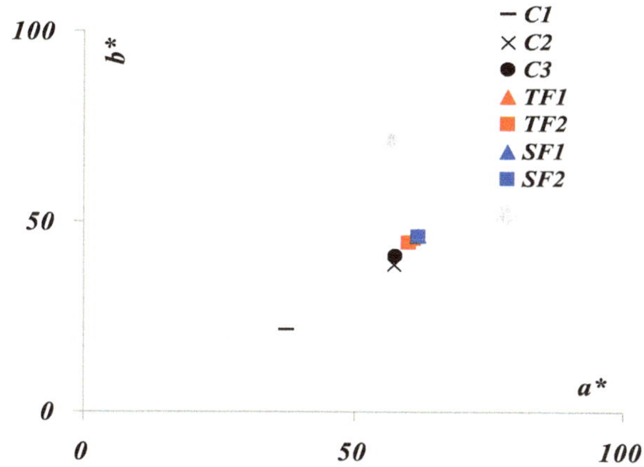

Fig. 7 Color strength (K/S) and chroma (C^*) outcomes of synthetic red ink marble-printed silk fabric samples without mordant

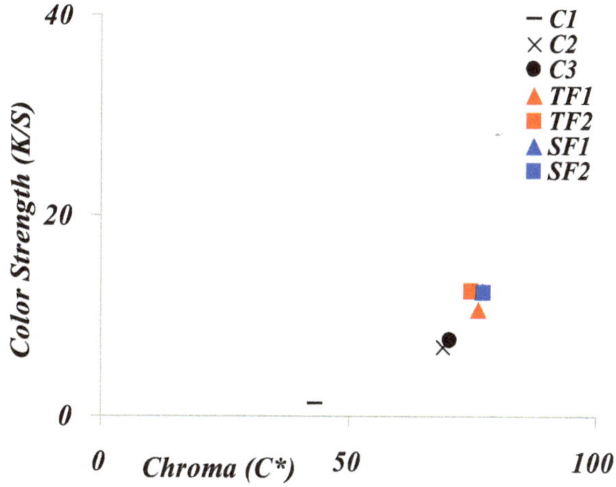

Fig. 8 Lightness (L^*) and chroma (C^*) values of synthetic red ink marble-printed silk fabric samples without mordant

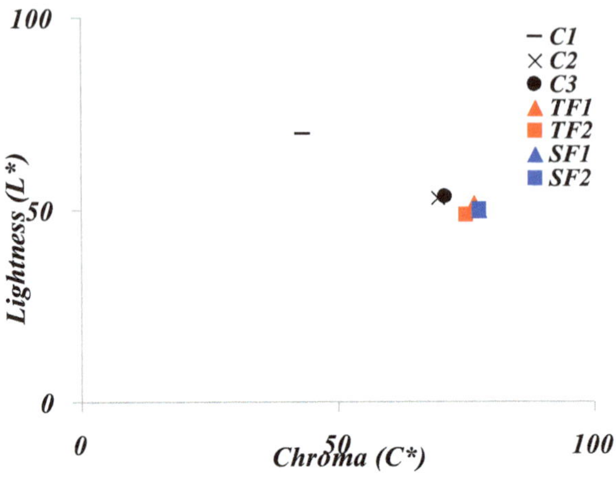

Table 3 Colorimetric properties of synthetic red ink marble-printed silk fabric samples without mordant

Without mordant						
Fabrics	*K/S*	*L**	*a**	*b**	*h*°	*C**
C1	1.2	69.4	37.51	21.6	29.9	43.26
C2	6.9	52.98	57.47	38.7	33.9	69.27
C3	7.5	53.29	57.61	40.8	35.3	70.62
TF1	10.6	51.59	61.05	45.7	36.8	76.27
TF2	12.5	48.74	60.12	44.5	36.5	74.77
SF1	12.5	49.78	61.85	46.3	36.8	77.23
SF2	12.3	50.08	61.91	46.2	36.7	77.26

(Table 3 and Figs. 5, 6, 7, and 8). Like silk fabrics marble-printed with organic natural red ocher without using mordant (Table 2 and Figs. 1, 2, 3, and 4), all marble-printed silk fabrics with synthetic red ink (without mordant inclusion) displayed red color shades (Table 3 and Fig. 6). So, all $h°$ (hue angle) values of marble-printed silk fabrics with synthetic red ink (without mordant inclusion) were smaller than 90°, and this indicates that the color shades of these fabrics are in reddish tones (Table 3). Both the thermofixation and steamfixation processes after marble printing with synthetic red ink without using mordant caused higher color strength, darker colors, higher chroma, and redder and yellower color shades when compared with nonfixed marble-printed silk fabrics with synthetic red ink (Table 3 and Figs. 5, 6, 7, and 8). Here, TF2, SF1, and SF2 silk fabrics marble-printed with synthetic red ink displayed the highest color strength values of 12.5, 12.5, and 12.3, respectively (Table 3). So thermofixation at 150 °C for 4 minutes and steamfixation at 102 °C for 15 and 30 minutes after the fixation processes caused the highest color strength levels on silk fabrics marble-printed with synthetic red ink (without any mordant) (Table 3 and Fig. 7).

3.1.2 Colorimetric Properties of Marble-Printed (Ebru) Silk Fabric Samples Premordanted with Potassium Bichromate

3.1.2.1 Colorimetric Properties of Organic Natural Red Ocher Marble-Printed Silk Fabric Samples Premordanted With Potassium Bichromate

The colorimetric data obtained from marble-printed silk fabrics premordanted with potassium bichromate are shown in Figs. 9, 10, 11, and 12 and Table 4.

When the reflectance curves of the silk fabrics premordanted with potassium bichromate and then marble-printed with organic natural red ocher were examined, the reflectance curves of the silk fabrics were close to each other with one exception (control fabric C1) (Fig. 9). All the treated silk fabrics displayed red color shades. So, the $h°$ (hue angle) values of all marble-printed silk fabrics were smaller than 90°, and this indicates that the color shades of these fabrics are in reddish tones (Table 4). This finding is in line with the reflectance curves (Fig. 9 and Table 4). When the a^**(redness)* – b^**(yellowness)* values (Fig. 10 and Table 4) of silk fabrics premordanted with potassium bichromate and then marble-printed with organic natural red ocher were examined, it was observed that nonfixed marble-printed silk fabrics (C1 and C2) exhibited the lowest $a^* - b^*$ values. Both a^* and b^* values were increased after both thermofixation and steamfixation in comparison with non-fixed counterparts leading to yellower and redder appearances (Fig. 10 and Table 4). When the *K/S-Chroma* graph (Fig. 11) of organic natural red ocher marble-printed silk fabrics premordanted with potassium bichromate were examined, it was observed that the C^* values of marble-printed and then thermofixed (TF1 and TF2) and marble-printed and then steamfixed (SF1 and SF2) silk fabrics were higher than nonfixed marble-printed silk fabrics (C1, C2, C3) (Table 4 and Figs. 11 and 12). The

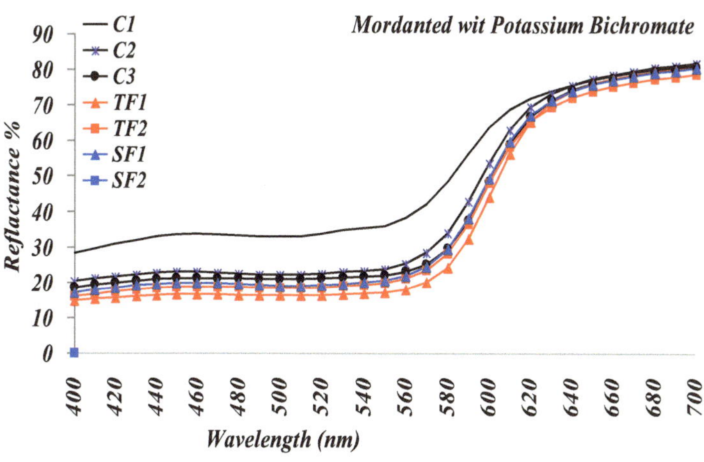

Fig. 9 Reflectance curves of organic natural red ocher marble-printed silk fabric samples premordanted with potassium bichromate

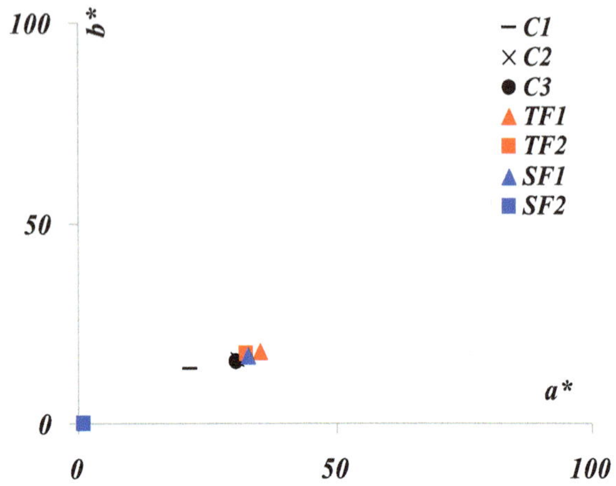

Fig. 10 Yellowness (b^*) and redness (a^*) values of organic natural red ocher marble-printed silk fabric samples premordanted with potassium bichromate

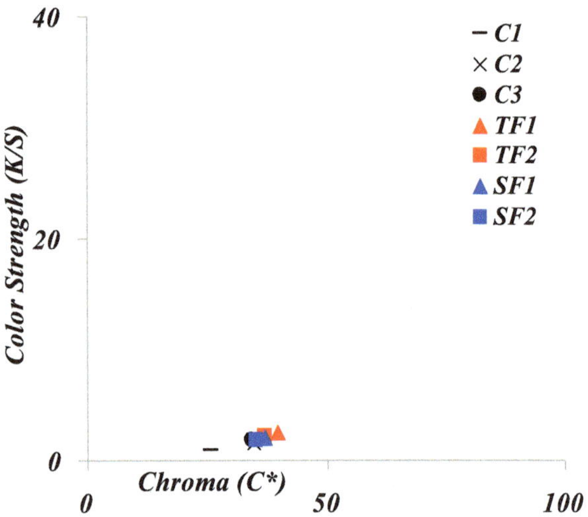

Fig. 11 Color strength (K/S) and chroma (C^*) values of organic natural red ocher marble-printed silk fabric samples premordanted with potassium bichromate

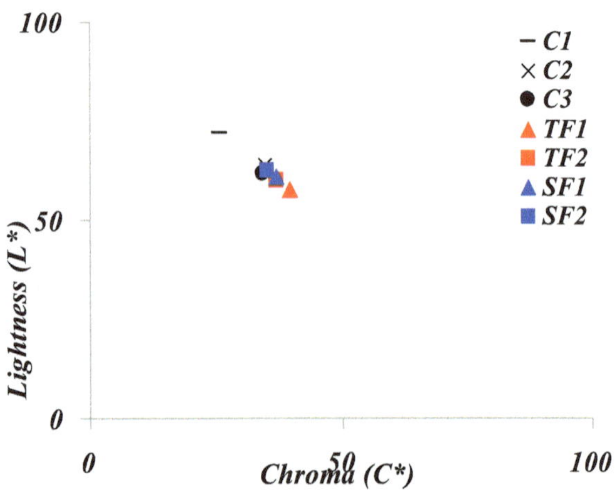

Fig. 12 Lightness (L^*) and chroma (C^*) values of organic natural red ocher marble-printed silk fabric samples premordanted with potassium bichromate

Table 4 Colorimetric properties of organic natural red ocher marble-printed silk fabric samples premordanted with potassium bichromate

Premordanted with potassium bichromate						
Fabrics	*K*/*S*	*L**	*a**	*b**	*h*°	*C**
C1	0.9	71.86	21.77	13.7	32.13	25.7
C2	1.6	63.75	30.79	16.1	27.54	34.7
C3	1.8	61.64	30.54	15.6	27.01	34.3
TF1	2.5	57.68	35.36	17.9	26.87	39.6
TF2	2.2	60.01	32.63	17.5	28.15	37
SF1	2	60.89	32.94	16.9	27.11	37
SF2	1.8	62.47	30.9	16.7	28.32	35.1

colorimetric results of thermofixed (TF1 and TF2) and steamfixed (SF1 and SF2) silk fabrics premordanted with potassium bichromate and then marble-printed with organic natural red ocher were comparable and close to each other. Overall, as can be seen from Figs. 9, 10, 11, and 12 and Table 4, silk fabrics premordanted with potassium bichromate and then marble-printed with organic natural red ocher subjected to thermofixation (TF1 and TF2) and steamfixation (SF1) after-treatments exhibited higher color strength (*K*/*S*), higher chroma (*C**), lower lightness (*L**), and higher *a** (redder appearance) and *b** (yellower appearance) values when compared with nonfixed marble-printed silk fabrics (C1, C2, C3). In other words, both thermofixation (TF1 and TF2) or steamfixation (SF1) after premordanting with potassium bichromate and then marble printing with organic natural red ocher caused higher color strength, darker colors, higher chroma, and redder and yellower color shades when compared with nonfixed silk fabrics but premordanted with potassium bichromate and marble-printed with organic natural red ocher.

3.1.2.2 Colorimetric Properties of Synthetic Red Ink Marble-Printed Silk Fabric Samples Premordanted With Potassium Bichromate

The colorimetric data of synthetic red ink marble-printed silk fabric samples premordanted with potassium bichromate are displayed in Figs. 13, 14, 15, and 16 and Table 5.

From Tables 4 and 5, silk fabrics premordanted with potassium bichromate and then marble-printed with synthetic red ink revealed significantly higher color strength, chroma (*C**), and *a** and *b** values than silk fabrics premordanted with potassium bichromate and then marble-printed with organic natural red ocher, leading to darker color shades. Silk fabrics premordanted with potassium bichromate and then marble-printed with synthetic red ink and then subjected to thermofixation (TF1 and TF2) after-treatments displayed higher color strength (*K*/*S*), higher chroma (*C**), lower lightness (*L**), and higher *a** (redder appearance) and *b** (yellower appearance) values when compared with their steamfixed (SF1 and SF2) and nonfixed (C1, C2, C3) counterparts (Table 5 and Figs. 13, 14, 15, and 16). Similar to

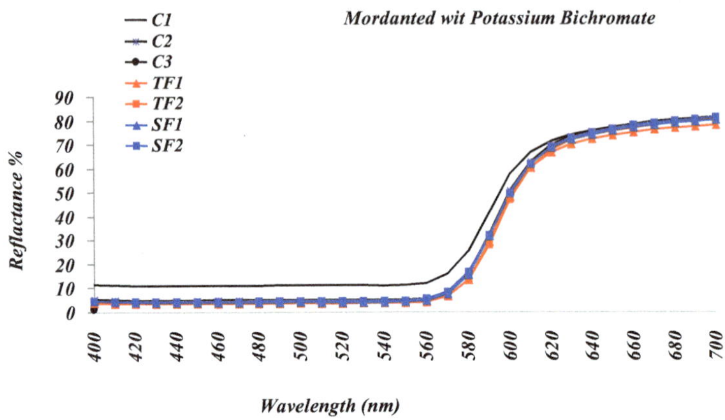

Fig. 13 Reflectance curves of silk fabric samples marble-printed with synthetic red ink and premordanted with potassium bichromate

Fig. 14 Yellowness (b^*) and redness (a^*) of silk fabrics marble-printed with synthetic red ink and premordanted with potassium bichromate

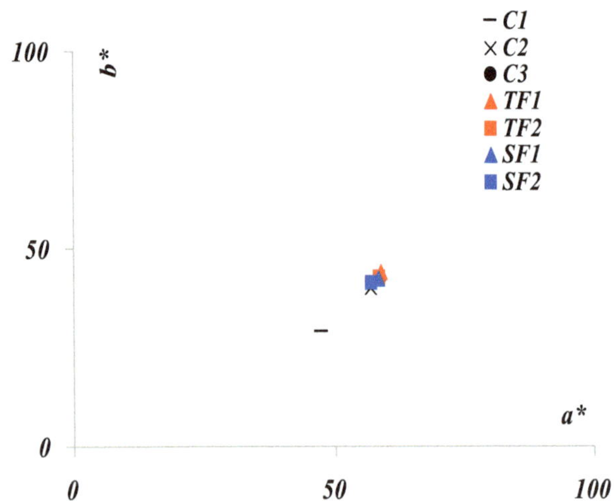

silk fabrics premordanted with potassium bichromate and then marble-printed with organic natural red ocher (Table 4 and Figs. 9, 10, 11, and 12), all silk fabrics premordanted with potassium bichromate and then marble-printed with synthetic red ink displayed red color shades (Table 4 and Fig. 14). So, all $h°$ (hue angle) values of silk fabrics premordanted with potassium bichromate and then marble-printed with synthetic red ink were smaller than 90°; this indicates that the color shades of these fabrics are in reddish tones (Table 5). Both thermofixation processes (TF1 and TF2) after premordanting with potassium bichromate and then marble printing with synthetic red ink caused higher color strength, darker colors, higher chroma, and redder and yellower color shades when compared with their steamfixed (SF1 and SF2) and nonfixed (C1, C2, C3) counterparts (Table 5 and Figs. 13, 14, 15, and 16). Here,

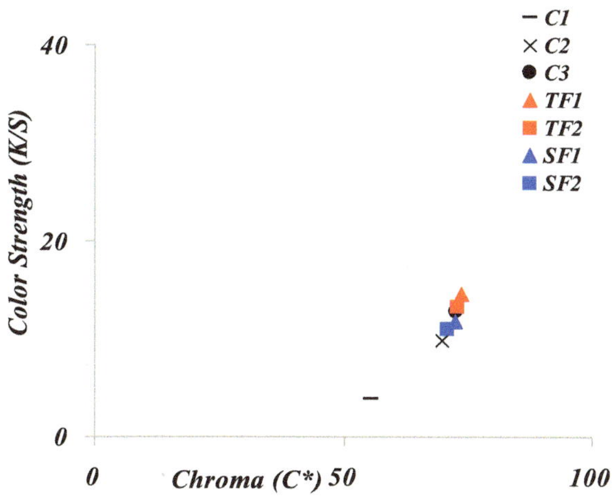

Fig. 15 Color strength (*K*/*S*) and chroma (*C**) values of silk fabric samples marble-printed with synthetic red ink and premordanted with potassium bichromate

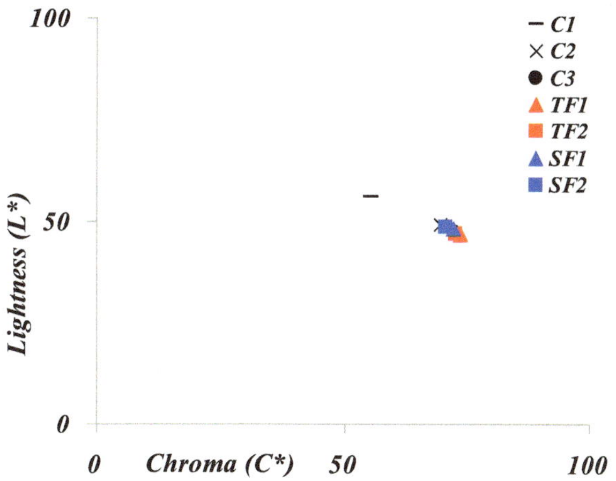

Fig. 16 Lightness (*L**) and chroma (*C**) values of silk fabric samples marble-printed with synthetic red ink and premordanted with potassium bichromate

Table 5 Colorimetric properties of synthetic red ink marble-printed silk fabric samples premordanted with potassium bichromate

Premordanted with potassium bichromate						
Fabrics	*K*/*S*	*L**	*a**	*b**	*h*°	*C**
C1	3.8	55.86	47.26	28.8	31.3	55.32
C2	9.8	49.18	56.84	40	35.1	69.5
C3	12.8	47.14	58.73	42.5	35.9	72.46
TF1	14.5	46.7	58.85	43.8	36.7	73.38
TF2	13.3	46.99	58.57	42.7	36.1	72.48
SF1	11.8	48.1	58.41	42.3	35.9	72.11
SF2	11	48.78	57.08	41.4	35.9	70.49

TF1- and TF2-treated silk fabrics premordanted with potassium bichromate and then marble-printed with synthetic red ink displayed the highest color strength values of 14.5 and 13.3, respectively (Table 5). So thermofixation at 130 °C for 4 minutes (TF1) and thermofixation at 150 °C for 4 minutes (TF2) resulted in the highest color strength levels among silk fabrics premordanted with potassium bichromate and then marble-printed with synthetic red ink (Table 5 and Fig. 15).

3.1.3 Colorimetric Properties of Marble-Printed (Ebru) Silk Fabric Samples Premordanted with Ferrous Sulfate (Iron (II) Sulfate)

3.1.3.1 Colorimetric Properties of Organic Natural Red Ocher Marble-Printed Silk Fabric Substrates Premordanted with Ferrous Sulfate (Iron (II) Sulfate)

The colorimetric properties of silk fabrics premordanted with ferrous sulfate (iron (II) sulfate) and then marble-printed with organic natural red ocher are displayed in Figs. 17, 18, 19, and 20 and Table 6.

The reflectance curves of silk fabrics premordanted with ferrous sulfate and then marble-printed with organic natural red ocher were close to each other with one exception (control fabric C1) (Fig. 17). All treated silk fabrics displayed red color shades. Therefore, all $h°$ (hue angle) values of all marble-printed silk fabrics were smaller than 90°; this means that the color shades of these fabrics are in reddish tones (Table 6). This finding is in parallel with the reflectance curves (Fig. 17 and Table 6). a^* values were increased after both thermofixation (TF1 and TF2) and SF1 steamfixation in comparison with nonfixed counterparts, leading to redder appearances (Fig. 18 and Table 6). When the *K/S-Chroma* graph (Fig. 19) of organic natural red ocher marble-printed silk fabrics premordanted with ferrous sulfate were examined, it was monitored that the chroma (C^*) values of marble-printed and then

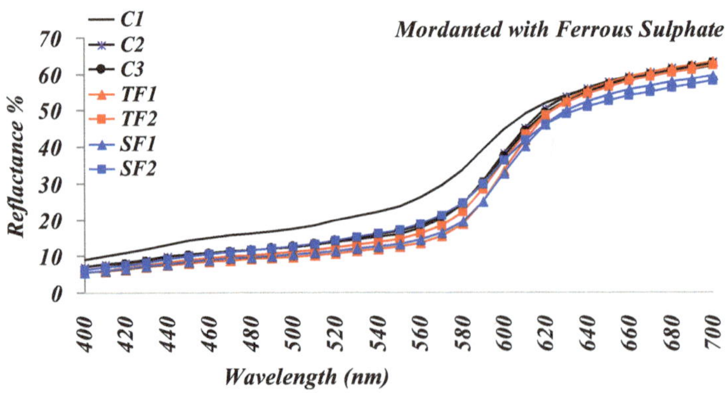

Fig. 17 Reflectance curves of organic natural red ocher marble-printed silk fabric samples premordanted with ferrous sulfate

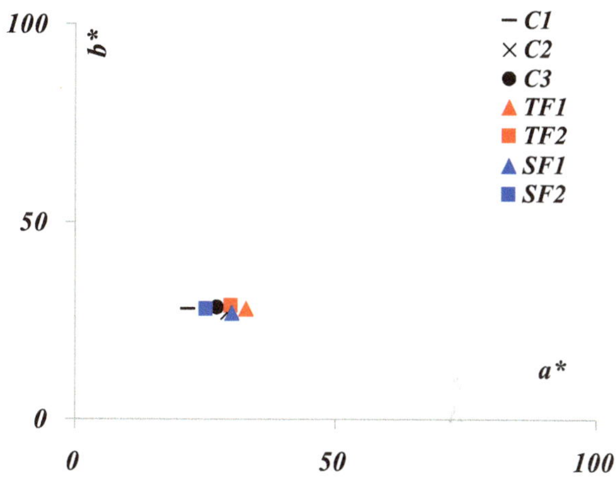

Fig. 18 Yellowness (b^*) and redness (a^*) values of organic natural red ocher marble-printed silk fabric samples premordanted with ferrous sulfate

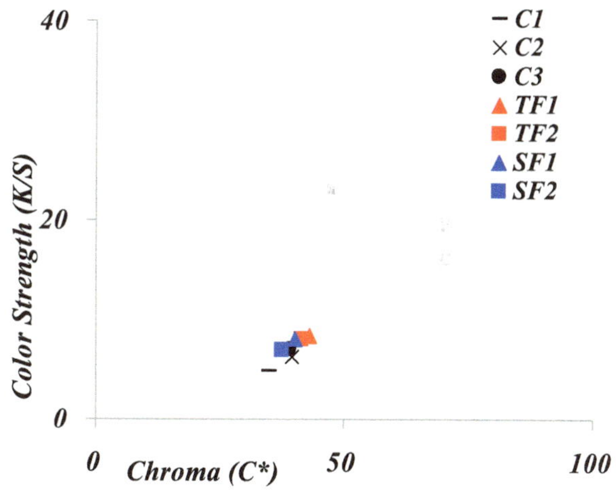

Fig. 19 Color strength (K/S) and chroma (C^*) outcomes of organic natural red ocher marble-printed silk fabric substrates premordanted with ferrous sulfate

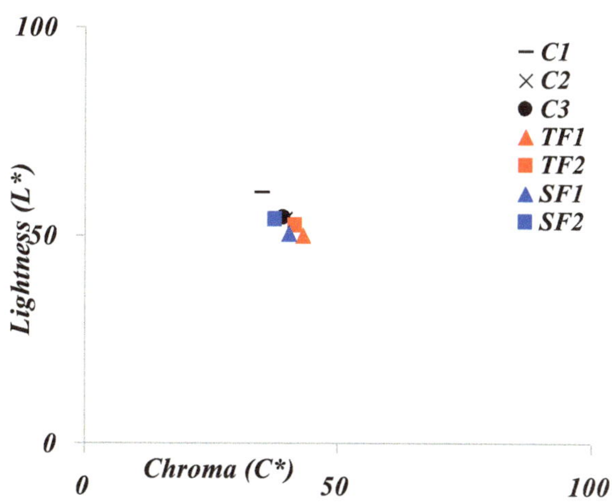

Fig. 20 Lightness (L^*) and Chroma (C^*) values of organic natural red ocher marble printed silk fabric substrates premordanted with ferrous sulfate

Table 6 Colorimetric properties of organic natural red ocher marble-printed silk fabric samples premordanted with ferrous sulfate

Premordanted with ferrous sulfate						
Fabrics	*K/S*	*L**	*a**	*b**	*h°*	*C**
C1	4.7	60.14	21.78	27.6	51.68	35.1
C2	6.2	53.97	29.2	26.9	42.62	39.7
C3	6.9	54.09	27.35	28.3	45.93	39.3
TF1	8.3	49.99	32.87	27.9	40.32	43.1
TF2	8	52.26	29.96	28.8	43.84	41.5
SF1	8	50.25	30.05	26.9	41.81	40.3
SF2	7	53.76	25.22	27.9	47.93	37.6

thermofixed (TF1 and TF2) and marble-printed and then SF1 steamfixed silk fabrics were higher than nonfixed marble-printed silk fabrics (C1, C2, C3) (Table 2 and Figs. 3 and 4). The colorimetric results of thermofixed (TF1 and TF2) and SF1 steamfixed silk fabrics premordanted with ferrous sulfate and then marble-printed with organic natural red ocher were comparable and very close to each other. Overall, as can be seen from Figs. 17, 18, 19, and 20 and Table 6, silk fabrics premordanted with ferrous sulfate and then marble-printed with organic natural red ocher that were subjected to thermofixation (TF1 and TF2) and SF1 steamfixation after-treatments exhibited higher color strength (*K/S*), higher chroma (*C**), lower lightness (*L**), and higher *a** (redder appearance) values when compared with nonfixed marble-printed silk fabrics (C1, C2, C3). In other words, both thermofixation (TF1 and TF2) and steamfixation (SF1) after premordanting with ferrous sulfate and then marble printing with organic natural red ocher caused higher color strength, darker colors, higher chroma, and redder color shades when compared with nonfixed silk fabrics but premordanted with ferrous sulfate and then marble-printed with organic natural red ocher.

3.1.3.2 Colorimetric Properties of Synthetic Red Ink Marble-Printed Silk Fabric Samples Premordanted With Ferrous Sulfate (Iron (II) Sulfate)

The colorimetric data of synthetic red ink marble-printed silk fabric samples premordanted with ferrous sulfate (*iron (II) sulfate*) are presented in Figs. 21, 22, 23, and 24 and Table 7.

From Tables 6 and 7, silk fabrics premordanted with ferrous sulfate and marble-printed with synthetic red ink demonstrated higher color strength (*K/S*) than silk fabrics premordanted with ferrous sulfate and then marble-printed with organic natural red ocher. Silk fabrics premordanted with ferrous sulfate and then marble-printed with synthetic red ink and then subjected to thermofixation (TF1 and TF2) and steamfixation (SF1 and SF2) displayed higher color strength (*K/S*) and lower lightness (*L**) when compared with their nonfixed (C1, C2, C3) counterparts (Table 7 and Figs. 21, 22, 23, and 24). Like silk fabrics premordanted with ferrous sulfate and then marble-printed with organic natural red ocher (Table 6 and Figs. 17, 18, 19, and

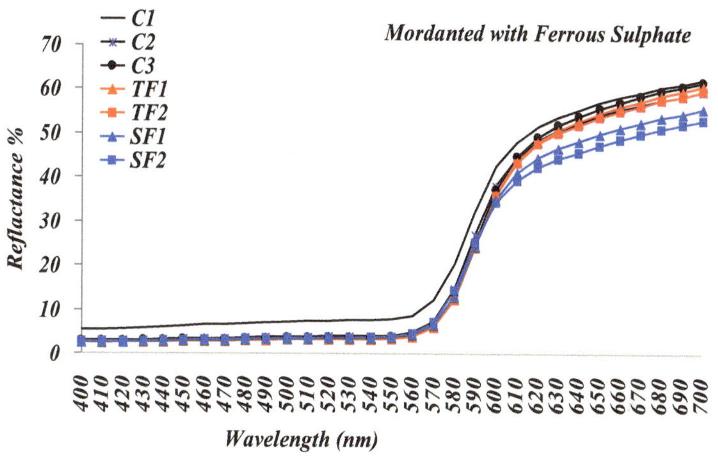

Fig. 21 Reflectance curves of synthetic red ink marble-printed silk fabric substrates premordanted with ferrous sulfate

Fig. 22 Yellowness (b^*) and redness (a^*) values of synthetic red ink marble-printed silk fabric substrates premordanted with ferrous sulfate

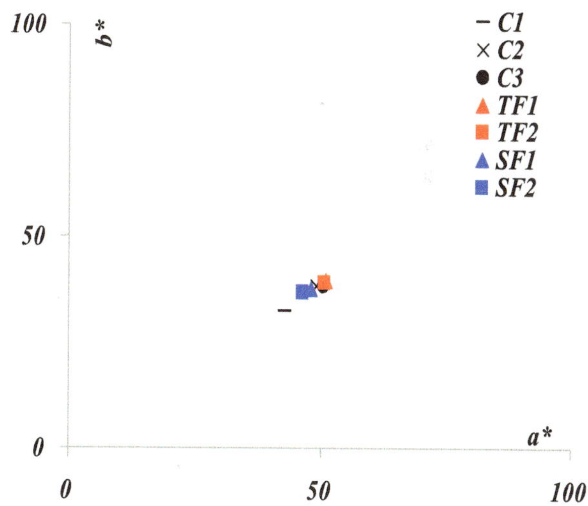

20), all silk fabrics premordanted with ferrous sulfate and then marble-printed with synthetic red ink displayed red color shades (Table 7 and Fig. 22). So, all $h°$ (hue angle) values of silk fabrics premordanted with ferrous sulfate and then marble-printed with synthetic red ink were smaller than 90°; this indicates that the color shades of these fabrics are in reddish tones (Table 7). Both thermofixation (TF1 and TF2) and steamfixation (SF1 and SF2) processes after premordanting with ferrous sulfate and then marble printing with synthetic red ink led to higher color strength when compared with their nonfixed (C1, C2, C3) counterparts (Table 7 and Fig. 23). Here, TF1-, TF2-, SF1-, and SF2-treated silk fabrics premordanted with ferrous sulfate and then marble-printed with synthetic red ink displayed the highest color strength values of 19.7, 19, 18.2, and 17.7, respectively (Table 7).

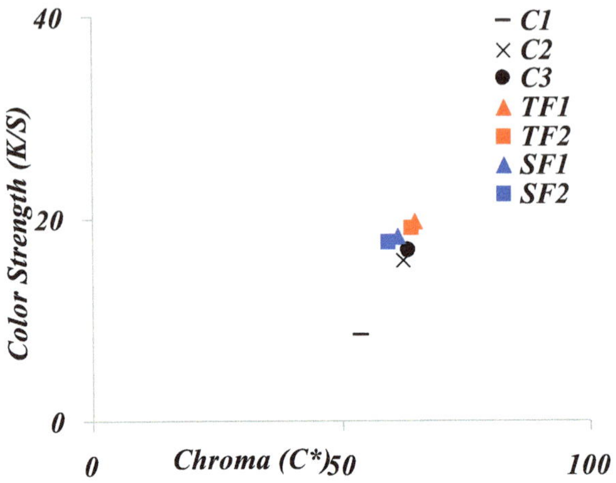

Fig. 23 Color strength (*K*/*S*) and chroma (*C**) values of synthetic red ink marble-printed silk fabric samples premordanted with ferrous sulfate

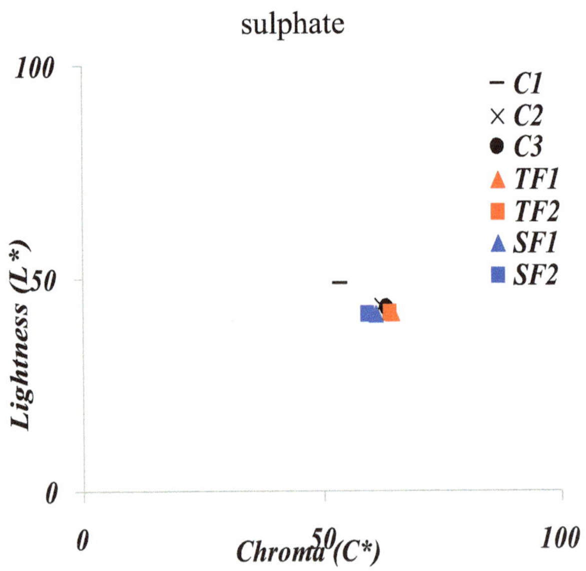

Fig. 24 Lightness (*L**) and chroma (*C**) values of synthetic red ink marble-printed silk fabric samples premordanted with ferrous sulfate

Table 7 Colorimetric properties of synthetic red ink marble-printed silk fabric samples premordanted with ferrous sulfate

Premordanted with ferrous sulfate						
Fabrics	*K*/*S*	*L**	*a**	*b**	*h*°	*C**
C1	8.5	48.63	42.86	32.1	36.8	53.55
C2	15.8	43.38	49.01	38.2	37.9	62.12
C3	16.8	42.75	50.35	38.2	37.2	63.18
TF1	19.7	41.79	50.98	39.5	37.8	64.47
TF2	19	41.72	50.6	39	37.6	63.86
SF1	18.2	41.56	47.99	37.6	38.1	60.96
SF2	17.7	41.55	46.31	36.9	38.6	59.21

3.1.4 Colorimetric Properties of Marble-Printed (Ebru) Silk Fabric Samples Premordanted with Gallnut

3.1.4.1 Colorimetric Properties of Organic Natural Red Ocher Marble-Printed Silk Fabric Samples Premordanted with Gallnut (Natural Mordant)

The colorimetric properties of silk fabrics premordanted with gallnut (natural mordant) and then marble-printed with organic natural red ocher are displayed in Figs. 25, 26, 27, and 28 and Table 8.

All silk fabrics premordanted with gallnut (natural mordant) and then marble-printed with organic natural red ocher displayed red color shades (Table 8). a^* (redness) values were increased after both thermofixation (TF1 and TF2) and steamfixation (SF1 and SF2) in comparison with nonfixed counterparts, leading to

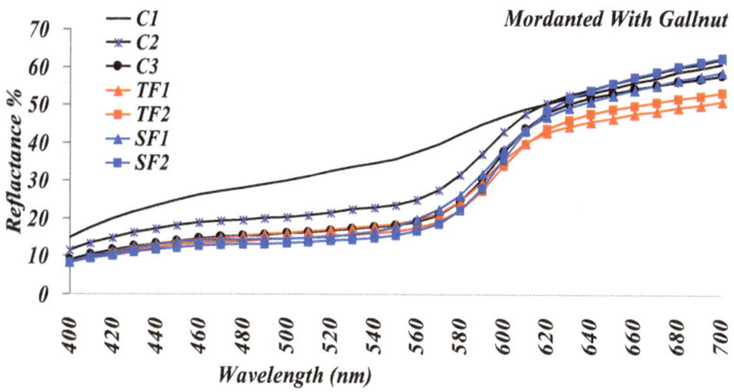

Fig. 25 Reflectance curves of organic natural red ocher marble-printed silk fabric samples premordanted with gallnut

Fig. 26 Yellowness (b^*) and redness (a^*) values of organic natural red ocher marble-printed silk fabric samples premordanted with gallnut

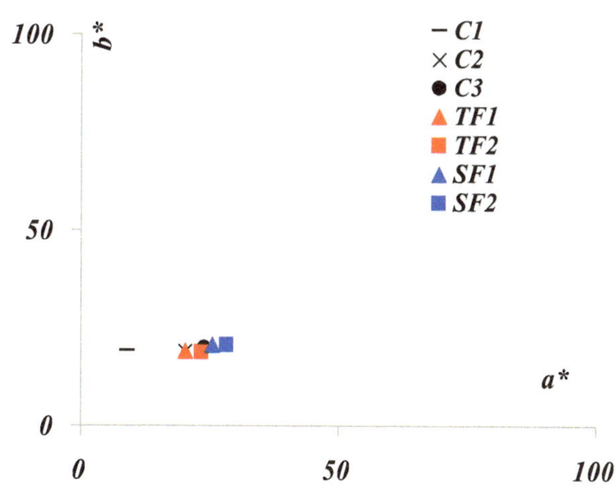

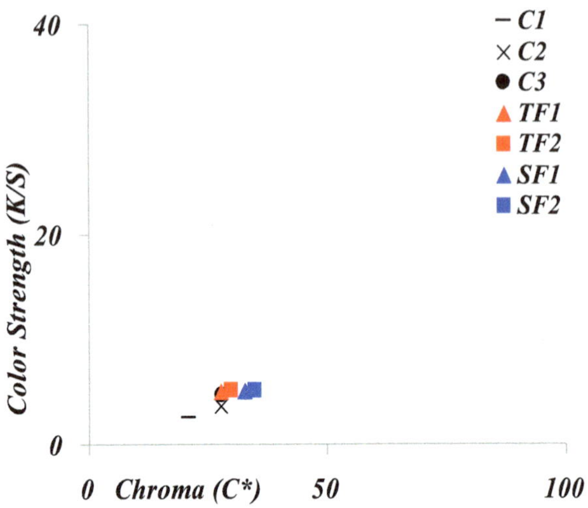

Fig. 27 Color strength (*K/S*) and chroma (*C**) values of organic natural red ocher marble-printed silk fabric samples premordanted with gallnut

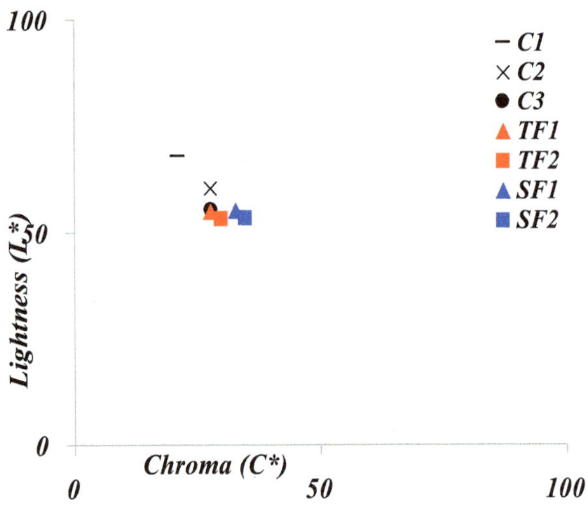

Fig. 28 Lightness (*L**) and chroma (*C**) values of organic natural red ocher marble-printed silk fabric samples premordanted with gallnut

Table 8 Colorimetric properties of organic natural red ocher marble-printed silk fabric samples premordanted with gallnut

Premordanted with gallnut						
Fabrics	*K/S*	*L**	*a**	*b**	*h*°	*C**
C1	2.5	67.75	8.94	19	64.81	21
C2	3.6	60.36	20.2	19.1	43.37	27.8
C3	4.6	55.19	20.01	19.5	44.27	28
TF1	4.9	54.82	20.22	19.1	43.39	27.8
TF2	5.1	53.21	23.25	18.9	39.08	30
SF1	5	55.21	25.51	20.6	38.91	32.8
SF2	5.1	53.27	28.09	20.7	36.35	34.9

redder appearances (Fig. 26 and Table 8). The colorimetric results of thermofixed (TF1 and TF2) and steamfixed (SF1 and SF2) silk fabrics premordanted with gallnut and then marble-printed with organic natural red ocher were comparable and very close to each other (Table 8 and Figs. 25, 26, 27, and 28). Overall, as can be seen from Figs. 25, 26, 27, and 28 and Table 8, silk fabrics premordanted with gallnut and then marble-printed with organic natural red ocher subjected to thermofixation (TF1 and TF2) and steamfixation (SF1 and SF2) displayed higher color strength (*K*/*S*) and higher a^* (redder appearance) values when compared with their nonfixed counterparts (C1, C2, C3).

3.1.4.2 Colorimetric Properties of Synthetic Red Ink Marble-Printed Silk Fabric Samples Premordanted with Gallnut (Natural Mordant)

The colorimetric properties of synthetic red ink marble-printed silk fabric samples premordanted with gallnut are exhibited in Figs. 29, 30, 31, and 32 and Table 9.

From Tables 8 and 9, silk fabrics premordanted with gallnut and then marble-printed with synthetic red ink demonstrated higher color strength (*K*/*S*) than silk fabrics premordanted with gallnut and then marble-printed with organic natural red ocher. Silk fabrics premordanted with gallnut and then marble-printed with synthetic red ink and then subjected to thermofixation (TF1 and TF2) and steamfixation (SF1) displayed higher color strength (*K*/*S*) and lower lightness (L^*) compared with their nonfixed (C1, C2, C3) counterparts (Table 9 and Figs. 31 and 32). Similar to silk fabrics premordanted with gallnut and then marble-printed with organic natural red ocher (Table 8 and Figs. 25, 26, 27, and 28), all silk fabrics premordanted with gallnut and then marble-printed with synthetic red ink displayed red color shades

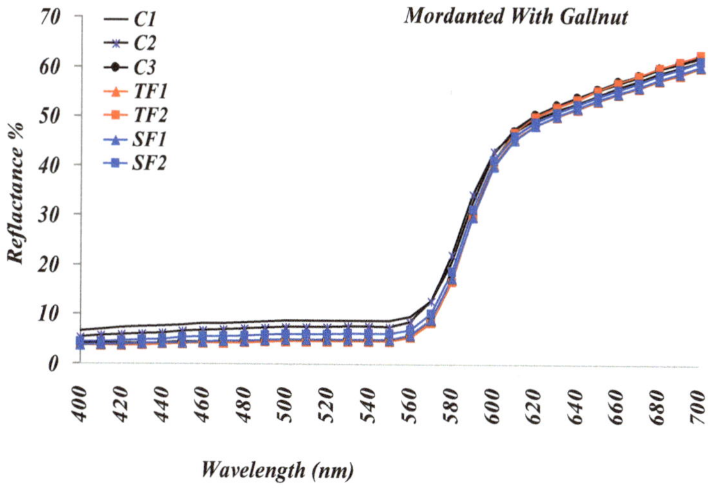

Fig. 29 Reflectance values of synthetic red ink samples mordanted with gallnut

Fig. 30 Yellowness (b^*) and redness (a^*) synthetic red ink values of samples mordanted with gallnut

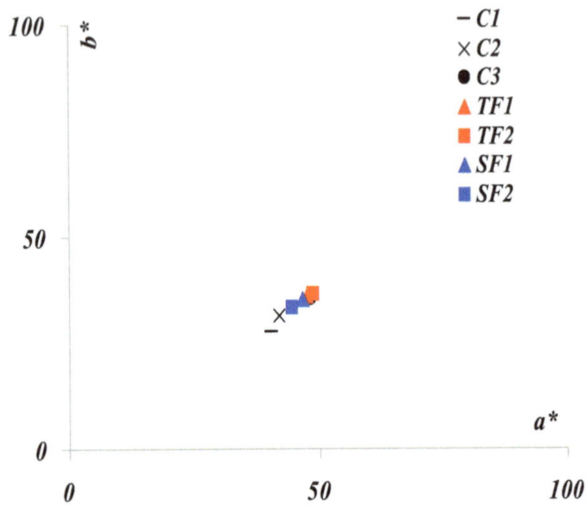

Fig. 31 Color strength (K/S) and chroma (C^*) synthetic red ink outcomes of samples with gallnut

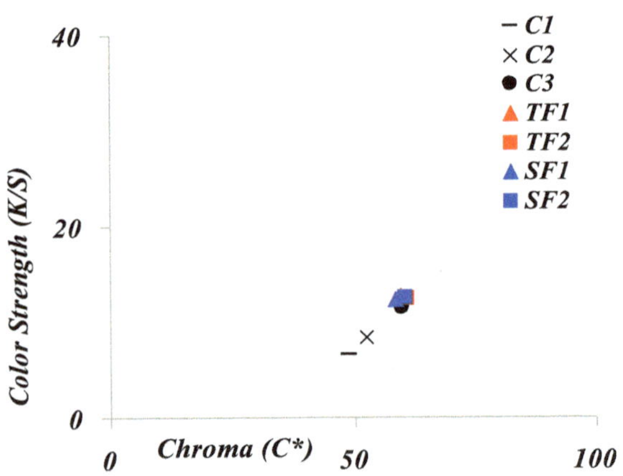

(Table 9 and Fig. 30). Both thermofixation (TF1 and TF2) and SF1 steamfixation after premordanting with gallnut and then marble printing with synthetic red ink led to higher color strength (K/S) when compared with their nonfixed (C1, C2, C3) counterparts (Table 9 and Fig. 31). Here, TF1-, TF2-, and SF1-treated silk fabrics premordanted with gallnut and then marble-printed with synthetic red ink displayed the highest color strength values of 12.7, 12.4, and 12.3, respectively (Table 9).

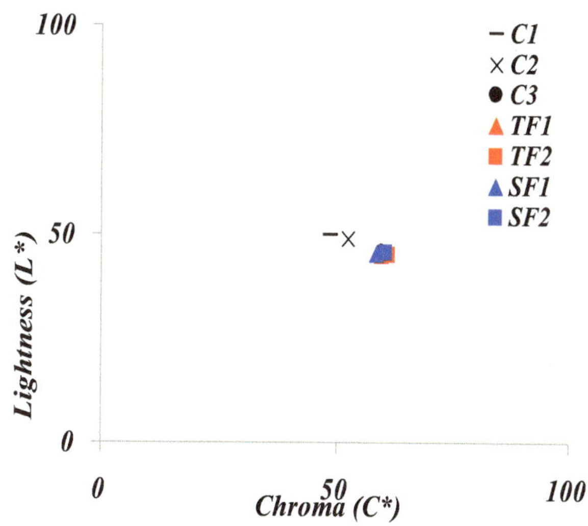

Fig. 32 Lightness (*L**) and chroma (*C**) synthetic red ink values of samples mordanted with gallnut

Table 9 Colorimetric properties of synthetic red ink marble-printed silk fabric samples premordanted with gallnut

Premordanted with gallnut						
Fabrics	*K/S*	*L**	*a**	*b**	*h*°	*C**
C1	6.6	49.59	40.27	27.5	34.3	48.74
C2	8.4	48.84	41.87	31.4	36.9	52.33
C3	11.5	45.43	48.07	35.5	36.4	59.76
TF1	12.7	44.89	47.37	35.9	37.1	59.42
TF2	12.4	45.03	48.64	36.5	36.9	60.82
SF1	12.3	45.18	46.6	35.2	37.1	58.4
SF2	10.5	46.73	44.55	33.3	36.8	55.64

3.2 *Color Fastness Properties of Marble-Printed (Ebru) Silk Fabric Samples*

3.2.1 Color Fastness Properties of Marble-Printed Silk Fabric Samples Without Mordant

Color fastness (washing, light, rub, water, and perspiration fastness) properties of unmordanted silk fabrics printed with both colorants are shown in Tables 10, 11, 12, 13, and 14.

As can be seen from Table 10, the washing fastness outcomes of all marble-printed silk substrates were between 4–5 and 5 gray scale levels, which are commercially acceptable levels. All dry rub fastness outcomes were in the range of 2–3 to 4 gray-scale levels. Moreover, wet rub fastness outcomes were generally slightly lower than dry rub fastness values. Especially wet rubbing fastness results of fabrics

Table 10 Washing (WF), light (LF), and rubbing (dry rub and wet rub) fastness properties of unmordanted organic natural red ocher dye marble-printed silk fabric samples

Without mordant										
			Rub fastness		Washing fastness					
Fabrics	*K/S*	LF	Dry	Wet	WO	PC	PES	N6.6	CO	AC
C1	0.4	3	4	4–5	5	5	5	5	5	5
C2	1	5	3–4	3–4	5	5	5	5	5	5
C3	1.2	6	2–3	1–2	5	5	5	4–5	4–5	4–5
TF1	1.7	6	2–3	1–2	5	4–5	4–5	4–5	4–5	4–5
TF2	1.8	6	2–3	2	5	4–5	4–5	4–5	4–5	4–5
SF1	1.6	6	2–3	1–2	5	4–5	4–5	4–5	4–5	4–5
SF2	1.4	6	3–4	3	5	4–5	4–5	4–5	4–5	4–5

Table 11 Light (LF), rub (dry and wet), and wash fastness characteristics of unmordanted synthetic red ink marble-printed silk fabric samples

Without mordant										
			Rub fastness		Washing fastness					
Fabrics	*K/S*	LF	Dry	Wet	WO	PC	PES	N6.6	CO	AC
C1	1.2	3	4–5	4–5	5	5	5	5	5	5
C2	6.9	5	4–5	3–4	5	5	5	5	4–5	5
C3	7.5	5	4–5	3–4	5	5	5	5	5	5
TF1	10.6	5	4–5	3–4	5	5	5	5	5	5
TF2	12.5	6	4–5	3–4	5	5	5	5	4–5	5
SF1	12.5	5	4–5	3–4	5	5	5	5	5	5
SF2	12.3	6	4–5	4	5	5	5	5	5	5

Table 12 Water fastness properties of organic natural red ocher and synthetic red ink marble-printed samples without mordant

Water fastness														
	Without mordant (organic natural red ocher)							Without mordant (synthetic red ink)						
	K/S	WO	PC	PES	N6.6	CO	AC	*K/S*	WO	PC	PES	N6.6	CO	AC
C1	0.4	5	5	5	5	5	5	1.2	5	5	5	5	5	5
C2	1	5	5	5	5	5	5	6.9	5	5	5	5	5	5
C3	1.2	5	5	5	5	5	5	7.5	5	5	5	5	5	5
TF1	1.7	5	5	5	5	5	5	10.6	5	5	5	5	5	5
TF2	1.8	5	5	5	5	5	5	12.5	5	5	5	5	5	5
SF1	1.6	5	5	5	5	5	5	12.5	5	5	5	5	5	5
SF2	1.4	5	5	5	5	5	5	12.3	5	5	5	5	5	5

(fabrics after treated with immediate rinsing, 24 hours storage at ambient conditions than rinsing and steamfixed at 102 °C for 30 minutes than rinsing) exhibited 1–2 and 4–5 wet rub fastness levels, respectively. All light fastness values were in the range of 3–6 levels. The light fastness value of fabrics after being treated with

Table 13 Acidic and alkaline perspiration fastness characteristics of organic natural red ocher marble-printed silk fabric samples without mordant existence

Without mordant													
		Acidic perspiration fastness						Alkaline perspiration fastness					
Fabrics	*K/S*	WO	PC	PES	N6.6	CO	AC	WO	PC	PES	N6.6	CO	AC
C1	0.4	5	5	5	5	5	5	5	5	5	5	5	5
C2	1	5	5	5	5	5	5	5	5	5	5	5	5
C3	1.2	5	5	5	4/5	4/5	4/5	5	5	5	5	5	5
TF1	1.7	5	5	5	4/5	4/5	4/5	5	5	5	5	5	5
TF2	1.8	4/5	4/5	4/5	4/5	4/5	4/5	5	5	5	5	5	5
SF1	1.6	5	5	5	4/5	4/5	4/5	5	5	5	5	5	5
SF2	1.4	5	5	5	5	5	5	5	5	5	5	5	5

Table 14 Acidic and alkaline perspiration fastness properties of synthetic red ink marble-printed silk fabric samples without mordant

Without mordant													
		Acidic perspiration fastness						Alkaline perspiration fastness					
Fabrics	*K/S*	WO	PC	PES	N6.6	CO	AC	WO	PC	PES	N6.6	CO	AC
C1	1.2	5	5	5	5	5	5	5	5	5	5	5	5
C2	6.9	5	5	5	5	5	5	5	5	5	5	5	5
C3	7.5	5	5	5	5	5	5	5	5	5	5	5	5
TF1	10.6	5	5	5	5	5	5	5	5	5	5	5	5
TF2	12.5	5	5	5	5	5	5	5	5	5	5	5	5
SF1	12.5	5	5	5	5	5	5	5	5	5	5	5	5
SF2	12.3	5	5	5	5	5	5	5	5	5	5	5	5

immediate rinsing was the lowest, which was 3. C3, TF1, TF2, SF1, and SF2 led to quite high light fastness level of 6 in accordance with blue wool scale.

As shown in Table 11, the wash fastness outcomes of all marble-printed samples were between 4–5 and 5 gray-scale levels, which are commercially acceptable levels. All rub fastness outcomes were in the range of 3–4 and 4–5 gray-scale levels. Wet rub fastness results were generally slightly lower than those of dry rub fastness. All light fastness outcomes were in the range of 3–6 levels. The lowest light fastness values were shown on the unmordanted and immediately washed printed sample (C1) with a 3 rating.

Water fastness properties of silk fabrics printed without mordant are shown in Table 12. As displayed in Table 12, the water fastness values of the printed samples were 5 gray-scale levels and no staining whatsoever, leading to excellent water fastness results. Acidic and alkaline perspiration fastness characteristics of unmordanted silk fabrics printed with organic natural red ocher are demonstrated on Table 13.

From Table 13, all perspiration fastness outcomes of organic natural red ocher marble-printed samples were between 4/5 and 5 gray-scale levels, which are commercially acceptable. Acidic and alkaline perspiration fastness characteristics of unmordanted silk fabrics printed with synthetic red dye are presented in Table 14. The acidic and alkaline perspiration fastness outcomes of printed silk substrates were 5 gray-scale levels and no staining whatsoever, leading to excellent fastness results (Table 14).

3.2.2 Color Fastness Properties of Marble-Printed Silk Fabric Samples Premordanted with Potassium Bichromate

The color fastness properties of silk fabrics premordanted with potassium bichromate mordant and then printed with both colorants are shown in Tables 15, 16, 17, 18, and 19.

The washing fastness outcomes of all marble-printed silk fabric samples premordanted with potassium bichromate were between 4–5 and 5 gray-scale levels (Table 15), which were commercially acceptable (Table 15). All rubbing fastness values were in the range of 1–2 to 4–5 gray-scale levels. The wet rub fastness outcomes of samples were generally slightly lower than those of dry rub fastness (Table 15). All light fastness outcomes were in the range of 3–6 levels. The light

Table 15 Light (LF), rub (dry and wet), and wash fastness characteristics of organic natural red ocher marble-printed silk fabric substrates premordanted with potassium bichromate

Premordanted with potassium bichromate										
			Rub fastness		Washing fastness					
Fabrics	*K/S*	LF	Dry	Wet	WO	PC	PES	N6.6	CO	AC
C1	0.9	3	4–5	4–5	5	5	5	5	5	5
C2	1.6	5	4	3–4	5	5	5	5	5	5
C3	1.8	5	2–3	1–2	5	5	5	5	5	5
TF1	2.5	6	3	1–2	5	4–5	4–5	4–5	4–5	4–5
TF2	2.2	5	3–4	2	5	4–5	4–5	4–5	4–5	4–5
SF1	2	6	2–3	1–2	5	4–5	4–5	4–5	4–5	4–5
SF2	1.8	6	2–3	1–2	5	5	5	5	5	5

Table 16 Water fastness properties of organic natural red ocher marble-printed silk fabric samples premordanted with potassium bichromate

Water fastness							
	Premordanted with potassium bichromate						
	K/S	WO	PC	PES	N6.6	CO	AC
C1	0.9	5	5	5	5	5	5
C2	1.6	5	5	5	5	5	5
C3	1.8	5	5	5	5	4/5	4/5
TF1	2.5	4/5	4/5	4/5	4/5	4/5	4/5
TF2	2.2	5	5	5	5	4/5	4/5
SF1	2	5	4/5	4/5	4/5	4/5	4/5
SF2	1.8	5	5	5	5	4/5	5

Table 17 Acidic and alkaline perspiration fastness properties of organic natural red ocher marble-printed silk fabric samples premordanted with potassium bichromate

Premordanted with potassium bichromate													
		Acidic perspiration fastness						Alkaline perspiration fastness					
Fabrics	*K/S*	WO	PC	PES	N6.6	CO	AC	WO	PC	PES	N6.6	CO	AC
C1	0.9	5	5	5	5	5	5	5	5	5	5	5	5
C2	1.6	5	5	5	5	5	5	5	5	5	5	5	5
C3	1.8	5	5	5	5	5	5	5	5	5	5	5	5
TF1	2.5	5	5	5	5	5	5	5	5	5	5	5	5
TF2	2.2	5	5	5	5	5	5	5	5	5	5	5	5
SF1	2	5	5	5	5	5	5	5	5	5	5	5	5
SF2	1.8	5	5	5	5	5	5	5	5	5	4/5	4/5	4/5

Table 18 Light (LF), rub (dry and wet), and wash fastness characteristics of synthetic red ink marble-printed silk fabric samples premordanted with potassium bichromate (LF = light fastness)

Premordanted with potassium bichromate											
			Rub fastness		Washing fastness						
Fabrics	*K/S*	LF	Dry	Wet	WO	PC	PES	N6.6	CO	AC	
C1	3.8	6	4–5	4–5	5	5	5	5	5	5	
C2	9.8	6	4–5	2–3	5	5	5	5	5	5	
C3	12.8	6	3–4	2–3	5	5	5	5	5	5	
TF1	14.5	6	3–4	2–3	5	5	5	5	4–5	5	
TF2	13.3	6	3–4	2–3	5	5	5	5	5	5	
SF1	11.8	6	3–4	3	5	5	5	5	5	5	
SF2	11	6	3–4	3	5	5	5	5	5	5	

Table 19 Water fastness properties of synthetic red ink marble-printed silk fabric samples premordanted with potassium bichromate

Water fastness							
	Premordanted with potassium bichromate						
	K/S	WO	PC	PES	N6.6	CO	AC
C1	3.8	5	5	5	5	5	5
C2	9.8	5	5	5	5	5	5
C3	12.8	5	5	5	5	5	5
TF1	14.5	5	5	5	5	5	5
TF2	13.3	5	5	5	5	5	5
SF1	11.8	5	5	5	5	5	5
SF2	11	5	5	5	5	5	5

fastness value of fabrics after treatment with immediate rinsing was the lowest, which was 3. The rest of the samples premordanted with potassium bichromate showed quite high light fastness values. The water fastness properties of silk fabrics marble-printed with organic natural red ocher and premordanted with potassium

Table 20 Acidic and alkaline perspiration fastness properties of synthetic red ink marble-printed silk fabric samples premordanted with potassium bichromate

Premordanted with potassium bichromate													
		Acidic perspiration fastness						Alkaline perspiration fastness					
Fabrics	*K/S*	WO	PC	PES	N6.6	CO	AC	WO	PC	PES	N6.6	CO	AC
C1	3.8	5	5	5	5	5	5	5	5	5	5	5	5
C2	9.8	5	5	5	5	5	5	5	5	5	5	5	5
C3	12.8	5	5	5	5	5	5	5	5	5	5	5	5
TF1	14.5	5	5	5	5	5	5	5	5	5	5	5	5
TF2	13.3	5	5	5	5	5	5	5	5	5	5	5	5
SF1	11.8	5	5	5	5	5	5	5	5	5	5	5	5
SF2	11	5	5	5	5	5	5	5	5	5	5	5	5

bichromate are shown in Table 16. The water fastness values of organic natural red ocher marble-printed silk fabric samples premordanted with potassium bichromate were between 4–5 and 5 gray-scale levels, which are commercially acceptable (Table 16).

Acidic and alkaline perspiration fastness properties of silk fabrics premordanted with potassium bichromate and then marble-printed are shown in Table 17. As can be seen from Table 17, all the perspiration fastness values of organic natural red ocher marble-printed silk fabric samples premordanted with potassium bichromate are between 4–5 and 5 gray-scale levels which are commercially acceptable.

The washing fastness outcomes of all silk fabric substrates premordanted with potassium bichromate and then marble-printed with synthetic red ink were between 4–5 and 5 gray-scale levels (Table 18). All rub fastness outcomes were in the range of 2–3 and 4–5 gray-scale levels. The wet rub fastness outcomes were slightly lower than the dry rub fastness values (Table 18). All light fastness values were very good with a 6 rating.

Water fastness properties of silk fabrics marble-printed with synthetic red and then premordanted with potassium bichromate are shown in Table 19. The water fastness values of silk fabric samples marble-printed with synthetic red ink and then premordanted with potassium bichromate were excellent with a 5 gray-scale level with no staining whatsoever (Table 19).

As can be seen from Table 20, the acidic and alkaline perspiration fastness values of synthetic red ink marble-printed silk fabric samples premordanted with potassium bichromate exhibited excellent fastness levels, 5 gray-scale level.

3.2.3 Color Fastness Properties of Marble-Printed Silk Fabric Substrates Premordanted with Ferrous Sulfate (Iron (II) Sulfate)

The color fastness properties of silk fabrics premordanted with ferrous sulfate (iron (II) sulfate) and then printed with both colorants are shown in Tables 21, 22, 23, 24, 25, and 26.

Table 21 Light (LF), rub (dry and wet), and wash fastness characteristics of organic natural red ocher marble-printed substrates premordanted with ferrous sulfate

Premordanted with ferrous sulfate										
			Rub fastness		Washing fastness					
Fabrics	*K/S*	LF	Dry	Wet	WO	PC	PES	N6.6	CO	AC
C1	4.7	6	4	3–4	5	4/5	4/5	4/5	4–5	4
C2	6.2	6	2–3	1–2	4/5	4–5	4–5	4–5	4	3–4
C3	6.9	6	3	2	5	4/5	5	5	4/5	5
TF1	8.3	6	2–3	1	4/5	4–5	4–5	4–5	4	3–4
TF2	8	6	3	1–2	5	4/5	4/5	4/5	4–5	4
SF1	8	6	2–3	1	5	4/5	4/5	4/5	4–5	4–5
SF2	7	6	4	2	5	5	5	5	5	5

Table 22 Water fastness properties of organic natural red ocher marble-printed silk fabric samples premordanted with ferrous sulfate

Water fastness							
	Premordanted with ferrous sulfate						
	K/S	WO	PC	PES	N6.6	CO	AC
C1	4.7	5	5	5	5	5	5
C2	6.2	5	4/5	4/5	4/5	4/5	4/5
C3	6.9	4/5	4/5	4	4	4	4
TF1	8.3	4/5	4/5	4/5	4/5	4/5	4/5
TF2	8	4/5	4/5	4/5	4/5	4/5	4/5
SF1	8	4/5	4/5	4/5	4/5	4/5	4/5
SF2	7	4/5	4/5	4/5	4/5	4/5	4/5

Table 23 Acidic and alkaline perspiration fastness properties of organic natural red ocher marble-printed samples premordanted with ferrous sulfate

Premordanted with ferrous sulfate													
		Acidic perspiration fastness						Alkaline perspiration fastness					
Fabrics	*K/S*	WO	PC	PES	N6.6	CO	AC	WO	PC	PES	N6.6	CO	AC
C1	4.7	4/5	5	5	5	5	5	5	5	5	5	5	5
C2	6.2	5	5	5	5	5	5	5	5	5	5	5	5
C3	6.9	5	5	5	5	5	5	4/5	4/5	4/5	4/5	4/5	4/5
TF1	8.3	4/5	5	5	5	4/5	4/5	4/5	4/5	4/5	4/5	4/5	4/5
TF2	8	5	5	5	4/5	4/5	4/5	5	5	5	4/5	4/5	4/5
SF1	8	4/5	4/5	4/5	4/5	4/5	4/5	4/5	4/5	4/5	4/5	4/5	4/5
SF2	7	4/5	4/5	4/5	4/5	4/5	4/5	5	5	5	4/5	4/5	4/5

The washing fastness values of all marble-printed silk fiber fabric substrates with premordanted ferrous sulfate were between 3–4 and 5 gray-scale level (Table 21). All rubbing fastness values were in the range of 1–4 gray-scale levels (Table 21). The wet rub fastness outcomes of the samples were slightly lower than those of the

Table 24 Light (LF), rub (dry and wet), and wash fastness characteristics of synthetic red ink marble-printed silk fabric samples premordanted with ferrous sulfate

Premordanted with ferrous sulfate											
			Rub fastness		Washing fastness						
Fabrics	*K*/*S*	LF	Dry	Wet	WO	PC	PES	N6.6	CO	AC
C1	8.5	6	4–5	4–5	5	5	4–5	5	4–5	4–5
C2	15.8	6	3–4	3–4	5	5	5	5	5	5
C3	16.8	6	4	3–4	5	5	5	5	5	5
TF1	19.7	6	4	4	5	5	5	5	5	5
TF2	19	6	4–5	4	5	5	5	5	5	5
SF1	18.2	6	4–5	4	5	5	5	5	5	5
SF2	17.7	6	4–5	4	5	5	5	5	5	5

Table 25 Water fastness properties of synthetic red ink marble-printed silk fabric samples premordanted with ferrous sulfate

Water fastness							
	Premordanted with ferrous sulfate						
	K/*S*	WO	PC	PES	N6.6	CO	AC
C1	8.5	5	5	5	5	5	5
C2	15.8	5	5	5	5	5	5
C3	16.8	5	5	5	5	5	5
TF1	19.7	5	5	5	5	5	5
TF2	19	5	5	5	5	5	5
SF1	8	4/5	4/5	4/5	4/5	4/5	4/5
SF2	7	4/5	4/5	4/5	4/5	4/5	4/5

Table 26 Acidic and alkaline perspiration fastness properties of synthetic red ink marble-printed silk fabric samples premordanted with ferrous sulfate

Premordanted with ferrous sulfate													
		Acidic perspiration fastness						Alkaline perspiration fastness					
Fabrics	*K*/*S*	WO	PC	PES	N6.6	CO	AC	WO	PC	PES	N6.6	CO	AC
C1	8.5	5	5	5	5	5	5	5	5	5	5	5	5
C2	15.8	5	5	5	5	5	5	5	5	5	5	5	5
C3	16.8	5	5	5	5	5	5	5	5	5	5	5	5
TF1	19.7	5	5	5	5	5	5	5	5	5	5	5	5
TF2	19	5	5	5	5	5	5	5	5	5	5	5	5
SF1	18.2	5	5	5	5	5	5	5	5	5	5	5	5
SF2	17.7	5	5	5	5	5	5	5	5	5	5	5	5

dry rub fastness. All light fastness values were quite high, with 6 blue wool scale rating.

The water fastness properties of organic natural red ocher marble-printed silk fabric samples premordanted with ferrous sulfate are shown in Table 22. From

Table 22, the water fastness values of organic natural red ocher marble-printed silk fabric samples premordanted with ferrous sulfate were between 4 and 5 gray-scale levels, which are commercially acceptable.

The acidic and alkaline perspiration fastness characteristics of silk fabrics premordanted with ferrous sulfate and then marble-printed with organic natural red ocher are shown in Table 23. As can be seen from Table 23, all perspiration fastness values of organic natural red ocher marble-printed silk fabric samples premordanted with ferrous sulfate are between 4–5 and 5 gray-scale levels, which are commercially acceptable.

Washing fastness outcomes of synthetic red ink marble printed silk fabric samples premordanted with ferrous sulfate were quite high, with 4–5 and 5 gray-scale levels (Table 24). All rub fastness outcomes were in the range of 3–4 and 4–5 gray-scale levels (Table 24). The wet rub fastness outcomes were generally slightly lower than the dry rub fastness. All light fastness values were quite high with a 6 level.

As presented in Table 25, the water fastness values of synthetic red ink marble-printed silk fabric substrates premordanted with ferrous sulfate were in the range of 4–5 and 5 gray-scale levels, which are commercially acceptable.

As shown in Table 26, the acidic and alkaline perspiration fastness values of synthetic red ink marble-printed silk fabric samples premordanted with ferrous sulfate exhibited 5 gray-scale levels, which are commercially acceptable.

3.2.4 Color Fastness Properties of Marble-Printed Silk Fabric Samples Premordanted with Gallnut (Natural Mordant)

The color fastness characteristics of marble-printed silk fabric substrates premordanted with gallnut (natural mordant) are shown in Tables 27, 28, 29, 30, and 31.

As seen from Table 27, the washing fastness values of organic natural red ocher marble-printed silk fabric samples premordanted with gallnut were between 4–5 and 5 gray-scale rating. All the washing fastness outcomes are commercial acceptable (Table 27). All rubbing fastness values were in the range of 1–2 to 4 gray-scale

Table 27 Light (LF), rub (dry and wet), and wash fastness characteristics of organic natural red ocher marble-printed silk fabric samples premordanted with gallnut

Premordanted with gallnut										
			Rub fastness		Washing fastness					
Fabrics	*K/S*	LF	Dry	Wet	WO	PC	PES	N6.6	CO	AC
C1	2.5	5	4	3–4	5	5	5	5	5	5
C2	3.6	6	3–4	2–3	5	5	5	4–5	4–5	4–5
C3	4.6	6	2–3	1–2	5	5	5	4–5	4–5	4–5
TF1	4.9	6	2–3	1–2	5	5	5	4–5	4–5	4–5
TF2	5.1	6	3	1–2	5	5	4–5	4–5	4–5	4–5
SF1	5	6	3	1–2	5	5	4–5	4–5	4–5	4–5
SF2	5.1	6	3	1–2	5	5	5	5	4–5	5

Table 28 Water fastness properties of organic natural red ocher marble-printed silk fabric samples premordanted with gallnut

Water fastness							
	Premordanted with gallnut						
	K/S	WO	PC	PES	N6.6	CO	AC
C1	2.5	5	5	5	5	4/5	5
C2	3.6	5	5	5	5	4/5	5
C3	4.6	5	5	5	5	4/5	5
TF1	4.9	5	5	5	5	4/5	5
TF2	5.1	5	5	5	5	4/5	5
SF1	5	5	5	5	5	4/5	5
SF2	5.1	4/5	4/5	4/5	4/5	4/5	4/5

Table 29 Acidic and alkaline perspiration fastness properties of organic natural red ocher marble-printed silk fabric samples premordanted with gallnut

Premordanted with gallnut													
		Acidic perspiration fastness						Alkaline perspiration fastness					
Fabrics	*K/S*	WO	PC	PES	N6.6	CO	AC	WO	PC	PES	N6.6	CO	AC
C1	2.5	4/5	5	5	5	4/5	5	5	5	5	5	4/5	5
C2	3.6	4/5	5	5	5	4/5	5	5	5	5	5	5	5
C3	4.6	4/5	5	5	5	4/5	4/5	4/5	5	5	5	4/5	5
TF1	4.9	4/5	5	5	5	4/5	4/5	5	5	5	5	4	4/5
TF2	5.1	4/5	5	5	5	4/5	4/5	5	5	5	5	4	4/5
SF1	5	4/5	5	5	5	4/5	5	4/5	5	5	5	4/5	5
SF2	5.1	4/5	4/5	5	5	4/5	4/5	4/5	5	5	5	4/5	5

Table 30 Light (LF), rub (dry and wet), and wash fastness characteristics of synthetic red ink marble-printed silk fabric samples premordanted with gallnut

Premordanted with gallnut										
			Rub fastness		Washing fastness					
Fabrics	*K/S*	LF	Dry	Wet	WO	PC	PES	N6.6	CO	AC
C1	6.6	4	4–5	4–5	5	5	5	5	5	5
C2	8.4	5	4–5	4	5	5	5	5	5	5
C3	11.5	5	4–5	3	5	5	5	5	5	5
TF1	12.7	6	4–5	4	5	5	5	5	5	5
TF2	12.4	6	4–5	4	5	5	5	5	5	5
SF1	12.3	6	4–5	4–5	5	5	5	5	5	5
SF2	10.5	6	4–5	4	5	5	5	5	5	5

levels. The wet rub fastness outcomes were slightly lower than the dry rub fastness (Table 27). All light fastness values were in the range of 5–6 levels. The light fastness values of fabrics after treatment with immediate rinsing were the lowest, with a 5 rating (Table 27). All the rest displayed quite high light fastness level, which was 6 (Table 27).

Table 31 Water fastness properties of synthetic red ink marble-printed silk fabric samples premordanted with gallnut

Water fastness							
	Premordanted with gallnut						
	K/S	WO	PC	PES	N6.6	CO	AC
C1	6.6	4/5	4/5	4/5	4/5	4/5	4/5
C2	8.4	5	5	5	5	4/5	5
C3	11.5	5	5	5	5	5	5
TF1	12.7	5	5	5	5	5	5
TF2	12.4	5	5	5	5	4/5	5
SF1	12.3	5	5	5	5	5	5
SF2	10.5	5	5	5	5	5	5

The water fastness properties of organic natural red ocher marble-printed silk fabric samples premordanted with gallnut are presented in Table 28. The water fastness values of organic natural red ocher marble-printed silk fabric samples premordanted with gallnut were between 4–5 and 5 gray-scale levels, which are commercially acceptable (Table 28).

The acidic and alkaline perspiration fastness properties of organic natural red ocher marble-printed silk fabric samples premordanted with gallnut are presented in Table 29. All perspiration fastness values of organic natural red ocher marble-printed silk fabric samples premordanted with gallnut were between 4–5 and 5 gray-scale levels, which are commercially acceptable (Table 29).

Silk fabrics premordanted with gallnut (natural mordant) were successfully marble-printed (Ebru) with organic natural red ocher. Furthermore, silk fabrics premordanted with gallnut (natural mordant) and then marble-printed with organic natural red ocher displayed quite high light, washing, water, alkaline, and acidic perspiration results. In this study, this type of marbling process makes the most contribution to sustainability. First, the fabric used in this study was made of silk fiber, which is a natural and sustainable fiber. Apart from this, here, a natural, renewable, and sustainable gallnut was utilized as a mordant via the premordanting technique. In addition, organic natural red ocher was used here for the coloration process of the silk fabric made using the marbling technique (Ebru).

The washing fastness values of synthetic red ink marble-printed silk fabric samples premordanted with gallnut were excellent with 5 gray-scale levels (Table 30). All rub fastness outcomes were in the range of 3 and 4–5 gray-scale levels (Table 30). The wet rub fastness outcomes were generally slightly lower than the dry rub fastness (Table 30). All light fastness outcomes were in the range of 4–6 levels. The lowest light fastness value was observed on the immediately washed printed sample with a 4 rating. The water fastness values of synthetic red ink marble-printed silk fabric samples premordanted with gallnut were in the range of 4–5 and 5, leading to quite high and commercially acceptable fastness levels (Table 31).

Table 32 Acidic and alkaline perspiration fastness properties of synthetic red ink marble-printed silk fabric samples premordanted with gallnut

Premordanted with gallnut													
		Acidic perspiration fastness						Alkaline perspiration fastness					
Fabrics	*K/S*	WO	PC	PES	N6.6	CO	AC	WO	PC	PES	N6.6	CO	AC
C1	6.6	4/5	5	5	5	4/5	5	5	5	5	5	5	5
C2	8.4	4/5	5	5	5	4/5	5	5	5	5	5	4/5	5
C3	11.5	4/5	5	5	5	5	5	5	5	5	5	4/5	5
TF1	12.7	4/5	5	5	5	5	5	4/5	5	5	5	4	5
TF2	12.4	4/5	5	5	5	4/5	4/5	5	5	5	5	4/5	5
SF1	12.3	4/5	5	5	5	5	5	5	5	5	5	4/5	5
SF2	10.5	4/5	5	5	5	4/5	4/5	4/5	5	5	5	4/5	5

The acidic and alkaline perspiration fastness values of synthetic red ink marble-printed silk fabric samples premordanted with gallnut were in the range of 4 and 5 gray-scale levels, which are commercially acceptable (Table 32).

Overall, the marble printed silk fabrics generally presented good to excellent washing, light, perspiration (acidic and alkaline), and water fastness results.

4 Comparison of the Colorimetric and Color Fastness Properties of Organic Natural Red Ocher Pigment and Synthetic Red Ink Marble-Printed (Ebru) Silk Fabrics

Color strength (*K/S*) comparison for premordanted, marble-printed (with organic natural red ocher pigment or synthetic red ink), and fixed (thermofixed and steam-fixed) silk fabric samples are presented in Fig. 33.

When the color strength values of unmordanted or mordanted and then marble-printed and then fixed silk fabric samples were examined (Fig. 33), the highest color yield (*K/S*) values were observed in ferrous sulfate (iron (II) sulfate) premordanted silk fabric for both colorant types (organic natural red ocher pigment and synthetic red ink).

In this study, in the case of organic natural red ocher pigment use, the highest color strength (*K/S*) value of 8.3 was observed in silk fabric premordanted with ferrous sulfate (iron (II) sulfate) and then marble-printed with organic natural red ocher pigment and then thermofixed at 130 °C for 4 minutes (TF1 process) (Fig. 33). Similarly, in the case of synthetic red ink use, the highest color strength (*K/S*) value of 19.7 was observed in silk fabric premordanted with ferrous sulfate (iron (II) sulfate) and then marble-printed with synthetic red ink and then thermofixed at 130 °C for 4 minutes (TF1 process) (Fig. 33).

Similar to the case of premordanting with ferrous sulfate (iron (II) sulfate), when the premordanted silk fabrics with potassium bichromate mordant were examined, the color strength (*K/S*) outcomes (2.5 with organic natural red ocher pigment and

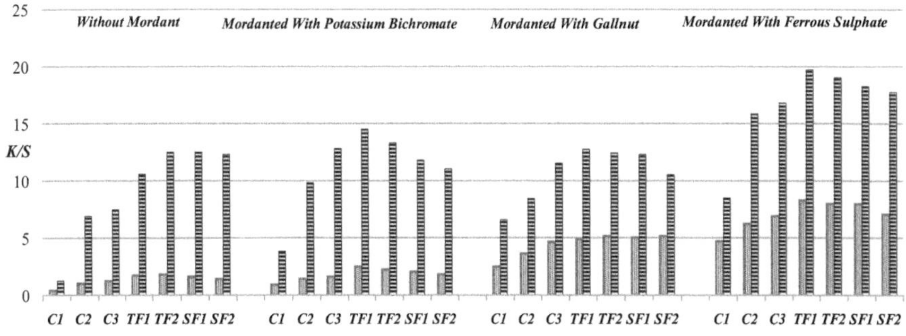

Fig. 33 Comparison of the color strength (*K*/*S*) values of premordanted, marble-printed (with organic natural red ocher pigment or synthetic red ink), and fixed (thermofixed and steamfixed) silk fabric samples

14.5 with synthetic red ink) were obtained on silk fabrics thermofixed at 130 °C for 4 minutes (TF1 process) (Fig. 33).

When the silk fabrics premordanted with gallnut natural mordant were examined (Fig. 33), in the case of the use of organic natural red ocher pigment, the highest color strength (*K*/*S*) outcomes were observed in thermofixation for 4 minutes at 130 °C (TF1) (*K*/*S* 4.9) and 150 °C (TF2) (*K*/*S* 5.1) and steamfixation for 15 minutes (SF1) (*K*/*S* 5.0) and 30 minutes (SF2) (*K*/*S* 5.1) at 102 °C. This type of marbling process makes the most contribution to sustainability. Apart from the natural, renewable, and sustainable silk fiber fabric usage, here, a natural, renewable, and sustainable gallnut was also used as a mordant via the premordanting technique. In addition, organic natural red ocher was used here for the coloration process of the silk fabric made using the marbling technique (Ebru). In the case of synthetic red ink, the highest color strength value was obtained in 4 minutes at 130 °C (TF1 process).

5 Conclusion

Today, sustainability is one of the most important issues in terms of protecting the environment and the world. The more we can reduce the possible damage to ecology in the production processes, the better for our planet. For this reason, it is very important to integrate sustainable and renewable natural products into production processes and to increase their utilization. Coloration is one of the processes that can create the most critical ecological troubles from the viewpoint of textile industry. The successful application of natural and natural-based products in textile product coloration will contribute to the sustainability efforts on our planet. As is known, many different techniques and colorants can be used in textile coloring. In this research study, the evaluation of colorimetric and color fastness characteristics of marble-printed (Ebru) silk fabrics using various mordants and different after-treatments was performed. Ebru (Turkish art of marbling) is a technique by which very different designs can be transferred to paper and fabrics. Marble printing (Ebru) is a traditional art that involves decoration of paper or fabrics. In this study, the "Ebru" marbling method is used for printing silk fabrics. Here, unmordanted and premordanted silk fabrics were marble-printed with organic natural red ocher and synthetic red ink for comparison. Here, silk fabrics were firstly premordanted with one natural mordant (gallnut) and two chemical mordants [potassium bichromate and iron (II) sulfate] for comparison. Moreover, unmordanted silk fabrics were used as control fabrics. After the premordanting process, premordanted samples were printed using the marbling method (Ebru) using organic natural red ocher and synthetic red ink (Pebeo Marbling Vermillion 02 Red ink). Then, marble-printed (Ebru) silk fabric samples were posttreated to undergo various steamfixation (at 102 °C for 15 minutes and at 102 °C for 30 minutes) and thermofixation (at 130 °C for 4 minutes and at 150 °C for 4 minutes) processes. Therefore, in this study, there are several aspects that contribute to sustainability efforts. The fabric utilized here is made of silk fiber, which is a natural, renewable, and sustainable fiber. Apart from this, a natural, renewable, and sustainable gallnut was used as a mordant via the premordanting technique. In addition, organic natural red ocher was used in the coloration process of the silk fabric made using the marbling style (Ebru). Here, silk fabrics premordanted with gallnut (natural mordant) were successfully marble printed (Ebru) with organic natural red ocher colorant. Furthermore, silk fabrics premordanted with gallnut (natural mordant) and then marble-printed with organic natural red ocher also displayed quite high light, washing, water, alkaline, and acidic perspiration results. In this study, this type of marbling process makes the most contribution in terms of sustainability.

In the case of organic natural red ocher pigment use, the highest color strength (*K*/*S*) value of 8.3 was observed in ferrous sulfate (iron (II) sulfate) premordanted and then marble-printed with organic natural red ocher pigment and then thermofixed at 130 °C for 4 minutes (TF1 process) silk fabric. Similarly, in the case of synthetic red ink utilization, the highest color strength (*K*/*S*) value of 19.7 was observed in silk fabric premordanted with ferrous sulfate (iron (II) sulfate) and then marble-printed with synthetic red ink and then thermofixed at 130 °C for 4 minutes (TF1 process). Similar to the case of premordanting with ferrous sulfate (iron (II) sulfate), when the premordanted silk fabrics with potassium bichromate mordant were examined, the color strength (*K*/*S*) values (2.5 with organic natural red ocher pigment and 14.5 with synthetic red ink) were obtained on silk fabrics thermofixed at 130 °C for 4 minutes (TF1 process). In conclusion, silk fabrics premordanted with ferrous sulfate (iron (II) sulfate) and then marble-printed exhibited the highest color strength (*K*/*S*) values. The highest *K*/*S* value was 19.7, which was observed in silk fabrics premordanted with ferrous sulfate (iron (II) sulfate) and marble-printed and then thermofixed at 130 °C for 4 minutes. Overall, the utilization of iron mordant generally led to the highest color strength values leading to darker color shades on silk fabrics. Thermofixation processes after marble printing generally led to the highest color strength values. Marble-printed silk fabric samples generally displayed good to excellent washing, light, perspiration (acidic and alkaline), and water fastness results. As a result of this study, it has been revealed that the premordanting process, the type of mordant used in premordanting, the fixation method applied after marble printing, and the application conditions of these fixation methods greatly affect the final colorimetric and color fastness characteristics of marble-printed silk fiber fabrics.

References

1. Babaoğlu, A. (2007). *Ebru İstanbul* (Geleneksel Sanatlar Serisi II). İstanbul Büyükşehir Belediyesi Kültür A.Ş.
2. Funda, T. (2007). *Türklerin Ebru Sanatı* (Kültür ve Turizm Bakanlığı Kütüphaneler ve Yayımlar Genel Müdürlüğü, Sanat Eserleri Dizisi 458). Semih Ofset Matbaacılık.
3. Barutçugil, H. (2007). *Türklerin Ebru Sanatı* (Kültür ve Turizm Bakanlığı Kütüphaneler ve Yayımlar Genel Müdürlüğü, Sanat Eserleri Dizisi 458). Semih Ofset Matbaacılık.
4. Tozun, H., & Uzunca, G. (2015). Ebru sanatının tekstilde kullanımı. *İnönü Üniversitesi Sanat ve Tasarım Dergisi, 5*, 93–99.
5. Benli, H., Yilmaz, M., & Bahtiyari, İ. (2014). Geleneksel Türk Ebru Sanatında Doğal Boyarmaddelerin Kullanılabilirliği. *Akdeniz Sanat Dergisi, 5*(10), 21–22.
6. Adıbelli, T. S. (2006). *Dekoratif Ürünlerde Ebru* [Doctoral dissertation, Selçuk Üniversitesi Sosyal Bilimler Enstitüsü].
7. Barutçugil, H. (2001). *Suyun Rüyası Ebru, Yaşayan Gelenek* (p. 68). Ebristan Yayınları.
8. Sönmez, G. (2007). *Geleneksekden Günümüze Ebru* (p. 13). İnkılâp Kitabevi Yayın Sanayi ve Ticaret A.Ş.
9. Başer, F. (2006). *Türk Ebru Sanatı*. Gözen Kitap ve Yayınevi.

Natural Dyeing of PA 6, PTT, PBT, PLA, Silk, and Soybean Textile Fibers with Black Mulberry Fruit Extract by Conventional and Microwave Dyeing Techniques

Ece Kalayci, Arzu Yavas, and Ozan Avinc

Abstract With the spread of sustainable, environmentally friendly, and conscious consumption habits, natural dyes, especially those of plant origin, have gained much popularity in recent years. In all industries where dyeing processes are used, such as the textile industry, food industry, medicine, etc., the demand for natural dyes is increasing. In this present study, dyeing processes were applied to various natural, synthetic, and regenerated fibers using black mulberry fruit extract. Here, 100% polyamide 6 (nylon 6) was used as a synthetic fiber; 100% PLA as a first renewable, sustainable, and biodegradable synthetic fiber; 100% silk fiber as a natural protein fiber; 100% soybean fiber as a regenerated protein fiber; and 100% PTT and 100% PBT fibers as elastic polyester synthetic fibers for dyeing with black mulberry fruit extract. Conventional dyeing with black mulberry fruit extract was performed in an HT-IR exhaustion dyeing machine at 100 °C and pH 4 with 1:50 liquor ratio for 60 minutes. Microwave dyeing was performed in a microwave oven at 900 watts and pH 4 with 1:50 liquor ratio for 1, 3, and 5 minutes. A simultaneous mordanting technique was performed for both dyeing techniques using gallnut as biomordant at 20% concentration. The colorimetric and color fastness properties (rub [dry and wet], perspiration [alkaline and acidic], wash, water, and seawater fastness) were then explored. The color yields of microwave dyeings were lower than those of conventional dyeings, while the color fastness levels of both dyeings were similar and quite high. Overall, natural dyeing of studied textile fibers with black mulberry fruit extract resulted in light and light-to-medium shades with good to quite high and satisfactory color fastness properties.

Keywords Black mulberry · Morus nigra · Natural dye · Microwave dyeing · Biomordant · Natural dyeing · Gallnut · PLA · Soybean · PTT · PBT · Silk · PA 6

E. Kalayci · A. Yavas · O. Avinc (✉)
Textile Engineering Department, Engineering Faculty, Pamukkale University, Denizli, Türkiye
e-mail: oavinc@pau.edu.tr

S. S. Muthu (ed.), *Natural Dyes and Sustainability*, Sustainable Textiles: Production, Processing, Manufacturing & Chemistry,
https://doi.org/10.1007/978-3-031-47471-2_14

1 Introduction

Environmental effects play a major role in sustainable and conscious production. The textile sector is one of the industries with the highest environmental impact. Coloration processes such as dyeing and printing, especially in textile finishing, are among the areas with the greatest environmental damage in terms of clean water consumption and the intensity of chemical use [1–6]. To produce just 1 kg of textile products, the water requirement can range from 95 to 400 liters [7]. The textile sector, which meets most of its energy needs from solid fuels such as petroleum derivatives or coal, consumes 19% of total industrial energy [7].

When you consider that more than 100 billion tons of garments are produced each year, it becomes clearer what a serious environmental impact the textile industry causes each year [8]. Textile production, which is one of the most polluting industries not only from the standpoint of water and energy consumption but also from the standpoint of emission levels, generates 1.2 billion tons of CO_2 annually [9–11]. More than 60% of textiles are utilized in the garment industry, and it was declared that about 5% of global emissions come from the fashion industry. Synthetic fibers have had a higher manufacture rate in recent years. Polyester fibers are the most widely utilized fibers, and at the start of the twenty-first century, they took first place as the most used fiber, surpassing cotton [12–16]. Emissions from polyester and other synthetic materials are very high because they are manufactured from fossil fuels. In 2015, polyester manufacture caused 706 billion kg of CO_2 [10].

In addition to dyes, textile finishing chemicals with different structures are often used, for example, to provide fragrance; for their fire-resistant, water-repellent, antibacterial, and antiviral properties; or to prevent bad odors [17–19]. The Global Textile Chemicals Market Analysis report published by Research and Markets [5] states that global demand for textile chemicals, which was 1.13 million tons in 2020, is estimated to attain 1.90 million tons by 2030 [20].

Global warming, climate change, increasing air pollution, soil pollution, water pollution, accumulation of microplastics, and chemicals are putting our future and future generations in grave danger day by day. The abundant resources that have been available to humanity for millions of years are becoming increasingly scarce and polluted. As a result of growing consumer awareness and restrictions and controls by states, authorities, and large companies, industrial production steps are becoming more and more environmentally friendly and responsible [1, 21–23].

For these reasons, academic and industrial studies have accelerated in recent years to introduce sustainable and more environmentally friendly textile finishing processes. Clean energy and raw material resources [24–32] and methods and technologies that use less energy, water, and chemicals [33–36] do not pollute the environment, in addition to those that use biodegradable, recyclable, and reusable raw materials and auxiliaries; issues such as wastewater management, carbon footprint, water footprint, and life cycle analysis are carefully studied [37–43].

The utilization of natural resources in textile dyeing processes with the highest environmental burden is as old as textile history. Based on information from archeological excavations, it is clear that the utilization of various natural resources in textile coloring processes dates back to ancient times [44–46]. Although synthetic dyes have dominated the need for industrial commercially utilized dyes in recent years due to enhanced consumption, the usage of natural products in the coloring of textile products has become popular again due to the awareness of sustainable and responsible production/consumption.

Using natural, environmentally friendly, and biodegradable raw material sources is the first step for environmentally friendly and responsible product design [1, 47]. The next step is to minimize the ecological impact of the process steps involved in manufacturing the product. In this regard, many new methods and technologies have been developed in recent years and have gained importance in the industry. According to the expected characteristics of the final product and the limitations of the raw material used, these methods are those that

- Use/recover and reuse fewer clean water resources.
- Consume less energy; use clean energy; minimize process steps and production times.
- Utilize natural or biodegradable chemicals or excipients with minimal environmental impact on production.
- Choose chemicals and excipients in the smallest possible quantities.

Although the utilization of microwave energy in the textile sector is not an industrially widespread technology today, its potential for use in textiles is one of the current areas of research due to its extremely low processing times and indirect energy consumption [48–50]. The use of microwaves, which are high-frequency radio waves, has been proven in many different industrial fields since the second half of the twentieth century [51]. The use of microwave technology in the textile industry is mainly related to wet processes and includes areas such as heating, drying, condensation, dyeing, fixation in printing, and disinfection of wool fabrics [51, 52].

In this study, natural dyeing of PA 6, PTT, PBT, PLA, silk, and soybean textile fibers with black mulberry fruit extract by conventional and microwave dyeing techniques was explored. In the below part, information about natural dye sources and natural dyeing in the textile sector was given.

1.1 Natural Dye Sources and Natural Dyeing in the Textile Industry

Mankind's interest in colors dates back to ancient times, as evidenced by archaeological research. Colors have been derived from various sources found in nature, including plants, animals, and minerals [53–56]. However, with increasing consumption, natural resources became insufficient, leading to the invention of synthetic dyes toward the end of the nineteenth century [57, 58].

Synthetic dyes have found extensive use in many sectors, with their variety and properties increasing over time. However, as scientific advancements have revealed risks to the environment and human health associated with certain chemical structures, the concept of sustainability has gained prominence.

As an outcome of research on the utilization of synthetic dyes in recent years, some chemical structures have been found to pose a risk to the ecology and human health. In this context, the utilization of some synthetic dyes has been restricted, and the sensitivity is increasing day by day, and new dyes and auxiliary chemicals are added to this list [59–61].

Natural dyes, being extracted from limited natural resources, have inherent limitations in terms of production and usage amounts. Their reproducibility in dyeing is generally low. This is due to variations in plants used as dye sources, which can be influenced by factors such as growth location, climate, water availability, and soil conditions [53–55, 62]. For instance, a dye obtained from plant leaves can vary in structure and color due to environmental factors. While it is unlikely that natural dyes can fully meet the industry's demand for synthetic dyes, they have gained popularity in recent years for niche production and design products due to their biodegradability and lesser harm or potential harm to the environment and human health [39, 62–64]. Although the color spectrum of natural dyes is not as wide as that of synthetic dyes, and there is a vast range of plant, animal, and even mineral sources from which natural dyes can be derived. In Turkey alone, more than 150 plant species could be utilized as natural dye sources, depending on the desired color, different natural sources, and whether plant, animal, or mineral can be chosen.

Yellow, greenish-yellow, and even red colors can be achieved through the presence of carthamine and corocetin in the petals of safflower (*Carthamus tinctorius* L.), while indican and isatin in the leaves of indigo (*Isatis tinctoria* L.) provide dark blue, blue, and green colors. The leaves of hibiscus (*Malva sylvestris* L.) impart a green color, the flower of poppy (*Papaver somniferum* L.) a purple color, the flower of Norouz (*Linaria* sp.) a red color, and the leaves of mulberry (*Morus* sp.) a yellow color [62]. This process is referred to as mordanting [62, 65, 66]. The type of mordant, the mordanting method, and the processing conditions play a crucial function in the color yield and characteristics obtained through natural dyeing. As an outcome of the change of chemical components in the mordant structure and the change of conditions, different complex structures can be formed with the dye, and different color shades and color fastness performances can be obtained in different variations [64, 67]. Water-soluble metal salts such as alum, iron alum, and copper alum, which contain heavy metals like aluminum, iron, copper, tin, and chromium, are commonly used mordants in traditional natural dyeing [63, 65, 67–69]. However, due to the hazardous influences of heavy metals on the ecology and human health, the utilization of many metal-containing mordants such as copper sulfate, potassium bichromate, and tin-containing structures in natural dyeing has been restricted and/or not been preferred. Compounds like potassium aluminum sulfate (potassium alum), aluminum sulfate (alum), aluminum acetate, and ferrous sulfate are preferred as more environmentally friendly metal mordants. However, the use of metal salts may not align with the sustainable and eco-friendly goals of natural dyeing.

Consequently, the exploration of biomaterials from natural sources as potential alternatives to metal mordants has been extensively studied in recent years [63, 67, 70–73]. In addition to coloration, natural dyeing processes can also impart numerous functional properties to textile surfaces, such as UV protection, antibacterial, antifungal, insect repellent, etc., thanks to the inherent properties of the natural resources used [69, 74–81].

As aforementioned, in this present study, natural dyeing of PA 6, PTT, PBT, PLA, silk, and soybean textile fibers with black mulberry fruit extract by conventional and microwave dyeing techniques was examined. Therefore, in the next section of this book chapter, information about black mulberry and its use in natural dyeing is given.

1.2 Black Mulberry and Its Use for Natural Dyeing

Black mulberry, scientifically known as *Morus nigra*, is a flowering plant species belonging to the Moraceae family [82–85]. It is native to a vast region ranging from Western Europe to Southwest Asia. Black mulberry is a tree species that can reach a height of up to 12 m and a width of 15 m. Its ripe fruits, which are edible, have a rich aroma and are dark purple, almost black in color. These fruits measure around 2–3 cm in length (Fig. 1) [83, 86, 87].

In the literature, various studies have been conducted in fields such as textiles, energy, and biomedicine to examine the coloring properties of black mulberry fruit. These studies have demonstrated that black mulberry could be utilized as an alternative to synthetic dyes for tissue staining in biomedical applications, including animal and human tissues such as nervous tissue [88–90]. For instance, Gedikli 2006

Fig. 1 (**a**) A black mulberry tree with branches in December (in Denizli, Türkiye). (**b**, **c**) A black mulberry tree with leaves and fruits in May (in Mersin, Türkiye)

[91] investigated the use of natural dyes, including black mulberry (*Morus nigra*), walnut (*Juglans regia*), madder (*Rubia tinctorum*), barberry (*Berberis crataegina*), and alder (*Alnus glutinosa*), for staining blood proteins (albumin, globulin, hemoglobin, fibrinogen) in electrophoresis gels [91, 92].

In another study by Ozturk et al. 2011 [93], wool fabrics were naturally dyed with black mulberry extract using various mordants such as clay, citric acid, tartaric acid, ferrous sulfate, sodium dichromate, copper sulfate, alum, and oxalic acid [93]. The color properties and fastness values were examined after the dyeing processes. The highest color yield was obtained for samples dyed with sodium dichromate as a mordant. When dyed with sodium dichromate, the wash fastness was found to be commercially acceptable, but the light fastness values were slightly lower compared to other mordants [93].

Lee 2010 [94] investigated the natural dyeing performance of Sangju silk samples with mulberry fruit extract to which cooper (Cu) complexes were added. It was found that the cooper metal complex acted as a mordant, allowing successful natural dyeing and achieving good color yield without the need for an additional mordanting process [94].

It was stated that another outstanding property of black mulberry fruit is its antibacterial structure. Trak and Arslan 2021 [95] investigated the usability, antibacterial activity, and dye degradation of dried black mulberry fruit extract in the synthesis of silver nanoparticles [28]. The outcomes suggest that silver nanoparticles synthesized in dried black mulberry fruit extract have the potential to be used in various fields [95].

In the cosmetics industry, products that are both environmentally friendly and harmless to human health are of great significance. For this reason, the availability of natural sources is also a frequent research topic in the cosmetics industry. In the research conducted by Boga et al. 2013 [96], the usability of organic structures in hair dyes with a green approach was studied. As a result of the research, the usability of anthocyanin pigments extracted from the fruit of the black mulberry tree was determined in hair dyes [96].

Another use of black mulberry fruit as a natural dye is also found in studies on solar cells [97–100]. In particular, the studies conducted in the last 10 years show the usability of black mulberry extract in dye-sensitive organic solar cells [98].

In this present study, natural dyeing process utilizing the black mulberry fruit extract was applied to various natural, regenerated, and synthetic fibers. Here, 100% polyamide 6 (nylon 6) as a synthetic fiber; 100% PLA as a first renewable, sustainable, and biodegradable synthetic fiber; 100% silk fiber as a natural protein fiber; 100% soybean fiber as a regenerated protein fiber; and 100% PTT and 100% PBT fibers as elastic polyester synthetic fibers were used for the dyeing with black mulberry extract. As is known, silk fiber is a natural protein fiber obtained from mulberry silkworms, and soybean fiber is a regenerated protein fiber made from soybeans. PLA is the first synthetic fiber obtained from natural sources, that is, naturally sourced (usually corn), sustainable, renewable, and biodegradable. In this context, the use of PLA fibers in production as well as silk and soybean fibers and finding life forms in different products are very important in terms of sustainability.

The PBT and PTT fibers used in this study are elastic polyester fibers and can be utilized instead of polyurethane fibers in many different textile applications. What is more in here, the 1,3-propanediol (Bio-PDO) monomer is synthesized from corn-derived glucose using an engineered *Escherichia coli* bacterium, and this monomer is utilized for the creation of PTT fibers [101]. Here, in the fermentation process, yeast ferment glucose to glycerol and microbes ferment glycerol to impart Bio-PDO [101].

The conventional natural dyeing process using black mulberry fruit extract for all studied textile fibers was carried out in a HT laboratory dyeing machine in company with 20% natural gallnut mordant at 100 °C and pH 4 with 1:50 liquor ratio for 60 minutes. Here, in terms of sustainability, gallnut (oak gall), an important natural mordant, was utilized as a mordant during the natural dyeing process, along with black mulberry natural dye. According to the results of conventional natural dyeing, silk and PA 6 fibers, which exhibited high color strength (K/S) values, were also dyed with black mulberry natural dye along with 20% gallnut natural mordant by microwave dyeing technique at 900 watts and pH 4 with 1:50 liquor ratio for various durations (1, 3, and 5 minutes) for comparison. Finally, the colorimetric properties and the various color fastness properties such as alkaline and acid perspiration, dry and wet rub fastness, and water and seawater fastness were investigated for the fabric samples dyed with natural dyes. These tests aimed to evaluate the color durability and resistance of the naturally dyed fabrics.

2 Materials and Methods

2.1 Material

Different types of fiber fabrics were selected to examine the dyeability performance of black mulberry fruit extract. Here, 100% polyamide 6 (nylon 6) as a synthetic fiber; 100% PLA as a first renewable, sustainable, and biodegradable synthetic fiber; 100% silk fiber as a natural protein fiber; 100% soybean fiber as a regenerated protein fiber; and 100% PTT and 100% PBT fibers as elastic polyester synthetic fibers were used for the dyeing with black mulberry extract. Fiber and fabric types and their dyeing types (conventional dyeing and/or microwave dyeing) were shown in Table 1. Here, in this study, natural dye application was implemented in two steps. In the first step, all selected fibers were dyed with black mulberry extract using conventional dyeing technique (Table 1). Conventional natural dyeing process using black mulberry fruit extract for all studied textile fibers is carried out in a HT laboratory dyeing machine in company with 20% natural gallnut mordant at 100 °C and pH 4 with 1:50 liquor ratio for 60 minutes. According to the results of conventional natural dyeing, silk and PA 6 fibers, which exhibited high color strength (K/S) values, were also dyed with black mulberry natural dye along with 20% gallnut natural mordant by microwave dyeing technique at 900 watts and pH 4 with 1:50 liquor ratio for various durations (1, 3, and 5 minutes) for comparison. Therefore, in

Table 1 A brief summary of the materials, methods, and process steps used in this study

		First step	Second step
Fiber type	Fabric type	Conventional dyeing	Microwave dyeing
PTT	Knitted	✓	
PBT	Knitted	✓	
PLA	Knitted	✓	
Soybean	Knitted	✓	
Polyamide 6	Woven	✓	✓
Silk	Woven	✓	✓

Table 2 Conditions of conventional dyeing and microwave dyeing

Conventional dyeing	Microwave dyeing
1:50 liquor ratio	1:50 liquor ratio
20% gallnut mordant	20% gallnut mordant
100 °C 60 minutes (4 °C/min temperature rise)	900 watts 1, 3, and 5 minutes
pH 4	pH 4
5-min rinsing	5-min rinsing

this second step, PA 6 and silk fibers were dyed with black mulberry natural dye according to the microwave dyeing method (Table 1).

2.2 *Natural Dyeing Process*

The utilized black mulberry fruits were obtained from the Bergama/Izmir region of Türkiye. Collected black mulberry fruit has been juiced, and it was used as a natural dye source to color selected fibers. The natural pH value of black mulberry extract was measured as pH 4. Therefore, both conventional and microwave dyeing processes were carried out at pH 4, which is the own pH of the black mulberry extract. Conventional dyeing with black mulberry fruit juice was performed in an ATAC LAB DYE HT IR laboratory type exhaustion dyeing machine at pH 4 and 100 °C with 1:50 liquor ratio for 60 minutes. Microwave dyeing was performed in a microwave oven (Arçelik MD74s Microwave) at 900 watts and pH 4 with 1:50 liquor ratio for 1, 3, and 5 minutes (Table 2). A simultaneous mordanting method was performed for both dyeing techniques by using gallnut powder as biomordant at 20% concentration. Natural gallnut powder was obtained from the Siirt region of Türkiye. In the case of 6-minute microwave dyeing, the dyeing liquor level decreases due to evaporation, causing unlevel dyeing and even damage to some parts of the fabrics. Therefore, microwave dyeing duration was limited to a maximum of 5 minutes. After both conventional and microwave natural dyeing processes, fabric samples were rinsed and washed with warm water and cold water for 5 minutes, respectively.

2.3 Colorimetric Measurements

After conventional dyeing and microwave dyeing processes, the color characteristics and color coordinates of the dyed fabric substrates were measured with the CIE Lab system using a DataColor SpectraFlash 600 spectrophotometer (DataColor International, Lawrenceville, NJ, USA) at illuminant D65 from a 10° standard observer. The color yield (K/S values) was calculated from the reflectance values at the corresponding wavelength of maximum absorption (λ_{max}) for both the conventional and microwave-dyed fabric substrates.

2.4 Color Fastness Determination

Rub fastness (wet and dry), alkali perspiration fastness, acidic perspiration fastness, and wash fastness characteristics were assessed after natural dyeing processes to determine dyeing performances of studied fabric samples. Wash fastness to domestic laundering (C06) was performed in accordance with ISO 105:C06 test in a M228 Rotawash machine (SDL ATLAS, UK). The test (C06/A2S) was carried out at 40 °C accompanied with the existence of sodium perborate. Seawater fastness and water fastness tests were performed in accordance with ISO 105-E02 and ISO 105-E01 protocols, respectively. Both alkaline and acid perspiration fastness testing was obtained in accordance with the ISO 105-E04 (color fastness to perspiration) protocol. Both dry and wet rub fastness testing was achieved following the ISO 105:X12 protocol. The evaluation of color fastness of the dyed fabrics to wash and to perspiration (alkaline and acid) was determined via staining of multifiber strip and color change (fading) of samples. Dry and wet rub fastness characteristics were obtained with the usage of ISO gray scales via level of staining of the cotton rubbing fabric.

3 Results and Discussion

3.1 Colorimetric and Color Fastness Results of Conventional Dyeing (First Step)

In the first step of the study, as aforementioned, conventional dyeing of PTT, PBT, PLA, PA, silk, and soybean fiber fabric samples was carried out with black mulberry extract at pH 4 and 100 °C with a liquor ratio of 1:50 in an HT IR laboratory type dyeing machine for 60 minutes.

3.1.1 Colorimetric Properties of Conventional Dyeings

Colorimetric properties (K/S, L^*, a^*, b^*, C^*, and h^o) and visual appearances of conventional dyeings are displayed in Table 3. The highest K/S value with 12.3 belonged to soybean fiber (Table 3). When both the K/S values and the appearances of the dyed fabrics were investigated, it is obvious that silk (K/S 8.73) and polyamide (K/S 7.85) fibers follow the soybean fiber, respectively. As expected, soybean, silk, and polyamide fibers exhibited the lowest lightness (L^*) values. Again, these three fibers exhibited the highest a^* values among all the fibers studied leading to redder appearance when compared with other three fibers (Table 3). These outcomes were in line with the respective measured K/S values (Table 3). What is more, silk and polyamide 6 fibers dyed with black mulberry fruit extract exhibited the highest chroma (C^*) values with 12.9 and 14, respectively (Table 3).

It is the right spot to point out that the dyeing process performed at the boiling point at 100 °C caused the soybean fiber fabric to lose its softness and harden due to high temperature sensitivity. Furthermore, silk and polyamide 6 fibers, which do not experience any problems in the handle while exhibiting high color efficiency and

Table 3 Colorimetric properties of conventionally dyed substrates with black mulberry fruit extract

Dyeing type	Fiber type	Dyed fabric appearances	K/S	L^*	a^*	b^*	C^*	h^o
Conventional	PTT	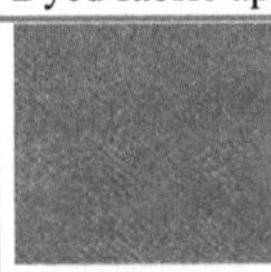	1.68	60.2	4.01	7.13	8.18	60.7
	PBT	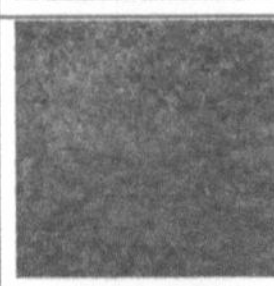	1.94	57.1	4.00	6.56	7.69	58.6
	PLA		1.63	57.9	4.36	5.36	6.91	50.9
	Polyamide 6		7.85	41.6	8.52	11.1	14.0	52.4
	Silk		8.73	35.4	9.57	8.71	12.9	42.3
	Soybean		12.3	28.7	8.28	6.22	10.4	36.9

highest chroma (C^*), are dyed in the darkest tones with black mulberry. As aforementioned, in the second stage of the study, microwave dyeing method is used. Soybean fibers were not included in the second step due to their heat sensitivity, as the microwave dyeing step was also performed at boiling temperature. Therefore, in the second step, only silk and polyamide fibers which exhibited high color yield, have the highest chroma, and did not experience any handle deterioration problems were utilized for microwave dyeing.

After the conventional dyeing methods in the first step, color fastness characteristics of all studied fibers dyed conventionally with black mulberry fruit extract were investigated. The rub (wet and dry) fastness and perspiration (acidic and alkaline) fastness results of studied fibers dyed conventionally with black mulberry fruit extract are displayed in Table 4 and Table 5, respectively.

3.1.2 Rub Fastness Properties of Conventional Dyeings

Rub fastness characteristics of fibers dyed conventionally with black mulberry fruit extract are shown in Table 4. When examining the cotton staining levels, it can be seen that the best fastness values with 5 grayscale rating (no staining whatsoever) obtained for both wet and dry rub fastness belonged to dyed polyamide 6 and soybean fiber samples (Table 4). They are followed by the dyed silk fiber samples with $4/\underline{5}$ grayscale points. While the PLA fiber samples display commercially acceptable rub fastness, the elastic polyester PTT and PBT fiber samples exhibited moderate to commercially acceptable good rub fastness values (Table 4). Overall, soybean, silk, polyamide 6, and PLA dyed conventionally with black mulberry fruit extract displayed commercially acceptable (equal to or more than four grayscale rating) rub fastness levels.

Table 4 Rub fastness properties of fibers dyed conventionally with black mulberry fruit extract

		Rub fastness[a] (cotton staining)	
Dyeing type	Fiber type	Dry	Wet
Conventional	PTT	3/4	4
	PBT	3	4
	PLA	4	4/5
	Polyamide 6	5	5
	Silk	$4/\underline{5}$	$4/\underline{5}$
	Soybean	5	5

[a]The lower line in the evaluation means a sensitivity of 0.25 points. For example, $4/\underline{5}$ is better than 4/5 and less than 5 grayscale rating

Table 5 Perspiration fastness properties of fibers dyed conventionally with black mulberry fruit extract

				Multifiber staining[a]					
Dyeing type			Color change	Diacetate	Cotton	Polyamide	Polyester	Acrylic	Wool
Conventional	Alkali perspiration	Polyamide 6	5	5	5	5	5	5	5
		Silk	5	5	4/$\underline{5}$	5	5	5	5
		Soybean	5	5	4/5	4/$\underline{5}$	5	5	5
		PTT	5	4/$\underline{5}$	4/$\underline{5}$	4/$\underline{5}$	5	4/$\underline{5}$	5
		PBT	5	5	4/$\underline{5}$	4/$\underline{5}$	4/$\underline{5}$	4/$\underline{5}$	5
		PLA	5	5	5	5	5	5	5
	Acidic perspiration	Polyamide 6	5	5	5	5	5	5	5
		Silk	5	5	4/5	5	5	5	5
		Soybean	5	5	4/$\underline{5}$	4/$\underline{5}$	5	5	5
		PTT	5	4/$\underline{5}$	4/$\underline{5}$	4/$\underline{5}$	5	5	5
		PBT	5	5	4/$\underline{5}$	4/$\underline{5}$	5	5	5
		PLA	5	5	5	4/$\underline{5}$	5	5	5

[a]The lower line in the evaluation means a sensitivity of 0.25 points. For example, 4/$\underline{5}$ is better than 4/5 and less than 5 grayscale rating

3.1.3 Perspiration Fastness Properties of Conventional Dyeings

Alkaline and acidic perspiration fastness properties of studied fibers dyed conventionally with black mulberry fruit extract are displayed in Table 5. All dyed fibers did not experience color change after both perspiration fastness leading to 5 grayscale rating for color change (Table 5). All dyed studied fibers showed fastness values of 4–5 or higher in both perspiration fastness tests leading to commercially acceptable perspiration levels (Table 5). Only dyed polyamide 6 fiber showed no staining whatsoever in all fiber components of the adjacent multifiber after both perspiration fastness tests leading to 5 grayscale rating.

3.2 Colorimetric and Color Fastness Results of Microwave Dyeing (Second Step)

As mentioned before, in the first stage, all fibers were subjected to conventional dyeing with black mulberry fruit extract. At this second stage, the fibers that showed the best performance in the first stage were dyed with black mulberry fruit extract under microwave dyeing conditions. As aforementioned, soybean fibers were not involved in the second step owing to their heat sensitivity leading to stiffer and brittle handle generation, as the microwave dyeing step was also performed at boiling temperature. So, in this second step, only silk and polyamide fibers which exhibited high color yield and the highest chroma and did not experience any handle deterioration problems were used for microwave dyeing.

3.2.1 Colorimetric Properties of Microwave Dyeings

Colorimetric properties (K/S, L^*, a^*, b^*, C^*, and h^o) and visual appearances of microwave dyeings are exhibited in Table 6. Color yields of microwave dyeings were lower than those of conventional dyeings (Table 6). In the case for microwave dyeings, the highest K/S values belonged to 3- and 5-minute microwave dyeings (Table 6). In the case of microwave dyeings, the highest K/S value for silk fibers was achieved for 3-minute microwave dyeing process, whereas the highest K/S value for the polyamide 6 fiber samples was achieved for 5-minute microwave dyeing process. In both fibers, the results of the 3- and 5-minute microwave dyeings were close and comparable. For this reason, 3-minute microwave dyeing in both fibers can be considered ideal in terms of similar color results and lower energy consumption.

Color fastness properties of silk and polyamide 6 fibers dyed using microwave with black mulberry fruit extract were investigated. The rub (wet and dry) fastness and perspiration (acidic and alkaline) fastness results of studied fibers dyed using microwave with black mulberry fruit extract are exhibited in Tables 7 and 8, respectively.

Table 6 Colorimetric properties of microwave-dyed samples with black mulberry fruit extract

Fiber type	Dyeing type and duration	Dyed fabric appearances	K/S	L^*	a^*	b^*	C^*	h^o
Silk	Conventional dyeing		8.73	35.4	9.57	8.71	12.9	42.3
	Microwave dyeing For 1 min		1.31	63.0	7.16	4.24	8.33	30.7
	Microwave dyeing For 3 min		2.96	48.8	12.01	2.883	12.36	13.52
	Microwave dyeing For 5 min		2.80	47.7	10.2	0.93	10.2	5.24
Polyamide 6	Conventional dyeing		7.85	41.6	8.52	11.1	14.0	52.4
	Microwave dyeing For 1 min		1.76	69.5	7.12	4.14	8.23	30.2
	Microwave dyeing For 3 min		2.84	61.1	6.47	4.73	8.02	36.2
	Microwave dyeing For 5 min		3.13	52.8	13.4	0.82	13.4	3.52

3.2.2 Rub Fastness Properties of Microwave Dyeings

Rub fastness properties of fibers dyed using microwave with black mulberry fruit extract are shown in Table 7. Silk and polyamide 6 fibers dyed both conventionally and using microwave displayed quite high and commercially acceptable wet and rub fastness levels. Besides, both wet and dry rubbing fastness values of all microwave-dyed silk and polyamide 6 fibers were five points according to grayscale rating without any staining whatsoever (Table 7).

Table 7 Rub fastness properties of microwave-dyed fibers with black mulberry fruit extract

Fiber type	Dyeing type	Dyeing duration (min)	Rub fastness[a] (cotton staining)	
			Dry	Wet
Silk	Conventional	60	$4/\underline{5}$	$4/\underline{5}$
	Microwave	1	5	5
	Microwave	3	5	5
	Microwave	5	5	5
Polyamide 6	Conventional	60	5	5
	Microwave	1	5	5
	Microwave	3	5	5
	Microwave	5	5	5

[a]The lower line in the evaluation means a sensitivity of 0.25 points. For example, $4/\underline{5}$ is better than 4/5 and less than 5 grayscale rating

3.2.3 Perspiration Fastness Properties of Microwave Dyeings

Alkaline and acidic perspiration fastness properties of silk and polyamide 6 fibers dyed using microwave with black mulberry fruit extract are displayed in Table 8. All conventionally and microwave-dyed fibers did not experience color change after both perspiration fastness leading to 5 grayscale rating for color change (Table 8). All conventionally and microwave-dyed studied fibers displayed fastness values of 4–5 or higher in both perspiration fastness tests leading to commercially acceptable perspiration levels (Table 8). Only conventionally and microwave-dyed polyamide 6 fiber showed no staining whatsoever in all fiber components of the adjacent multifiber after both perspiration fastness tests leading to 5 grayscale rating (Table 8).

In both silk and polyamide 6 fibers, the colorimetric and studied color fastness performances (Tables 6, 7, and 8) of the 3- and 5-minute microwave dyeings were very close to each other and comparable. For this reason, 3-minute microwave dyeing for both fibers can be considered ideal from the standpoint of similar color and color fastness results and lower energy consumption due to lower application duration. Therefore, in this last below phase of this study, a 3-minute microwave dyeing process was determined as the ideal processing time and applied in order to compare with the conventional dyeing process.

3.2.4 Wash, Water, and Seawater Fastness Properties of Both Conventional and Microwave Dyeings

Apart from rub (wet and dry) and perspiration (alkaline and acidic) fastness tests, wash, water, and seawater fastness properties of conventional and microwave silk fiber dyeings were discovered and displayed in Table 9. As could be noticed in Table 9, very high fastness levels were obtained for both conventional and microwave dyeings. All conventionally and microwave-dyed silk fibers displayed fastness values of 4–5 or higher in wash, water, and seawater fastness tests leading to

Table 8 Perspiration fastness properties of microwave-dyed fibers with black mulberry fruit extract

					Multifiber staining[a]					
	Fiber type	**Dyeing type**	**Dyeing duration (min)**	**Color change**	**Diacetate**	**Cotton**	**Polyamide**	**Polyester**	**Acrylic**	**Wool**
Alkali perspiration	Silk	Conventional	60	5	5	4/$\underline{5}$	5	5	5	5
		Microwave	1	5	5	5	5	5	5	5
		Microwave	3	5	5	4/$\underline{5}$	5	5	5	5
		Microwave	5	5	5	4/5	5	5	5	5
	Polyamide 6	Conventional	60	5	5	5	5	5	5	5
		Microwave	1	5	5	5	5	5	5	5
		Microwave	3	5	5	5	5	5	5	5
		Microwave	5	5	5	5	5	5	5	5
Acidic perspiration	Silk	Conventional	60	5	5	4/5	5	5	5	5
		Microwave	1	5	5	4/$\underline{5}$	5	5	5	5
		Microwave	3	5	5	4/5	4/$\underline{5}$	5	5	5
		Microwave	5	5	4/$\underline{5}$	4/5	4/$\underline{5}$	5	5	5
	Polyamide 6	Conventional	60	5	5	5	5	5	5	5
		Microwave	1	5	5	5	5	5	5	5
		Microwave	3	5	5	5	5	5	5	5
		Microwave	5	5	5	5	5	5	5	5

[a]The lower line in the evaluation means a sensitivity of 0.25 points. For example, 4/$\underline{5}$ is better than 4/5 and less than 5 grayscale rating

Table 9 Wash, water, and seawater fastness properties of conventional and microwave silk fiber dyeings

			Multifiber staining[a]					
Fastness type	Dyeing type[b]	Color change	Diacetate	Cotton	Polyamide	Polyester	Acrylic	Wool
Wash	Conventional	5	5	5	5	5	5	5
Wash	Microwave	5	5	5	5	5	5	5
Water	Conventional	5	5	4/5	5	5	5	5
Water	Microwave	5	5	4/5̲	5	5	5	5
Seawater	Conventional	5	5	4/5	4/5̲	5	5	5
Seawater	Microwave	5	5	4/5̲	4/5̲	5	5	5

[a]The lower line in the evaluation means a sensitivity of 0.25 points. For example, 4/5̲ is better than 4/5 and less than 5 grayscale rating
[b]Microwave dyeing was performed for 3 minutes

commercially acceptable fastness levels (Table 9). Conventionally and microwave-dyed silk fiber showed no staining whatsoever in all fiber components of the adjacent multifiber after wash fastness test leading to 5 grayscale rating.

4 Conclusion

In this study, the effectiveness of black mulberry fruit extract in sustainable natural dyeing for various fibers was investigated. The initial stage involved subjecting PTT, PBT, PLA, soybean, polyamide 6, and silk fibers to a conventional dyeing process using black mulberry fruit extract. Among these fibers, soybean fiber dyed conventionally with black mulberry fruit extract exhibited the highest K/S value of 12.3. However, the conventional dyeing process implemented at the boiling point of 100 °C caused the soybean fiber fabric to lose its softness, becoming hardened and brittle due to its sensitivity to high temperatures.

When considering both the K/S values and the appearance of the dyed fabrics, it is apparent that silk (K/S 8.73) and polyamide 6 (K/S 7.85) fibers followed soybean fiber in terms of color intensity. These three fibers displayed the highest a* values among all the fibers studied, resulting in a redder appearance compared to the other three fibers. Additionally, silk and polyamide 6 fibers dyed with black mulberry fruit extract exhibited the highest chroma (C*) values, measuring 12.9 and 14, respectively.

When examining the cotton staining levels, the best fastness values with 5 grayscale rating (no staining whatsoever) obtained for both wet and dry rub fastness belonged to conventionally dyed polyamide 6 and soybean fiber samples. They were followed by conventionally dyed silk fiber samples with 4/5̲ grayscale points. While the conventionally dyed PLA fiber samples display commercially acceptable rub fastness, the conventionally dyed elastic polyester PTT and PBT fiber samples exhibited moderate to commercially acceptable good rub fastness values.

Overall, soybean, silk, polyamide 6, and PLA fibers dyed conventionally with black mulberry fruit extract displayed commercially acceptable (equal to or more than 4 grayscale rating) rub fastness levels. All conventionally dyed fibers did not experience color change after both perspiration fastness tests, receiving a rating of 5 on the grayscale for color change. Furthermore, all conventionally dyed fibers demonstrated fastness values of 4–5 or higher in both perspiration fastness tests, indicating commercially acceptable perspiration levels. Only dyed polyamide 6 fiber showed no staining whatsoever in all fiber components of the adjacent multifiber after both perspiration fastness tests, receiving a rating of 5 on the grayscale.

In the second stage of the study, the fibers that performed best in the first stage were subjected to dyeing with black mulberry fruit extract using microwave dyeing conditions. Only silk and polyamide fibers, which exhibited high color yield and the highest chroma and did not experience any handle deterioration issues, were chosen for microwave dyeing. It was observed that the color yields of microwave dyeings were lower than those achieved through conventional dyeing. Among the microwave dyeing durations tested, the highest K/S values were obtained for the 3-minute and 5-minute dyeing durations.

Furthermore, both conventionally and microwave-dyed silk and polyamide 6 fibers exhibited high and commercially acceptable wet and rub fastness levels. In particular, all microwave-dyed silk and polyamide 6 fibers achieved a rating of five points on the grayscale for both wet and dry rubbing fastness, indicating no staining whatsoever. Similarly, all conventionally and microwave-dyed fibers did not undergo any color change after the perspiration fastness tests, receiving a rating of 5 on the grayscale for color change. Additionally, all conventionally and microwave-dyed fibers displayed fastness values of 4–5 or higher in the perspiration fastness tests, indicating commercially acceptable perspiration resistance. Notably, conventionally and microwave-dyed polyamide 6 fibers exhibited no staining whatsoever in all fiber components of the adjacent multifiber after the perspiration fastness tests, receiving a rating of 5 on the grayscale. In both silk and polyamide 6 fibers, the colorimetric properties and color fastness performances of the 3-minute and 5-minute microwave dyeings were very similar and comparable. Consequently, the 3-minute microwave dyeing duration can be considered ideal for both fibers due to its comparable color and color fastness results, as well as lower energy consumption owing to the shorter application duration. Furthermore, all conventionally and microwave-dyed silk fibers displayed fastness values of 4–5 or higher in wash, water, and seawater fastness tests, indicating commercially acceptable fastness levels. Conventionally and microwave-dyed silk fibers showed no staining whatsoever in all fiber components of the adjacent multifiber after the wash fastness test, achieving a rating of 5 on the grayscale.

In conclusion, this study demonstrates that textile fibers can be successfully colored to achieve light and light-to-medium shades using black mulberry fruit extract through the natural dyeing method. The resulting color fastness characteristics were found to be good to quite high, indicating the durability of the dyed fabrics. Furthermore, the utilization of microwave technology in natural dyeing was explored and shown to be feasible, offering the potential to reduce energy consumption by shortening the dyeing duration. While the current commercial utilization of microwave technology in the textile sector is limited, its application in sustainable and responsible production holds great promise for the future. Promoting the widespread use of natural and natural-based resources and implementing innovative and environmentally friendly finishing processes such as ozone, $ScCO_2$, microwave, ultrasound, UV, and laser, among others, will significantly contribute to the sustainability of the textile sector. Each effort and step taken in this direction adds value to the sustainability of our planet and fosters a more responsible and environmentally conscious textile sector.

Acknowledgments The authors would like to thank Mr. Ahmet Güven for his help in the supply of black mulberry fruit.

References

1. Kalayci, E., Avinc, O., Yavas, A., & Coskun, S. (2019). Responsible textile design and manufacturing: Environmentally conscious material selection. In A. Y. Alqahtani, E. Kongar, K. K. Pochampally, & S. M. Gupta (Eds.), *Responsible manufacturing: Issues pertaining to sustainability*. Taylor & Francis.
2. Lara, L., Cabral, I., & Cunha, J. (2022). Ecological approaches to textile dyeing: A review. *Sustainability, 14*(14), 8353.
3. Roy Choudhury, A. (2014). Environmental impacts of the textile industry and its assessment through life cycle assessment. In *Roadmap to sustainable textiles and clothing: Environmental and social aspects of textiles and clothing supply chain* (pp. 1–39). Springer.
4. Khan, S., & Malik, A. (2014). Environmental and health effects of textile industry wastewater. In *Environmental deterioration and human health: Natural and anthropogenic determinants* (pp. 55–71). Springer.
5. Slater, K. (2003). *Environmental impact of textiles: Production, processes and protection*. Elsevier.
6. Christie, R. (2007). *Environmental aspects of textile dyeing*. Elsevier.
7. Demirer, G., & Alkaya, E. (2018). Tekstil Sektöründe Temiz Üretim Rehberi Kaynak Kullanımında Verimlilik, Maliyetlerde Düşüş, Çevreye Uyum. WWF-Türkiye, İstanbul.
8. Igini, M. (2023). *10 Stunning Fast Fashion Waste Statistics*. https://earth.org/statistics-about-fast-fashion-waste/. Accessed 13/01/2022.
9. Parikh, J., Panda, M., Ganesh-Kumar, A., & Singh, V. (2009). CO2 emissions structure of Indian economy. *Energy, 34*(8), 1024–1031.
10. Kant, R. (2011). Textile dyeing industry an environmental hazard. *Natural Science, 4*(1), 22–26.
11. Gbolarumi, F., Wong, K., & Olohunde, S. (2021). Sustainability assessment in the textile and apparel industry: A review of recent studies. In *IOP conference series: Materials science and engineering* (Vol. 1051, no. 1, p. 012099). IOP Publishing.
12. Aizenshtein, E. (2006). World production and consumption of polyester fibres and thread. *Fibre Chemistry, 38*(3), 264–271.
13. Palacios-Mateo, C., van der Meer, Y., & Seide, G. (2021). Analysis of the polyester clothing value chain to identify key intervention points for sustainability. *Environmental Sciences Europe, 33*(1), 2.

14. Kumartasli, S., & Avinc, O. (2020). Important step in sustainability: Polyethylene terephthalate recycling and the recent developments. In *Sustainability in the textile and apparel industries* (p. 1). Springer.
15. Kumartasli, S., & Avinc, O. (2021). Recycled thermoplastics: Textile fiber production, scientific and recent commercial developments. In *Recent developments in plastic recycling* (pp. 169–192). Springer.
16. Kalayci, E., Avinc, O., & Yavas, A. (2017). Sustainable decisions and approaches in textile production. In *3rd international conference on computational mathematics and engineering sciences*. Girne/Turkish Republic of Northern Cyprus.
17. Yavas, A., et al. (2017). *Water-oil repellent treatment of viscose fabric with microwave irradiation*. Presented at the Asia and Africa Science Platform Establishment of Collaboration Research for Neo-Fiber Technology in Asia and Africa+ Seminar Series 10 and Annual Report Meeting for 'Neo-Fiber Technology' Project 2016–2017, Kyoto, Japan.
18. Horrocks, A. R. (2011). Flame retardant challenges for textiles and fibres: New chemistry versus innovatory solutions. *Polymer Degradation and Stability, 96*(3), 377–392. https://doi.org/10.1016/j.polymdegradstab.2010.03.036
19. Kalın, M. B. (2008). Tekstil Yüzeylerinin Yanmaya Karşı Dirençlerinin Arttırılması. Yüksek Lisans, Tekstil Mühendisliği Anabilim Dalı, KahramanMaraş Sütçü İmam Üniversitesi, KahramanMaraş.
20. News-Desk, A. R. *Global demand for textile chemicals expected to reach 1.90 million tonnes by 2030*. https://in.apparelresources.com/business-news/manufacturing/global-demand-textile-chemicals-expected-reach-1-90-million-tonnes-2030/. Accessed 13/01/2022
21. Ilgin, M. A., & Gupta, S. M. (2010). Environmentally conscious manufacturing and product recovery (ECMPRO): A review of the state of the art. *Journal of Environmental Management, 91*(3), 563–591. https://doi.org/10.1016/j.jenvman.2009.09.037
22. Thackeray, S. J., et al. (2020). *Civil disobedience movements such as School Strike for the Climate are raising public awareness of the climate change emergency*.
23. Zandalinas, S. I., Fritschi, F. B., & Mittler, R. (2021). Global warming, climate change, and environmental pollution: Recipe for a multifactorial stress combination disaster. *Trends in Plant Science, 26*(6), 588–599.
24. Kalayci, E., Yavas, A., & Avinc, O. (2018). The effects of different alkali treatments with different temperatures on the colorimetric properties of lignocellulosic raffia Fibers. In *International conference on Recent Advances in Engineering and Technology*, Havana, Cuba, 22nd–23rd November 2018.
25. Karthik, T., & Murugan, R. (2016). Milkweed—A potential sustainable natural fibre crop. In *Sustainable fibres for fashion industry* (pp. 111–146). Springer.
26. Unal, F., Avinc, O., & Yavas, A. (2020). Sustainable textile designs made from renewable biodegradable sustainable natural abaca fibers. In *Sustainability in the textile and apparel industries* (pp. 1–30). Springer.
27. Kalayci, E., Avinc, O. O., Bozkurt, A., & Yavas, A. (2016). Sustainable textile fibers obtained from agricultural wastes: Pineapple leaf fibers. *Sakarya University Journal of Science, 20*(2), 203–221.
28. Zheng, Y., Wang, J., Zhu, Y., & Wang, A. (2015). Research and application of kapok fiber as an absorbing material: A mini review. *Journal of Environmental Sciences, 27*, 21–32.
29. Unal, F., Yavas, A., & Avinc, O. (2020). Contributions to sustainable textile design with natural raffia palm fibers. In *Sustainability in the textile and apparel industries* (pp. 67–86). Springer.
30. Turkoglu, K. B., Kalayci, E., Avinc, O., & Yavas, A. (2019). Oleofilik buoyans özellikli kapok lifleri ve yenilikçi yaklaşımlar. *Düzce Üniversitesi Bilim ve Teknoloji Dergisi, 7*(1), 61–89.
31. Joseph, S., Oommen, Z., & Thomas, S. (2006). Environmental durability of banana-fiber-reinforced phenol formaldehyde composites. *Journal of Applied Polymer Science, 100*(3), 2521–2531.
32. Kalayci, E., Avinc, O., & Turkoglu, K. B. (2022). Importance of Asclepias Syriaca (milkweed) fibers in sustainable fashion and textile industry and its potential end-uses. In S. S. Muthu

(Ed.), *Sustainable approaches in textiles and fashion: Fibres, raw materials and product development* (pp. 1–21). Springer Nature Singapore.
33. Haggag, K., Ragheb, A., Nassar, S. H., Hashem, M., El Sayed, H., & El-Thalouth, I. A. (2014). *Microwave irradiation and its application in textile industries*. Science Publishing Group.
34. Unal, F., Avinc, O., Yavas, A., Eren, H. A., & Eren, S. (2020). Contribution of UV technology to sustainable textile production and design. In *Sustainability in the textile and apparel industries* (pp. 163–187). Springer.
35. Rauscher, H., Perucca, M., & Buyle, G. (2010). *Plasma technology for hyperfunctional surfaces: Food, biomedical and textile applications*. Wiley.
36. Odabaşoğlu, H. Y., Avinç, O. O., & Yavaş, A. (2013). Susuz Boyama. *Tekstil ve Mühendis, 20*(90), 62–79.
37. İnkaya, T., Eren, H. A., & Aniş, P. (2010). Pamuk Ağartılmasında Lakkaz/Mediatör Sistemlerinin Oksijen ve Ozon ile Kombine Edilmesi. *Pamukkale University Journal of Engineering Sciences, 14*(1), 77–82.
38. Rissanen, T. (2013). *Zero-waste fashion design*. Doctor of Philosophy – Design, University of Technology.
39. Sharma, A. (2013). Eco-friendly textiles: A boost to sustainability. *Asian Journal of Home Science, 8*(2), 768–771.
40. Chapagain, A., Hoekstra, A., Savenije, H., & Gautam, R. (2006). The water footprint of cotton consumption: An assessment of the impact of worldwide consumption of cotton products on the water resources in the cotton producing countries. *Ecological Economics, 60*(1), 186–203.
41. Scrimshaw, J. (2013). *ITMA Sürdürülebilirlik Bülteni*. World Textile Information Network Ltd.
42. Dutta, P. K., Dutta, J., & Tripathi, V. (2004). Chitin and chitosan: Chemistry, properties and applications. *Journal of Scientific and Industrial Research, 63*(1), 20–31.
43. Ahmad, S., Ashraf, M., Abid, S., Jabbar, M., Shafiq, F., & Siddique, A. (2022). Recent developments in laser fading of denim: A critical review. *Journal of Natural Fibers, 19*(15), 11621–11631.
44. Ahsan, R., Masood, A., Sherwani, R., & Khushbakhat, H. (2020). Extraction and application of natural dyes on natural fibers: An eco-friendly perspective. *Review of Education, Administration & LAW, 3*(1), 63–75.
45. Yıldırım, F. F., Yavas, A., & Avinc, O. (2020). Printing with sustainable natural dyes and pigments. In *Sustainability in the textile and apparel industries–production process sustainability* (pp. 1–35). Springer.
46. Yusuf, M., Shabbir, M., & Mohammad, F. (2017). Natural colorants: Historical, processing and sustainable prospects. *Natural Products and Bioprospecting, 7*, 123–145.
47. Gupta, S. M., & Lambert, A. J. D. F. (Eds.). (2007). *Environment conscious manufacturing*. CRC Press.
48. Kalayci, E., Yavas, A., Avinc, O., & Tonguc, T. (2019). *Dyeing performance comparison of soybean fibers with direct dyes with conventional and microwave technologies*. Presented at the Textile Science and Economy X, Zrenjanin, Serbia.
49. Kalayci, E., Yavas, A., Avinc, O., & Bektas, M. (2018). Dyeability of soybean fibers with acid dyes under microwave conditions. In *II. International scientific and vocational studies congress*, Nevşehir/Turkey, 5–8th June 2018.
50. Büyükakıncı, B. Y. (2012). Usage of microwave energy in Turkish textile production sector. *Energy Procedia, 14*(1), 424–431.
51. Özerdem, A., Tarakçioğlu, I., & Özgüney, A. (2008). Mikrodalga enerjisinin Reaktif Baskılı Pamuklu Kumaşların Fiksajında Kullanılabilirliği. *Tekstil ve Konfeksiyon, 18*(4), 289–296.
52. Kanık, M. (1995). Yüksek Frekanslı Isıtma Teknolojisi, Tekstil Sanayindeki Uygulamaları ve Bu Alandaki Son Gelişmeler. *Tekstil&Teknik, Ocak*, pp. 64–74.
53. Cardon, D. (2007). Natural dyes. In *Sources, tradition, technology and science* (p. 268). Archetype Publishing.

54. Saxena, S., & Raja, A. (2014). Natural dyes: Sources, chemistry, application and sustainability issues. In *Roadmap to sustainable textiles and clothing: Eco-friendly raw materials, technologies, and processing methods* (pp. 37–80). Springer.
55. Prabhu, K., & Bhute, A. S. (2012). Plant based natural dyes and mordants: A review. *Journal of Natural Product and Plant Resources, 2*(6), 649–664.
56. Samanta, A. K., & Agarwal, P. (2009). *Application of natural dyes on textiles*.
57. Tamburini, D., Shimada, C. M., & McCarthy, B. (2021). The molecular characterization of early synthetic dyes in E. Knecht et al's textile sample book "A Manual of Dyeing"(1893) by high performance liquid chromatography-Diode array detector-Mass spectrometry (HPLC-DAD-MS). *Dyes and Pigments, 190*, 109286.
58. Baker, R. M. (2011). *Nineteenth century synthetic textile dyes. Their history and identification on fabric*. University of Southampton.
59. Ali, H. (2010). Biodegradation of synthetic dyes—A review. *Water, Air, & Soil Pollution, 213*, 251–273.
60. Chew, Y. L., Xing, J., Lim, L. G. S., & Zhan, Z. (2023). *Development of LC/MS/MS method for screening and quantitation of 47 synthetic dyes under restricted substance list in textiles*.
61. Simion Beldean-Galea, M., Copaciu, F.-M., & Coman, M.-V. (2018). Chromatographic analysis of textile dyes. *Journal of AOAC International, 101*(5), 1353–1370.
62. Güngörmez, H. (2015). Natural dyes and salt. *Journal of the Institute of Science and Technology, 5*(1), 57–63.
63. Hosseinnezhad, M., Gharanjig, K., Jafari, R., & Imani, H. (2021). Green dyeing of woolen yarns with weld and madder natural dyes in the presences of biomordant. *Progress in Color, Colorants and Coatings, 14*(1), 35–45.
64. Akramov, K., & Mukhitdinova, J. (2020). The importance of obtaining natural dyes. *Asian Journal of Multidimensional Research (AJMR), 9*(5), 424–435.
65. Karadağ, R. (2008). *Doğal boyamacılık*. TC Kültür ve Turizm Bakanlığı.
66. Ma, X., Wei, Y., Wang, S., Zuo, X., & Shen, B. (2020). Sustainable ultrasound-assisted ultralow liquor ratio dyeing of cotton fabric with natural turmeric dye. *Textile Research Journal, 90*(5–6), 685–694.
67. Işmal, Ö. E. (2019). Doğal boya uygulamalarının değişen yüzü ve yenilikçi yaklaşımlar. *Yedi, 22*, 41–58.
68. Eyüpoğlu, Ş. (2020). Yabani Ekin Çiçeği Bitkisinden Elde Edilen Doğal Boya ile Keten Kumaşın Boyama Özelliklerinin İncelenmesi. *Düzce Üniversitesi Bilim ve Teknoloji Dergisi, 8*(1), 317–325.
69. Mohammad, F. (2015). Natural colorants in the presence of anchors so-called mordants as promising coloring and antimicrobial agents for textile materials. *ACS Sustainable Chemistry & Engineering, 3*(10), 2361–2375.
70. Vankar, P. S., Shanker, R., Mahanta, D., & Tiwari, S. (2008). Ecofriendly sonicator dyeing of cotton with Rubia cordifolia Linn. using biomordant. *Dyes and Pigments, 76*(1), 207–212.
71. İşmal, Ö. E., & Yıldırım, L. (2019). Metal mordants and biomordants. In *The impact and prospects of green chemistry for textile technology* (pp. 57–82). Elsevier.
72. Jahangiri, A., et al. (2018). Natural dyeing of wool by madder (rubia tinctorum l.) root extract using tannin-based biomordants: Colorimetric, fastness and tensile assay. *Fibers and Polymers, 19*, 2139–2148.
73. Yusuf, M., Khan, S. A., Shabbir, M., & Mohammad, F. (2017). Developing a shade range on wool by madder (Rubia cordifolia) root extract with gallnut (Quercus infectoria) as biomordant. *Journal of Natural Fibers, 14*(4), 597–607.
74. Jajpura, L., Rangi, A., & Khandual, A. (2020). Natural finishes, technologies and recent developments. In *Sustainable technologies for fashion and textiles* (pp. 209–229). Elsevier.
75. Joshi M., Ali, S. W., Purwar, R., & Rajendran, S. (2009). Ecofriendly antimicrobial finishing of textiles using bioactive agents based on natural products. *Indian Journal of Fibre & Textile Research*, 34 (September), 295–304.

76. Natarajan, G., Rajan, T. P., & Das, S. (2022). Application of sustainable textile finishing using natural biomolecules. *Journal of Natural Fibers, 19*(11), 4350–4367.
77. Sankaran, A., Kamboj, A., Samant, L., & Jose, S. (2021). Synthetic and natural UV protective agents for textile finishing. In *Innovative and emerging technologies for textile dyeing and finishing* (pp. 301–324). John Wiley and Sons.
78. Zaghloul, S., El-shafie, A., El-bisi, M., & Refaie, R. (2017). Herbal textile finishes–natural antibacterial finishes for cotton fabric. *Egyptian Journal of Chemistry, 60*(2), 161–180.
79. Kamel, M. Y., & Hassabo, A. G. (2021). Anti-microbial finishing for natural textile fabrics. *Journal of Textiles, Coloration and Polymer Science, 18*(2), 83–95.
80. Pandit, P., Gayatri, T., & Maiti, S. (2018). Green and sustainable textile materials using natural resources: : Processing and characterization. *Green and Sustainable Advanced Materials, 1*, 213–261.
81. Khurshid, M. F., Ayyoob, M., Asad, M., & Shah, S. N. H. (2015). Assessment of eco-friendly natural antimicrobial textile finish extracted from aloe vera and neem plants. *Fibres & Textiles in Eastern Europe, 6*(114), 120–123.
82. Koyuncu, F., Çetinbas, M., & Erdal, İ. (2014). Nutritional constituents of wild-grown black mulberry (Morus nigra L.). *Journal of Applied Botany and Food Quality, 87*, 93.
83. Koyuncu, F., Vural, E., Koyuncu, M., & Yildirim, F. (2004). Evaluation of black mulberry (Morus nigra L.) genotypes from lakes region, Turkey. *European Journal of Horticultural Science, 69*(3), 125–131.
84. Ahlawat, T., Patel, N., Agnihotri, R., Patel, C., & Tandel, Y. (2016). Black mulberry (Morus nigra). *Underutilized fruit crops: Importance and cultivation, 195*, 212.
85. Lim, S. H., & Choi, C.-I. (2019). Pharmacological properties of Morus nigra L.(black mulberry) as a promising nutraceutical resource. *Nutrients, 11*(2), 437.
86. Vikipedi. "Kara dut." Kara dut - Vikipedi (wikipedia.org). Accessed
87. Okatan, V., Polat, M., & Aşkin, M. A. (2016). Some physico-chemical characteristics of black mulberry (Morus nigra L.) in Bitlis. *Scientific Papers-Series B, Horticulture, 60*, 27–30.
88. Tousson, E., & Al-Behbehani, B. (2011). Black mulberries (Morus nigra) as a natural dye for animal tissues staining. *Animal Biology, 61*(1), 49–56.
89. Ebrahimi, M., & Parham, A. (2020). Using herbal dyes as an alternative staining method for sperm evaluation. *Veterinary Medicine and Science, 6*(3), 441–446.
90. Tousson, E. M., & Al-Behbehani, B. (2010). Black mulberries (Morus Nigra) as a natural dye for nervous tissues staining. *The Egyptian Journal of Experimental Biology (Zoology), 6*(1), 159–164.
91. Gedikli, F. (2006). "Ceviz (Juglans regia), karadut (Morus nigra), karamuk (Berberis crataegina), kök boya (Rubia tinctorum) ve kızılağaç (Alnus glutinosa)nın protein elektroforez jellerinin boyanmasında kullanılabilirliğinin araştırılması," Fen Bilimleri Enstitüsü.
92. Gedikli, F., & Gökçe, İ. (2015). Protein Elektroforez Jellerinin Boyanmasında Coomassie'ye Alternatif Doğal Boyalar. *Gaziosmanpaşa Bilimsel Araştırma Dergisi, 11*, 1–10.
93. Öztürk, Z., Koçak, E. D., Merdan, N., & Acar, K. (2011). Karadut (Morus Nigra) Meyvesinden Elde Edilen Boyalarla Boyanmış Yünlü Materyalin Yıkama ve Işık Haslıklarının İncelenmesi. *Akdeniz Sanat, 4*(8), 99–102.
94. Lee, K.-W. (2010). Analysis of anthocyanidin in mulberry of cu complex compound and natural dyeing on Sangju silk. *Journal of the Korean Society of Industry Convergence, 13*(4), 191–196.
95. Trak, D., & Arslan, Y. (2021). Synthesis of silver nanoparticles using dried black mulberry (Morus nigra L.) fruit extract and their antibacterial and effective dye degradation activities. *Inorganic and nano-metal chemistry* (pp. 1–13). Taylor & Francis.
96. Boga, C., et al. (2013). Investigation on the dyeing power of some organic natural compounds for a green approach to hair dyeing. *Dyes and Pigments, 97*(1), 9–18.
97. Özel, K. "ZnO tabakalı, TiO2 bazlı boya duyarlı organik güneş pillerinin sentezlenmesi, karakterizasyonu ve veriminin geliştirilmesi," Fen Bilimleri Enstitüsü.

98. Önen, T. (2016). *Yeni türdeki jellerin boya duyarlı güneş pillerinde kullanılması*. Fen Bilimleri Enstitüsü.
99. Yildiz, Z., Atilgan, A., Atli, A., Özel, K., Altinkaya, C., & Yildiz, A. (2019). Enhancement of efficiency of natural and organic dye sensitized solar cells using thin film TiO2 photoanodes fabricated by spin-coating. *Journal of Photochemistry and Photobiology A: Chemistry, 368*, 23–29.
100. Khalili, M., Abedi, M., & Amoli, H. S. (2017). Influence of saffron carotenoids and mulberry anthocyanins as natural sensitizers on performance of dye-sensitized solar cells. *Ionics, 23*(3), 779–787.
101. Karthik, T., & Rathinamoorthy, R. (2018). Sustainable biopolymers in textiles: An overview. In *Handbook of ecomaterials* (pp. 1–27). Springer.

Index

A

Adsorption, 13, 113, 118, 180, 184, 195, 270, 274–277, 279–281
Anthocyanin, 28, 31, 36, 37, 67–70, 138, 167, 175, 183, 219, 248–250, 280, 334

B

Biodegradable, 8, 14, 15, 20, 26, 27, 46, 48, 58, 131–133, 189, 205, 233, 235, 237, 266, 271, 278, 279, 330, 331, 334, 335
Biomordants, 104, 119, 120, 145, 336
Black mulberries, 330–347
By-product, 135, 138, 145, 237, 254, 267

C

Characterization, 64, 142, 227–228, 271
Cleaner production, 121
Color, 27, 46, 64, 79, 154, 194, 212, 233, 266, 288, 331
Coloration, 9, 10, 12, 17, 46–58, 64–67, 82, 86, 118, 154, 214, 222, 288, 323, 325, 326, 330, 333
Corrosion inhibitors, 266–281
Cosmetics, 8, 46, 48, 58, 64, 65, 67–69, 72, 80, 95–98, 104, 105, 129, 154, 250, 253, 266, 334

D

Denim, 10, 196, 197, 202, 204–208

E

Ebru technique, 288–326
Eco-friendly color alternatives, 98
Extraction, 2, 26, 46, 64, 81, 107, 131, 154, 195, 215, 232, 267
Extraction techniques, 10, 48–50, 92, 93, 108, 232, 236–239, 244, 246, 254

F

Fastness, 2, 9, 12, 13, 15–19, 26, 36, 46, 53–57, 65, 67, 71, 104, 116–120, 129, 135–137, 139, 141, 144, 146, 185, 194, 195, 200, 202, 203, 206–208, 212, 216, 219, 221, 223, 224, 226, 254, 255, 288–327, 332, 334, 335, 337–346
Food, 8, 25, 46, 64, 79, 104, 129, 154, 197, 212, 233, 266
Food industry, 8, 49, 52, 68, 239, 252, 267

G

Gallnut, 71, 288, 290, 291, 309–313, 321–326, 335, 336
Gallo-tannins, 175, 220, 225

H

Hazardous, 38, 48, 49, 80, 104, 105, 128, 129, 132, 154, 186, 206, 232, 233, 332

S. S. Muthu (ed.), *Natural Dyes and Sustainability*, Sustainable Textiles: Production, Processing, Manufacturing & Chemistry,
https://doi.org/10.1007/978-3-031-47471-2

I

Indigo, 4, 5, 7–11, 13, 15–18, 21, 47, 65, 67, 81, 89, 139, 142, 154, 157, 165, 166, 178, 180, 186, 189, 194–209, 212, 214, 218, 221, 226, 227, 254, 332

M

Madder, 7, 9, 13, 16, 21, 47, 65, 81, 96, 105, 119, 142, 212, 214, 217–219, 225, 334
Majith, 119
Marble printing, 291–292, 294, 296, 298, 301, 302, 306, 307, 312, 326, 327
Microbial dyes, 72
Microwave dyeing, 330–347
Microwave radiation, 11, 12, 109, 118, 120, 137, 238
Mordants, 2, 36, 46, 65, 104, 135, 182, 203, 212, 235, 267, 288, 332
Morus nigra, 278, 279, 333, 334
Mushroom-based biopigments, 79–98

N

Natural dyeing, 6, 11, 13–16, 18, 21, 46, 52, 71, 128–147, 200, 212–214, 227–228, 235, 330–347
Natural dyeing process, 6, 117, 138, 139, 142, 200, 222–227, 333–337
Natural dyes, 2, 36, 46, 64, 80, 104, 128, 154, 194, 211, 232, 266, 290, 331
Natural dyestuffs, 25–36, 39–42, 235, 254

O

Organic natural red ocher, 288, 290–302, 304–306, 309–311, 314–326

P

PA 6, 330–347
Packaging, 64, 65, 68–69, 72, 121, 128
PBT, 330–347
Pigments, 15, 18, 29, 31, 46–58, 64–70, 72, 80–83, 86–98, 104, 137, 138, 145, 196, 208, 212, 217, 219, 220, 223, 232–255, 266, 267, 278, 289, 290, 292, 324–326, 334
PLA, 38, 330–347
Prevention techniques, 270
PTT, 330–347

Q

Quercus robur (Oak), 112–113

R

Recycle, 19, 26, 36, 144, 246
Reduce, reuse and recycle (3R), 19, 27, 144
Renewable, 14, 20, 46, 48, 52, 97, 114, 132, 133, 205, 212, 215, 232–235, 237, 266, 271, 278, 279, 288, 291, 323, 325, 326, 334, 335
Reuse, 19, 26, 37, 39, 144, 145, 246, 331

S

Silk, 3, 37, 52, 65, 116, 128, 179, 196, 212, 254, 288, 331
Solar cell, 64, 65, 70–72, 334
Soybean, 31, 330–347
Supercritical fluids (SCFs), 36, 50, 93–95, 110, 222, 239–242, 246–253
Sustainability, 2, 4, 19–22, 26–27, 36–42, 46, 58, 121, 131, 132, 135–139, 142–147, 194, 204, 205, 208, 209, 232, 233, 237, 288, 291, 323, 325, 326, 332, 334, 335, 347
Sustainable, 2, 10–16, 18–21, 26, 36, 37, 40–42, 46, 58, 64, 65, 72, 97, 98, 117, 128, 131–135, 137, 144, 146, 147, 154–190, 195, 204–208, 212, 228, 232–237, 241, 255, 267, 288, 291, 323, 325, 326, 330–332, 334, 335, 345, 347
Sustainable Textile industry, 79–98, 131–135
Synthetic red ink, 288, 290–292, 294, 296–298, 301–304, 306–308, 311–315, 317, 318, 320–326

T

Textile applications, 53, 67, 335
Textile value chain, 128
Traditional textiles, 2–6, 15–22

U

Ultrasonic radiation, 117, 137

W

Walnut, 7, 65, 119–120, 250, 334
Waste valorization, 206, 208
Wastewater, 14, 20, 52–54, 104, 129, 133, 136, 138, 142, 145, 206, 207, 330

FSC
www.fsc.org
MIX
Papier | Fördert
gute Waldnutzung
FSC® C083411

Kolibri 360 GmbH
Ferdinand-Jühlke-Straße 7
99095 Erfurt, Deutschland
produktsicherheit@kolibri360.de